Symbiosis

Symbiosis

An Introduction to Biological Associations

Vernon Ahmadjian
Clark University

Surindar Paracer
Worcester State College

With a Foreword by
Lynn Margulis

Published for Clark University by
University Press of New England
Hanover and London, 1986

Printed in the United States of America

Library of Congress Cataloging-in-Publication Data

Ahmadjian, Vernon.
    Symbiosis : an introduction to biological associations.

    Includes bibliographies and index.
    1. Symbiosis.   I. Paracer, Surindar.   II. Title.
[DNLM: 1. Symbiosis.   QH 548 A286s]
QH548.A37   1986   574.5'2482   86–5471
ISBN 0–87451–371–5

Figure 3.2 A and B reproduced from J. C. Burnham, T. Hashimoto, and S. F. Conti. 1968. Electron microscopic observations on the penetration of *Bdellovibrio bacteriovorus* into gram-negative bacterial hosts. J. Bacteriology 96:1366–1381, by permission of J. C. Burnham and the American Society for Microbiology. **Figure 3.3** reproduced from H. W. Paerl and K. K. Gallucci. 1985. Role of chemotaxis in establishing a specific nitrogen-fixing cyanobacterial-bacterial association. Science 227:648, by permission of H. W. Paerl and the American Association for the Advancement of Science. **Figure 3.4 A** reproduced from L. B. Preer and A. Jurand. 1974. Bacteriological reviews 38:113–163, by permission of Louise B. Preer and the American Society for Microbiology. **Figure 3.4 B** reprinted from "The relation between virus-like particles and R bodies of *Paramecium aurelia*" by John R. Preer, Jr. and Arthur Jurand, *Genetical Research,* Vol. 12 (1968), pp 331–340, by permission of Cambridge University Press. © Cambridge University Press 1968. **Figure 3.5 A and B** reproduced from K. W. Jeon and I. J. Lorch. 1967. Unusual intra-cellular bacterial infection in large, free-living amoebae. Experimental Cell Research 48:236–240, by permission of K. W. Jeon and Academic Press. **Figure 4.2** adapted from F. B. Dazzo and D. H. Hubbell. 1975. Cross-reactive antigens and lectin as determinants of symbiotic specificity in the *Rhizobium*-clover association. Applied Microbiology 30:1017–1033, by permission of Frank B. Dazzo and Academic Press. **Figure 4.2 insert** reproduced from F. B. Dazzo and W. J. Brill. 1979. Bacterial polysaccharide which binds *Rhizobium trifolii* to clover root hairs. J. Bacteriology 137:1362–1373, by permission of Frank B. Dazzo and Academic Press. **Figure 4.3** reproduced from D. Baker, W. Newcomb, and J. G. Torrey. 1980. Characterization of an ineffective actinorhizal microsymbiont, *Frankia* sp. Eull (Actinomycetales). Can. J. Microbiology 26:1072–1089, by permission of William Newcomb and the Canadian Journal of Microbiology. **Figure 4.4 A and B** reproduced by permission of Steve R. Dunbar and Sandra K. Perkins. **Figure 6.4** reproduced from Raymond I. Carruthers, Dean L. Haynes, and Donald M. MacLeod. 1985. *Entomophthora muscae* mycosis in the onion fly. *Delia antiqua.* Journal of Invertebrate Pathology 45:81–93, by permission of Raymond I. Carruthers and Academic Press. **Figure 6.6 B** reproduced by permission of C. Gerald Van Dyke. **Figure 7.3 A** reprinted by permission from *Nature,* Vol. 289, No. 5794, pp. 169–172. Copyright © 1981 Macmillan Journals Limited. **Figures 7.5 A and C** reproduced from *Methods and Principles of Mycorrhizal Research,* by permission of Merton F. Brown and the American Phytopathological Society. **Figures 7.5 B, 7.6 B and C** reproduced by permission from the Canadian Journal of Botany, Vol. 59, pages 683–688. **Figure 7.6 A** reproduced by permission of USDA Forest Service. **Figure 8.4 A and B** reproduced by permission of Stanley L. Erlandsen. **Figure 8.6 A** reproduced from A. M. Fallis, ed. 1971. Ecology and Physiology of Parasites. Toronto University Press, Toronto, 75, by permission of Keith Vickerman and University of Toronto Press. **Figure 8.6 B** reproduced from J. E. Donelson and M. J. Turner. 1985. How the trypanosome changes its coat. Scientific American 252:45, by permission of Steven T. Brentano and Scientific American. **Figure 8.9 B** reproduced from Bannister et al. 1975. Structure and invasive behaviour of *Plasmodium knowlesi* merozoites *in vitro.* Parasitology 71:483–491, by permission of Lawrence Bannister and Cambridge University Press. **Figure 8.10 A, B, and C** reproduced by permission of Lawrence Bannister. **Figure 9.2** reproduced by permission of R. L. Pardy. **Figure 10.4 B and C** from Schistosoma Mansoni: The Parasite Surface in Relation to Host Immunity, by Diane J. McLaren, copyright 1980 by John Wiley and Sons Limited. Reproduced by permission of John Wiley and Sons Limited. **Figure 10.6 B and C** property of USDA reproduced by permission of James J. Peterson. **Figure 10.7** reproduced from W. R. Nickle (ed.) *Plant and Insect Nematodes,* Marcel Dekker, Inc., N.Y. 1984, by permission of W. R. Nickle and Marcel Dekker, Inc. **Figure 10.8 A and B** reproduced by permission of J. D. Eisenback. **Figure 11.3** reproduced by permission of M. C. F. Proctor. **Figure 12.2** reproduced by permission of R. N. Mariscal. **Guests and Hosts** by Dr. Milton Love reprinted with permission from Natural History, Vol. 88, No. 6; Copyright the American Museum of Natural History, 1979.

Contents

Foreword

Symbiosis, the association of organisms that are members of different species, is a pervasive fact of the living world. In spite of its importance to bacteriology, physiology, ecology, and evolution, however, mainstream biologists working in these subsciences have tended systematically to ignore symbiosis as a principle worthy of investigation. The authors of this book have returned our attention to the original broad nineteenth-century interpretation first envisioned by Anton de Bary, concerning themselves with all of the significant associations between members of different species. They have gracefully crossed the borders that separate zoology, botany, microbiology, ecology, parasitology, and the like. Because they use the symbiotic associations themselves as the focus, Ahmadjian and Paracer can take us to an extraordinary variety of ecological settings: the intestinal epithelium, the temperate forest, the coral reef, and the center of the plant cell. Furthermore, we learn to see such diverse subjects as cell photosynthesis, pollination ecology, and aspects of tropical medicine in a new light. It is a rewarding journey.

As we learn here, symbioses remain viable partnerships even when their values change. The value of any given association, measured by how much it increases or decreases the probability of survival of the partners, varies with genetic constitution, ecological setting, and other changing conditions. A sliding continuum, which may be viewed as a sort of rainbow spectrum of continuity, is revealed in which bacterial viruses, commensals, pathogens, parasites, and other familiar associates are recognized as symbionts: what may be thought of as their primary color in the continuum is more a consequence of the peculiarities of our perception than of the natural setting in which they are embedded.

The greatest virtue of this unprecedented textbook is the breadth of its coverage of the biological sciences. From glimpses of microbial predation to takeovers of entire colonies of ants by other ants, we begin to see the chemical, cellular, behavioral, and community levels of the ubiquitous symbiosis phenomenon. The advantage of having such well-expounded material collected in a single place for the curious student will, I hope, lead to the formalized study of symbiosis in the biology curriculum.

The case that an understanding of symbiosis is important for the budding biologist has been well made. For example, Ahmadjian and Paracer tell us that, in our favorite kingdom, perhaps half of the species are symbionts. Some animal phyla have only symbiotic members. The nucleated cell, that venerable unit of animal, plant, fungal, and protoctistan life, also seems to be a coevolved microbial symbiotic complex. This accessible book also gives us some notion of the long history of ideas about symbiotic associations, a history that is often ignored.

The Colorado biologist Ivan T. Wallin tried, in his book published in 1926 (see p. 66 for reference) and his other publications, to convince his early twentieth-century colleagues of his "principle of symbionticism." Wallin's fundamental assertions about symbionticism were not developed in great detail. With the exception of K. S. Mereschovsky of Kazan, a colleague with whom he never collaborated, Wallin was a very lonely voice in biological circles. Both men concluded, for example, that the mitochondria of animal and plant cells had originated as symbiotic bacteria, a theory we now accept as proved. Various reasons prevented this conclusion and other imaginative tenets from catching on in the professional biological circles of the time.[1]

Some of Wallin's ideas are still not generally accepted. He felt that symbioses (especially those between very different sorts of partners)

---

[1] Mehos, D. Ivan Wallin's principle of symbionticism. J. History of Biology, 1986 submitted.

provide a major source of inherited variation and as such an indispensable mechanism for the origins of new forms of plants and animals. Wallin's concept is still ignored in serious contemporary treatises on speciation. We continue to be taught the "mathematical proof" that the gradual accumulation of mutations suffices to account for the appearance of new species. Another of Wallin's claims is that former microbial symbionts play a direct role in the development and differentiation of animal and plant cells and tissues. This third insight has hardly ever even been mentioned in "polite biological society." In the end Wallin may be proved correct: the evolution of certain cell organelles, the origin of new species, and the phenomenon of cell and tissue differentiation may come to be viewed as applications of the principles of symbiosis. A new hard science of symbiology with its own postulates and derivations may be emerging.

Certainly the founding of such a biological subfield, "applied symbiology," and the continued development of principles of biological science depend on the availability of the kind of information that Ahmadjian and Paracer have so neatly presented here. The work of our authors in compiling this text will be deeply appreciated by students and instructors at many levels of the educational hierarchy. Indeed, now that we have *Symbiosis: An introduction to biological associations* we will wonder how we ever taught without it.

Lynn Margulis
May 1986

Preface

This introductory text is designed for students, instructors, and research workers who wish to learn about symbiosis. Our experience in teaching courses on symbiosis and "parasitism" has convinced us of the need for such a book. The rarity of academic courses on this subject and the general ignorance of many biologists about symbiosis underscores the need for suitable general texts. We hope that our book, which views symbiosis from a broad perspective, will help to meet these needs.

Symbiosis as a concept and a discipline has changed dramatically in recent years. It is no longer the exclusive domain of a relatively few scientists who study specialized associations but rather embraces all aspects of biology and all five kingdoms. From viruses and bacteria to protoctists, fungi, plants, and animals, symbiotic associations abound. No organism is immune from the direct or indirect influence of symbiosis, whose profound effect on biological evolution and species formation is being fully realized only now. Some of the greatest events in biological history, including the origin of the eukaryotic cell, as well as some of the most devastating diseases of mankind are the results of symbiotic relationships. New techniques of molecular biology have given impetus to research in symbiosis and are helping scientists to develop permanent vaccines against agents of diseases such as malaria and sleeping sickness. Molecular evidence has been crucial in the acceptance of the serial endosymbiosis theory for the origin of chloroplasts and mitochondria.

How one defines the term *symbiosis* has been a matter of confusion. For reasons that are not entirely clear to us, many scientists equate symbiosis with mutualism, that is, a relationship in which both partners benefit. This was not the intention of Anton de Bary, who coined the word *symbiosis*. De Bary used the word in a broad sense, to mean a living together of different types of organisms, and did not limit the word to only one type of relationship. In this book we follow de Bary's original purpose and consider commensalism, parasitism, and mutualism to be subcategories of symbiosis.

Apart from historical considerations, we feel that it is important for scientists to have a broad perspective of symbiosis. Until recently, most scientists in fields such as medical parasitology, microbiology, plant pathology, virology, and veterinary medicine have been concerned primarily with pathogenic relationships, but these relationships, in fact, constitute only a narrow segment of interorganismic associations. It is important for students and scientists to study nonpathogenic symbioses if they are to fully understand the mechanisms of pathogenic responses. The potential for biological management of pest organisms will be realized only if future scientists develop an appreciation of the functional relationships that exist among a broad range of associating organisms.

The vastness of the subject presented us with several problems, one of which was how to select examples to illustrate different symbiotic phenomena. We have tried to use examples that best represented particular types of symbiosis. Examples not included in this book are not necessarily those considered to be unimportant; rather, they are omitted because of space considerations. Another problem concerned the specialized terminology and information associated with different biological fields. We have assumed that few individuals are conversant with all facets of biology and, therefore, we have provided the reader with a brief introduction to each topic before describing specific details of a symbiosis.

There are many books on various aspects of symbiosis, from the molecular biology of parasites to pollination and insect-fungus relationships. One of our goals was to place together in one text the many facets of symbiosis, ranging from classical observations to the modern experimental approaches of molecular biology, to enable the reader to fully appreciate the vastness of the subject. Another goal was to develop in biologists a better understanding and recognition of symbiosis. Whether or not we accomplish the last goal depends on the reactions of our readers to this text. To that end, we would appreciate receiving comments, suggestions, and even criticism from readers on the utility of this book, the adequacy of the information presented, and the ways by which the text might be improved.

Many individuals have helped us in the task of writing this book. We are grateful to our colleagues who have read drafts of various chapters and offered constructive suggestions and criticisms. We particularly thank the following individuals: Professors James Kimbrough, Dave Mitchell, Charles Tarjan, and Grover Smart, all from the University of Florida, and Dr. Robert Esser, nematologist, Florida Department of Agriculture, and Professor James Sassar, North Carolina State University; also Professors William Johansen, Todd Livdahl, Donald Nelson, Margaret Comer, and John Brink, all from Clark University, and Ann Hirsch from Wellesley College. We also thank Professors Ray Welsch, University of Massachusetts Medical School, Warren Silver, University of South Florida, and Lynn Margulis,

Boston University, for their critical reviews, stimulating discussions and encouragement.

We express particular thanks to Selina Remmer for her insightful editorial suggestions on many of the chapters. We are deeply grateful to Mrs. Frances Ables, Elizabeth Rogers, Inis Cook, and Del-Marie Bachand for their typing of parts of the manuscript, to Kristin Meier for editorial comments, to Lori Dembowicz for computer assistance, to Mary Richard and Laurel Donahue for xeroxing, and to Mrs. Terry Reynolds and the staff of the Word Processing Center at Clark University. Our joint collaboration in writing this book is a tribute to the spirit of the Worcester Consortium for Higher Education, which encourages interinstitutional cooperation. One of us (SP) gratefully acknowledges President Philip Vairo and the Board of Trustees of Worcester State College for the grant of a sabbatical leave and extends his appreciation to the Department of Nematology and Plant Pathology at the University of Florida, Gainesville, for the use of library facilities and stimulating discussions he encountered during his sabbatical leave.

Finally, to our colleagues who sent us reprints, preprints, glossy prints, and words of encouragement we say, thank you.

Worcester, Massachusetts   *V.A.*
March 1986   *S.P.*

CHAPTER 1

Introduction

A. Concepts and Definitions

1. INTRODUCTION. *Symbiosis* is a permanent or long-lasting association between two or more different species of organisms. The partners of a symbiosis are called *symbionts* and they may benefit from, be harmed by, or not be affected by the association. Few phenomena in the world are as widespread as symbiosis. Thousands of associations exist, from fungal-algal unions called lichens to those between giant tube worms and sulfur-oxidizing bacteria. No organism is an island. Humans as well as tiny protoctists are symbiotic composites. The biological world is a web of organismal interrelationships at different levels of organization. It may be humbling to us, as humans, to realize that each of our cells contains organelles called mitochondria that may be transformed symbiotic bacteria, and that we cannot live normally without other types of beneficial bacteria that grow in our intestinal tract. Cells of plants contain organelles called chloroplasts, which are thought to have evolved from symbiotic photosynthetic bacteria. Every organism, large or small, forms associations with other organisms. Bacteria, which are symbiotic with higher forms of life, are themselves hosts to symbiotic viruses. Even some viruses are associated with obligate symbionts called satellite viruses. Without symbiotic relationships, life would be very different from what it is today.

Two types of organismic associations exist in nature: intraspecific (*intra* "within") and interspecific (*inter* "between"). *Intraspecific associations* are formed between members of the same species. Colonies, mate pairing, and parent-offspring relationships are examples of intraspecific associations. *Physalia,* the Portuguese man-of-war, is not an individual but a colony composed of a float and tentacles that are interdependent. The adult male of the blood fluke, *Schistosoma,* holds the female permanently within a narrow boatlike groove on its ventral surface. Scientists have suggested that the male produces a pheromone that controls the maintenance and maturation of the female. Another example of mate pairing is demonstrated by the nematode *Trichosomoides,* which lives in the bladder of rodents. The male spends its entire life in the uterus of the female. An example of an intraspecific plant association is the relationship between injured and healthy trees. Trees, such as sugar maple, that are injured by insects release volatile chemical substances that act as warning signals to neighboring healthy trees and stimulate them to synthesize chemical compounds that protect them from the attacking insects.

Interspecific associations, or symbioses, occur when two different species of organisms depend on each other for food, shelter, or protection. Early zoologists described interspecific associations primarily in terms of nutrition. For example, a *parasite* is an organism that obtains its food by living on or in another organism, the *host.*

The focus of this book is on interspecific relationships, their importance in the evolution of plants and animals, their significance in the earth's economy, and their role in the complexity of life. We consider only biological examples of symbiosis and examine in detail associations that are representative of the five kingdoms that constitute the living world. The concept of long-lasting associations has a wide range of application, from "symbiotic" stars to the complex psychological interactions between parents and their offspring. This book is an attempt to underscore the perspective that symbiosis is such a universal and important phenomenon that it should be an integral component of the education of biologists.

2. SUBDIVISIONS OF SYMBIOSIS. In this book, we use the word *symbiosis* in a broad sense, as originally intended by Anton de Bary in 1879, to refer to a living together of different organisms. Various types of symbiosis, whether beneficial or harmful, are described by the terms *commensalism, mutualism,* and *parasitism.* An association in which one symbiont benefits and the other is neither harmed nor benefited is a *commensalistic symbiosis.* An association in which both symbionts benefit is a *mutualistic symbiosis.* A relationship in which a symbiont receives nutrients at the expense of a host organism is a *parasitic symbiosis.* Let us examine each of these categories in more detail.

a. Commensalism. The term *commensal* was first used by P. -J. van Beneden in 1876 for associations in which one animal shared food caught by another animal. The two animals were considered to be "messmates" that ate from the same table. An example of a commensalistic relationship is that between silverfish and army ants. The silverfish live with the army ants, participate in their raids, and share their prey. They are of no harm or benefit to the ants. The term *commensalism* is used here in its broadest sense, indicating that the benefit to one of the symbionts may be nutritional or protective.

b. Mutualism. Mutualistic symbiosis appears to be the ideal type because both partners benefit from the relationship. The extent to which each symbiont benefits, however, may vary and generally is difficult to assess. The benefit a symbiont receives from an association should be considered in terms of its costs. As de Bary suggested, there probably is not an example of mutualism in which both partners benefit equally. Unfortunately, for most symbioses we do not fully understand the complex interactions that take place between the symbionts. In mutualism, the benefits may be nutritional or physical. In many associations there is a reciprocal exchange of nutrients. For example, in the symbioses of algae and invertebrates, the algae provide the animals with organic compounds, which are products of photosynthesis, while the animals provide the algae with waste products

such as nitrogenous compounds and carbon dioxide, which the algae use in photosynthesis. This type of nutrient interdependence allows the animals to colonize habitats that are unsuitable for their symbiont-free counterparts. Marine animals such as corals grow well in nutrient-poor tropical oceans because their symbiotic algae provide them with food and oxygen. Another benefit that may be obtained in a mutualistic association is protection. Some bacteria and unicellular algae are protected from a hostile environment by being within a host cell. Such protection has resulted in adaptations of the symbiont such as a greatly reduced or absent cell wall and a loss of sexual reproduction.

c. Parasitism. Parasitism is a one-sided symbiosis in which one of the symbionts benefits at the expense of the other. As in mutualism, the primary factor in parasitism is nutrition: the parasite uses the host as a source or supply of food. Parasitic symbioses vary in the degree to which the host is affected. Some parasites are so virulent that the host cells die shortly after the parasitism begins. Parasites that cause disease and possibly death of the host are called *pathogens.*

Mutualism, parasitism, and commensalism are the only categories of symbiosis considered in this text. Some scientists recognize other categories, such as *phoresis* and *inquilinism,* which are based on the transport or shelter of one of the symbionts. We consider such associations to be commensalistic. In phoretic relationships, one symbiont protects or transports another symbiont. For example, some insects deposit their eggs on the bodies of other animals, which transport them to new habitats. Some marine hydroids and sea anemones are attached to the shells of molluscs and crabs and are carried about by their hosts. In inquilinism, two or more animals of different species share a dwelling place. Some symbiologists consider the predation of one animal by another a type of symbiosis.

Terms such as *mutualism, parasitism,* and *commensalism* are used to conveniently categorize associations. But many relationships are not static and there may be frequent transitions from one type to another. Symbiotic associations may change because of environmental factors or internal influences caused by the development of the symbionts. A parasitic association could evolve into one of mutualism or commensalism. Indeed, it is difficult to conceive of two organ-

BOX ESSAY

Anton de Bary

PIONEER IN SYMBIOSIS

Anton de Bary, a German mycologist and professor at the University of Strasbourg, carried out pioneering studies in plant pathology during the mid-nineteenth century. He studied disease-causing bacteria and was the first to note that fungi cause plant diseases. De Bary studied sexual reproduction in many fungi and established the major groups of fungi, based on modes of reproduction, in early schemes of classification. Among his many studies of fungi, he discovered the complex life cycle of *Puccinia graminis,* the wheat rust fungus, describing the five different types of spores and confirming that the common barberry was the alternate host for the fungus.

De Bary also studied lichens and he deserves partial credit for the discovery that lichens are symbiotic associations. In 1866, in the first edition of his classic book *Morphologie und Physiologie der Pilze, Flechten und Myxomyceten,* de Bary stated that gelatinous lichens might be associations of blue-green algae and parasitic fungi. This idea was contrary to the contemporary view that lichens were single organisms, like mosses and liverworts. De Bary's view of lichens preceded by one year the work published by Simon Schwendener, the Swiss botanist, who is generally given credit for the discovery of the true nature of lichens. De Bary was cautious and limited his opinion to gelatinous lichens, whereas Schwendener had no doubt that all lichens were associations of fungi and algae.

De Bary coined the word *symbiosis* to describe the growing number of associations that were being discovered between different types of organisms. The word *symbiosis* first appeared in 1879 in a paper titled "Die Erscheinung der Symbiose" (The phenomenon of symbiosis). The paper was a printed speech that de Bary had given the previous year to a group of German naturalists and physicians. De Bary defined *symbiosis* in the broadest possible way, that is, as a living together of different organisms. He included in his definition all cases of intimate associations, including epiphytes growing on trees and insects pollinating flowers. According to de Bary, parasitism was also a form of symbiosis, as were mutualism and commensalism, terms that were coined earlier, in 1876, by Pierre-Joseph van Beneden, a professor at the University of Louvain, in his book *Animal Parasites and Messmates.*

During the next fifty years, confusion arose as to the exact meaning of symbiosis. The term became equated with mutualism and the two terms were used interchangeably to mean an association in which both symbionts benefit. The reason for this confusion is hard to understand, since de Bary clearly defined his terms and in his paper gave examples of parasitic as well as mutualistic symbioses. De Bary was studying lichens when he coined the word *symbiosis* and much of his original paper on symbiosis used lichens as examples. Later workers may have assumed that de Bary applied the term *symbiosis* only to associations such as lichens, which de Bary considered, for the most part, to be examples of mutualism. De Bary's earlier view of lichens as parasitic associations was restricted to ones with blue-green photobionts. At the present time, the original intention of de Bary is being honored by most scientists and symbiosis is once again being defined broadly.

isms becoming involved initially in a mutualistic association. Most mutualistic symbioses probably began as parasitic ones with one organism attempting to exploit another one. If we consider parasitism as an antagonistic relationship, then mutualism can be regarded as a standoff or a draw between two antagonists. For example, during the course of a parasitic relationship between two organisms defenses of the host may become strong enough to slow or stop the growth of the parasite. Over time, the virulence of the parasite lessens and a mutualistic relationship between parasite and host may begin. Two extreme examples of such a mutualistic evolution are mitochondria and chloroplasts in eukaryotic cells. These organelles are transformed bacteria that probably began as parasites in ancient prokaryotic cells. Conversely, a mutualistic association may degenerate into a pathogenic one if the defenses of the host weaken. For example, the common intestinal symbiont of humans *Escherichia coli* can become a pathogen of the body under abnormal conditions. Understanding how a parasitic or disease-causing association evolves into a mutualistic one can have important medical benefits for humans.

3. PARASITISM AND DISEASE. Because early workers misunderstood de Bary's definition of the term *symbiosis* and equated it with mutualism, parasitism has generally been considered a separate category rather than a subcategory of symbiosis. Recent research studies in cell biology, epidemiology, and evolutionary biology continue to use terms such as *parasitology, parasitic animals,* and *parasitism* without considering their position in the wider picture of symbiosis. Parasites have a poor public image because parasitologists have tended to link disease and parasitism. Many parasites, however, do not cause disease; they do not disrupt or seriously diminish the performance of their host even though they take nutrients from the host. Parasites should be viewed in the broader context of symbiosis and disease-causing parasites should be called pathogens. This view is similar to one held earlier by J. G. Horsfall and A. E. Dimond (1960), who argued that a pathogen and a parasite cannot be considered the same. In the nineteenth century P. -J. van Beneden similarly distinguished between a beast of prey and a parasite, the former killing its victim, the latter not. Equating symbiosis only with mutualism was a historical accident.

Recent developments in parasitology have emphasized aspects of host-symbiont relationships that involve cell membranes, molecules such as lectins, pheromones, and toxins, genetic mechanisms, nonspecific and specific immunity, and the ecological concepts of populations. Significant progress has been made in the axenic (pure) culture and maintenance of protoctist and metazoan symbionts *in vitro*. Let us now look at the details of host-pathogen relationships.

a. Pathogenic Relationships. Pathogenesis is a process during which vital functions of a host are disrupted or impaired because of the continuous presence of a pathogen. The pathogen causes disease, which manifests itself as a complex of symptoms that can be diagnosed by a trained specialist. Pathogens are, therefore, disease inducers and can be separated into two classes, abiotic and biotic. Air pollutants, asbestos, and nutritional imbalances and deficiencies are examples of abiotic pathogens. Organisms that induce disease in a host are called biotic pathogens. Many parasites are pathogenic to their hosts, however other parasites are not pathogenic.

The causes of disease have always fascinated people, beginning with ancient times. Early man thought that supernatural forces or evil spirits visited people and when they disobeyed God's will, diseases were sent to punish them. For centuries, man prayed to the gods and offered sacrifices to gain protection from disease for himself, his livestock, and his crops. It was not until the nineteenth century that scientific observations and studies produced the germ theory of disease. In 1807, Bénédict Prévost demonstrated that bunt disease of wheat was produced by a fungal pathogen. Prévost was the first scientist to prove experimentally the cause of a disease, but since he lived at a time when most people thought life developed spontaneously, his ideas did not gain wide acceptance. The role of fungi in producing plant diseases was not firmly established until the mid 1850s when Anton de Bary published a paper on smut and rust diseases of wheat and showed that fungi did not arise spontaneously. He, like his contemporaries, was not aware of Prévost's earlier discovery. Louis Pasteur, working on yeast, and Robert Koch, investigating anthrax and cholera, helped to close the chapter on the belief in the spontaneous generation of life and to establish the germ theory of disease. Bacteria, fungi, protozoans, and various worms were soon demonstrated to be causes of human, animal, and plant diseases. In 1883, Robert Koch developed four criteria to implicate a pathogen as the cause of a disease. The criteria are called *Koch's postulates of pathogenicity:*

The pathogen must be present in all host organisms that have the disease.

The pathogen must be isolated in axenic culture outside the host.

When a healthy host is inoculated with the isolated pathogen, disease symptoms must occur.

The pathogen must be reisolated from the experimentally diseased host and shown to be the same as the original isolate.

The science of animal and plant pathology has benefited immensely from the application of Koch's postulates. There are, however, many difficulties in meeting these criteria. First, many pathogens are obligate parasites and will not grow in chemically defined culture media. Only recently have advances in tissue culture techniques allowed some obligate parasites to be grown in axenic culture. Second, a disease may be a complex of more than one pathogen. Third, Koch's postulates assume that all diseases are induced by pathogenic organisms. Today, we know of many abiotic pathogens that can induce diseases in plants, animals, and humans.

Development of a disease follows a series of stages:

An infection stage, which involves contact of a susceptible host by a pathogen and its entry into the host. Cell surface interactions between pathogen and host are important in the infection process.

An incubation stage, which is the period between infection and the outbreak of disease symptoms. During this phase, the pathogen grows and multiplies. Pathogenic toxins and other metabolites cause physiological abnormalities in the host.

A defensive stage, when the host may develop responses that neutralize the pathogen. Pathogens that override host defense mechanisms are called *virulent.*

An immunity stage, in which a recovered host is immune to reinfection by the same pathogen.

Scientists who study diseases in a population are called epidemiologists. Some of these scientists find the concept of "parasitic infection" too broad and varied, and to simplify matters they have devised mathematical models to study patterns of disease development. Epidemiologists distinguish between two classes of parasites: microparasites and macroparasites. *Microparasites* include viruses, bacteria, and protoctists. They are microscopic, have a short generation time, and reproduce rapidly in the host. Microparasites tend to induce immunity to reinfection in individuals that have survived an earlier infection. The duration of a disease caused by a microparasite is short compared to the life-span of the host. *Macroparasites* include helminths and arthropods. They are large symbionts that have long generation times and multiply slowly in the host. The immune response in the host depends on the size of the "parasitic burden" and tends to be short-lived. Infection of a host from macroparasites may occur continually, in which case it is called *persistent.* There are many exceptions to the above characteristics of micro and macroparasites, but in general the definitions are useful for constructing theoretical models and for developing epidemiological concepts of parasitic pathogens.

b. The Pathogen. A *pathogen* is any factor that disrupts or impairs one or more vital functions in a host. In stricter terms, pathogens are cellular irritants that induce abnormal cell physiology and metabolism. The effect of a pathogen may be localized in a specific area, such as the respiratory system, or the effect may be *systemic,* with the disease developing in all parts of the host. Virulence is the relative ability of a pathogenic organism to induce a given amount of disease in a given host. Genetic constitution of the host and the predisposing environmental conditions are important factors in the development of a disease. Some scientists use *pathogenic fitness* as a synonym for *virulence,* but *pathogenic fitness* is more accurately the ability of an organism to infect, colonize, and reproduce in a susceptible host.

c. The Host. Disease induced by a pathogen in a *susceptible host* always produces symptoms. A *resistant host* is one in which a pathogen fails to develop and the host does not show disease symptoms. Host resistance or susceptibility to a particular pathogen is affected by factors such as age, sex, nutritional and hormonal levels, and genetic makeup. Vertebrate hosts have complex defense mechanisms. The entire reticuloendothelial system is engaged in the development of immune responses. Antibody production is the most specific host response to pathogenic antigens. There have been major breakthroughs in our understanding of immune responses. By means of recombinant DNA techniques, monoclonal antibodies can now be manufactured *in vitro.*

Parasitologists often describe a host in relation to the role it plays in the life cycle of a symbiont. A *definitive* or *final host* is one in which a symbiont reaches sexual maturity. An *intermediate host* is also necessary for the completion of a symbiont's life cycle. For example, in the life cycle of the sheep liver fluke, *Fasciola hepatica,* sheep are the definitive hosts and snails are the intermediate hosts; the fluke's larval stages develop in the snail. Some symbionts need more than one intermediate host. *Reservoir hosts* are species that harbor potential pathogenic symbionts without showing signs of disease. These species are the source of infection for other species. *Vectors* are carriers of transmittable pathogenic parasites. Many blood-sucking arthropods, such as horseflies, tsetse flies, mosquitoes, and fleas, are well-known vectors of viral, bacterial, and protoctistan pathogens.

B. Symbiosis in All Forms of Life

One of the earliest systems of classification consisted of two kingdoms, plants and animals. Although this classification lasted a long time, it was not satisfactory because some organisms such as *Euglena,* bacteria, and slime molds did not fit well in either kingdom. These odd groups were arbitrarily placed into either the plant or animal kingdom depending on the personal biases of scientists. As techniques in cell biology improved, and details of the cell were better understood, it became clear that the two-kingdom system of classification was too restrictive and would have to be expanded. In 1969, R. H. Whittaker proposed a five-kingdom classification, which met with initial resistance but now is accepted by most biologists. The five kingdoms proposed by Whittaker were the Monera, Protista, Fungi, Plantae, and Animalia. L. Margulis and K. V. Schwartz, in their recent book *Five Kingdoms,* have modified Whittaker's classifica-

tion. These scientists consider the kingdom Protista, which includes only single-cell organisms, as too narrowly defined and have replaced it with the kingdom Protoctista, a term used by early scientists to include multicellular as well as unicellular organisms. In this book we follow the classification of Margulis and Schwartz (Fig. 1.1).

The kingdom Monera consists of organisms that have prokaryotic cells. Organisms in the remaining four kingdoms have eukaryotic cells. The distinction between prokaryotic and eukaryotic cells is a fundamental one in the five-kingdom classification. Prokaryotes are the most primitive life form, whereas eukaryotes are more highly developed. It is presumed that eukaryotes evolved from prokaryotes but exactly how that might have happened is a much debated question in biology.

The kingdom Monera consists mostly of bacteria, the only living organisms with prokaryotic cells. Blue-green algae, long considered to be plants because they produce oxygen during photosynthesis, have prokaryotic cells and now are placed in the Monera and called cyanobacteria or blue-green bacteria. *Prochloron,* a newly discovered organism that lives as a symbiont in tropical tunicates, may be an intermediate form between cyanobacteria and eukaryotic algae. *Prochloron* has chlorophyll a and b and carotenoids, like green algae, and is bright green. It lacks the phycobiliproteins found in cyanobacteria, but because of the prokaryotic structure of its cells, *Prochloron* is placed in the kingdom Monera.

The kingdom Fungi includes organisms that lack chlorophyll and obtain nutrients by absorption from living or nonliving organic sources. Fungi excrete enzymes that break down organic matter into simpler, usable compounds and, like bacteria, are important decomposers in nature. Most fungi grow by means of filamentous threads called hyphae, although some, like yeasts, are unicellular. Fungal cell walls contain chitin, the material found in exoskeletons of many animals. Many fungi reproduce sexually, but the resulting diploid zygote undergoes meiosis either immediately or after a resting period to form haploid spores or offspring. There are over 100,000 species of fungi. They produce many spores and grow practically everywhere.

The kingdom Animalia consists of multicellular organisms that ingest their food. Reproduction consists of fertilization of an egg by a sperm

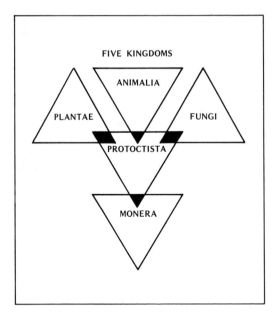

Figure 1-1. Model of the five-kingdom classification of biological organisms.

and the resulting zygote forms a ball of cells called a blastula from which the animal develops. Animal cells lack walls and most are diploid, the haploid condition occurring only in the sperm and eggs.

The kingdom Plantae consists mostly of photosynthetic, multicellular organisms. Some plants lack chlorophyll and absorb food from decaying matter or from other plants. Plant cells have walls that contain cellulose. Plants undergo sexual reproduction and develop from embryos, which are surrounded by a jacket of sterile tissues. Plants show alternation of generations in which a haploid generation (gametophyte) alternates with a diploid generation (sporophyte). In mosses and liverworts the haploid generation is dominant; in more advanced plants, such as gymnosperms and angiosperms, the diploid generation is dominant.

The kingdom Protoctista is the weakest link in the five-kingdom scheme of classification. Organisms included in this kingdom are ones that do not fit into the other four kingdoms. Many forms have chloroplasts and other forms obtain food by absorption and ingestion. Protoctists are considered to be the earliest types of eukaryotic organisms and include the algae, protozoa, slime molds, and water molds.

Symbiotic associations are common in all five

biological kingdoms, as well as among viruses. We consider viruses here because although they are noncellular, they have some basic characteristics of living organisms. Viruses contain genetic material that is transmittable. Their level of symbiotic interaction is molecular rather than organismal and generally more integrated than other symbioses. Viruses are obligate associates of many living forms and therefore are considered as symbionts. Viruses can form long-lasting associations with their hosts; for example, prophage viruses insert part of their chromosome into a chromosome of the host cell and replicate along with the host. The genetic colonization of host cells by viruses is a subtle type of symbiosis that has far-reaching implications, both positive and negative, for mankind. This natural form of genetic engineering is mirrored by man's attempts to create artificial symbioses by manipulating the genetic material of different organisms. Recombinant DNA technology strives to produce symbiotic hybrids that will be of use in medicine and agriculture.

Bacteria are common symbionts of plants, animals, and protoctists. Some bacteria are pathogenic but many have a useful role in the earth's ecology. Nitrogen-fixing bacteria, such as *Rhizobium, Frankia,* and cyanobacteria, are important in the nitrogen economy of the earth. Their ability to fix nitrogen, a necessary element for living organisms and one in short supply in natural substrates, makes them useful symbionts. Bacteria may play a greater role than is now realized in symbioses such as lichens, *Hydra-Chlorella,* and *Paramecium.* Undoubtedly, the importance of bacteria in symbiotic systems will be recognized as new and refined studies are conducted. The critical role that bacteria play as part of an animal's microbiota has been strikingly illustrated by studies on germ-free animals, which do not develop normally and die prematurely.

Photosynthetic protoctists such as algae appear to be ideal symbionts because they can manufacture food. In symbioses, these food-producing organisms provide nutrients for heterotrophic animals and fungi. Although such symbioses have occurred many times, the number of different photosynthetic protoctists involved is surprisingly low compared to the number of heterotrophs. For example, thousands of fungi form lichens but with only a few dozen types of algae. Similarly, in marine symbioses, many invertebrates contain symbiotic algae but the algae are mostly strains of one type of dinoflagellate. The driving force in the development of symbiotic associations between heterotroph and autotroph is the heterotroph, since it has the most to gain from symbiotic involvement. Associations between autotrophs and heterotrophs are difficult to classify. Many of the associations appear to be mutualistic because there usually is an exchange of metabolites between the symbionts. On closer examination of these symbiotic systems, however, we find that some of them, such as lichens and mycorrhizas, may better be considered as forms of controlled parasitism in which the heterotroph exploits the photobiont, or photosynthetic partner.

There are many symbiotic fungi and most of them are pathogens that cause virulent diseases of plants and animals. Fungi even parasitize other fungi. These mycoparasites are common and include different groups of fungi. Marine fungi form close associations, called *mycophycobioses,* with marine algae. Such associations probably have enabled ancient marine organisms to colonize the terrestrial environment. Presumably, the fungi protected the more sensitive algae from drying and helped them obtain water and minerals from the ground.

Most terrestrial plants form symbioses with mycorrhizal fungi. These root-fungus associations are vital for the well-being of plants. Many of the symbiotic associations that involve plants are parasitic, with the plants usually being hosts to fungi, monerans, protoctists, and animals. Some plants parasitize other plants. Most parasitic plants lack chlorophyll and obtain food from their hosts, but some, such as mistletoe, can manufacture food and also obtain nutrients from their hosts.

Animals, like plants, are hosts to many parasitic symbionts. They have developed a variety of responses, such as the immune response, to parasites and in some cases have developed mutualistic relationships that may have originated in parasitic forms. Insects and flowering plants have developed many close relationships during a long period of coevolution. In addition to parasitic symbioses, insects form mutualistic associations with plants. We consider pollination to be a type of symbiosis, as did de Bary. The contacts between insects and plants during pollination are transitory, but the repetitive nature of the contacts qualifies these associations as symbiotic. Insects also use plants as a place to deposit eggs or to hide from prey. Birds are com-

mon pollinators of plants and use them for shelter and nesting. Numerous symbioses between animals exist. Some birds commonly associate with large land mammals and not only remove insects from the animal's skin but also provide early warning of an impending threat from a predator. In the oceans, small fish and shrimp maintain cleaning stations and remove surface parasites from larger fish that periodically visit the stations.

C. Classification of Symbioses

Mortimer P. Starr has developed a classification of symbiotic associations in order to standardize the many conflicting terms that have been used to describe different symbioses. The following is a modification of Starr's system.

1. LOCATION OF THE SYMBIONTS. A symbiont may be inside or outside another symbiont. An *endosymbiont* is one that resides inside, whereas an *ectosymbiont* resides outside a host organism. The distinction between inside and outside may not be clear. We restrict the term *endosymbiont*[1] to those organisms that reside within the cells of other organisms. Thus, the extracellular microbes of the digestive tracts of animals are ectosymbionts.

Technically, very few symbionts lie within the host cytoplasm. Most endosymbionts are surrounded by the host plasma membrane, which separates them from the cytoplasm. Therefore, in a cellular sense, the symbionts are outside the host cell.

2. PERSISTENCE OF THE SYMBIOSIS. Most symbioses are persistent; that is, the symbionts remain together for a long time or their contacts are frequent. Associations in which one partner is an endosymbiont are usually the most persistent type. The symbionts remain together through all stages of their life cycles. Many endosymbionts have an arrested life cycle and generally remain in the vegetative stage. In some symbioses, the contacts between the partners are intermittent, as in the cleaning symbiosis of marine fishes or the pollination of flowers by insects and birds. In cases of parasitism, the sym-

---

[1] John O. Corliss has proposed that the word *xenosome* be used for endosymbionts.

biosis persists throughout the host's life, the duration of which depends on the virulence of the parasite.

3. DEPENDENCE ON THE SYMBIOSIS. *Obligate symbionts* are so highly adapted to a symbiotic existence that they cannot live outside it. *Facultative symbionts* can also live in the free-living condition. Sometimes, it is hard to tell whether a symbiont is obligate or facultative. For example, a symbiont may exist free-living but in specialized niches or in small populations that are difficult to identify. Algae such as *Trebouxia* are highly adapted to the lichen symbiosis but scattered colonies of these algae have been found free-living. Are such colonies truly nonsymbiotic, or are they established by individuals that escape from the symbiosis? Obligate symbionts depend on the symbiosis for nutrients. A symbiosis that involves two organisms, at least one of which obtains nutrients from the other, is called a *biotrophic symbiosis.* If one of the symbionts dies and the other uses it as a source of nutrients, the association is called a *necrotrophic* symbiosis. Organisms may obtain physical protection from a symbiotic association or receive some other benefit, such as the light produced by symbiotic bacteria in marine fishes.

4. SPECIFICITY OF THE SYMBIONTS. Symbionts may be highly specific to one organism, such as the bacterial symbionts of *Paramecium aurelia,* or they may associate with different organisms. The dinoflagellate *Symbiodinium microadriaticum* forms symbioses with many different marine animals. How specific an organism is may relate to the evolutionary stage of the symbiosis. Presumably, the more highly evolved a symbiosis is, the longer the symbionts have had to adapt to each other and the more specific is the association. An extreme example of specificity is seen in mitochondria, which are semi-independent, transformed organisms that cannot exist outside a eukaryotic cell.

5. SYMBIOTIC PRODUCTS. In some associations interaction of the symbionts results in the formation of new structures or chemical compounds. For example, a lichen thallus and many of its chemical compounds develop only as a result of the symbiosis. The fungal and algal partners cannot form these structures alone. Simi-

larly, legume root nodules and the red pigment leghemoglobin are products that develop from symbiotic associations of *Rhizobium* bacteria and legumes.

D. Who Studies Symbiosis?

Our research on the material to include in this book led us to many disciplines of biology. We found that each discipline has its own traditions and terminology and is concerned with specific aspects of symbiosis. There are many biological specialties within which active research on symbiosis is taking place.

Cell Biology and Biochemistry. The cell is a common habitat for intracellular symbionts such as viruses, bacteria, and protoctists. Special attention is being given to cell membrane structure and features of the membrane that change its permeability. Scientists are studying the physiological and morphological modifications of cell organelles and the integrated biochemical mechanisms between symbionts that result in the synthesis of unique metabolites and secondary products.

Immunology. Host immune response and mechanisms of antibody-antigen interactions are the focus of many studies. The concept of antigen sharing between host and symbiont (molecular mimicry) has fascinated immunologists interested in the evolution of the immune response. Immunologists are successfully using the techniques of hybridomas and recombinant DNA to mass-produce vaccines, interferon, and monoclonal antibodies.

Origin of Life and Evolutionary Biology. Lynn Margulis's theory of eukaryotic evolution through serial endosymbiosis of bacterial ancestors is accepted by most scientists. Molecular studies on chloroplasts and mitochondria support the relationship of these organelles to bacteria. Symbiotic associations are central to the evolution of herbivory in mammals, cellulose degradation by insects, nutrient uptake by land plants, and the rise of the angiosperms. The complexity of floral architecture among members of the orchid family demonstrates the coevolution of insects and flowering plants.

Genetics. Geneticists are concerned with the coevolution of the genetic mechanisms of host and symbionts and the nature of the equilibrium between the symbiont and host resistance. Increasingly, pathogenic disease is being recognized as a consequence of destabilized genetic equilibria between host and symbiont.

Ecology. Ecologists are exploring the population dynamics of symbionts and their hosts and the environmental conditions that influence the rapid growth of pathogenic symbionts. Mathematical models are being used to examine predator-prey interactions and plant-pollinator systems.

Epidemiology. Epidemiologists are studying patterns of disease prevalence and how disease is influenced by environmental factors. Scientists can predict disease outbreaks and identify factors that reduce the pathogenicity of a symbiosis.

Virology. Viruses interact with the host cell at the most elementary level. Even though viral infections that result in disease have received most of the attention from scientists, persistent, latent, and asymptomatic viruses have been discovered in a wide range of host organisms.

Bacteriology. Bacterial symbioses involving prokaryotes such as *Agrobacterium* and *Rhizobium* have provided new insights into the molecular and genetic mechanisms of cell regulation. Pathogenic bacteria cause diseases in humans, animals, and plants. Scientists are discovering secondary symbionts of pathogenic bacteria and evaluating their possible use as biological control agents.

Mycology. Fungal symbionts range from facultative to obligate in their mode of nutrition. Management of fungal diseases of crop plants is a continuing priority of modern agriculture. The role of some soil fungi in producing carcinogenic mycotoxins is being actively studied.

Plant Pathology. Scientists are studying plant diseases caused by biotic as well as abiotic pathogens. Interactions between higher plants and pathogenic symbionts such as viruses, bacteria, fungi, and nematodes are being examined.

Entomology. The role of insects as pathogens and carriers of pathogenic symbionts to humans and domesticated animals is being actively studied. Insects are the dominant class of animal life

on earth and they form many different types of symbiotic associations.

Parasitology. This broad subject deals with the study of protozoa, helminths, and arthropods. In recent years, advances have been made in the axenic culture and nutritional requirements of these symbionts, thereby providing new insights into the nature of host resistance and susceptibility. Helminthic symbionts include flukes, tapeworms, and roundworms, organisms that produce diseases in millions of people in the tropical and subtropical parts of the world. Host-symbiont relationships involving *Plasmodium* (a malarial parasite), *Schistosoma* (a blood fluke), *Trypanosoma* (sleeping sickness), and *Onchocerca* (a filarial worm causing blindness) have been targets of research by the World Health Organization and many other institutions. New techniques of molecular biology are being used to clone genes that code for surface antigens of malarial parasites in order to develop vaccines that will produce resistance to the disease.

Behavioral Biology. Social symbioses are common in the animal kingdom and are of interest to behavioral biologists. In these symbioses, interactions are considered in terms of a society rather than an individual. A complex taxonomy is associated with the different modes of social symbiosis.

E. Summary and Perspectives

Symbioses are interspecific associations that have played a significant role in the evolution of plants and animals and in shaping the earth's physical features. There are three main types of symbiosis: mutualism, parasitism, and commensalism. The boundary lines between these categories are not clear-cut and there are frequent transitions between them. The most common type of life style among living organisms is that of parasitism, and it appears, from an evolutionary point of view, that mutualism and commensalism arose from parasitism.

Parasites that cause disease are called pathogens and they may be localized or spread throughout a host. There are several types of host: intermediate host, final host, and reservoir host.

The most widely accepted classification of living organisms recognizes five kingdoms, Monera, Protoctista, Fungi, Plantae, and Animalia. The most poorly defined kingdom is the Protoctista, which includes single-cell as well as multicellular organisms that do not fit into the other kingdoms. All five kingdoms contain symbiotic associations. Viruses are also considered symbionts, since they have basic properties of living organisms and are commonly found in many life forms.

Because of the wide diversity of symbioses and the growing number of terms used to describe them, a classification system for symbiotic associations has been developed. This classification is based on several features: location of the symbionts, whether ectosymbionts or endosymbionts; persistence of the symbiosis; dependence on the symbiosis, whether obligate or facultative symbionts; and specificity of the symbionts. Biotrophic and necrotrophic symbionts are distinguished on the basis of whether nutrients are obtained from a living or dead partner.

Research on symbiosis is occurring in many disciplines of biology. There is a growing awareness of the fundamental importance of symbiosis as a unifying theme in biology, an awareness that organisms function only in relation to other organisms. Cellular and ecological approaches to mutualism are two vivid examples of expanding subject areas within the scope of symbiosis. The trend in the biological sciences to greater specialization should be balanced by broader studies of the relationships that exist among different associating organisms.

Review Questions

1. Define symbiosis.
2. Distinguish between intraspecific and interspecific associations.
3. List the three major types of symbiosis and define each type.
4. Distinguish between a parasite and a pathogen.
5. What are Koch's postulates of pathogenicity and what are the limitations of these criteria?
6. Describe the stages in the development of a disease.
7. What are microparasites and how do they differ from macroparasites?
8. What are reservoir hosts?
9. Distinguish between ectosymbiont and endosymbiont.

10. Why is it difficult to distinguish between an obligate and facultative symbiont?
11. Describe how a parasitic symbiosis might develop into one of mutualism.

Further Reading

Attenborough, D. 1979. Life on earth: A natural history. Little, Brown, Boston. 319 pp. (A beautifully illustrated survey of the major groups of living organisms on earth.)

Beringer, J. E., and A. W. B. Johnston. 1984. Concepts and terminology in plant-microbe interactions. In: Plant-microbe interactions: Molecular and genetic perspectives, vol. I, 3–18, ed. T. Kosuge and E. W. Nester. Macmillan, New York. (An introductory treatment of how viruses, bacteria, and fungi interact with plants.)

Corliss, J. O. 1984. The kingdom Protista and its 45 phyla. BioSystems 17:87–126.

Ewald, P. W. 1983. Host-parasite relations, vectors, and the evolution of disease severity. Ann. Rev. Ecol. Syst. 14:465–485.

Goff, L. J. 1982. Symbiosis and parasitism: Another viewpoint. BioScience 32:255–256. (Recommends that the word *symbiosis* be used in its original, broad sense.)

Lewin, R. A. 1982. Symbiosis and parasitism: Definitions and evaluations. BioScience 32:254–256. (Urges scientists to assign quantitative values to symbiotic interactions; recommends continuing the common practice of equating the term *symbiosis* with mutualism.)

Perry, N. 1983. Symbiosis: Close encounters of the natural kind. Blandford Press, Poole, Dorset. 128 pp. (A popular book with excellent color photographs.)

Sharp, J. D., Sr. 1980. Symbiosis: Creatures living together in nature. American Biology Teacher 42:290–292. (A popular article for biology teachers.)

Smith, D. C. 1979. From extracellular to intracellular: The establishment of a symbiosis. Proc. R. Soc. Lond. B. 204:115–130.

Smith, D. C., and A. Douglas. 1985. The biology of symbiosis. Edward Arnold, London (in press).

Starr, M. P. 1975. A generalized scheme for classifying organismic associations. In: Symbiosis: Symposia of the Soc. for Experimental Biology, no. 29, 1–20, ed. D. H. Jennings and D. L. Lee. Cambridge Univ. Press, Cambridge. (A detailed outline of the classification and terminology of symbiotic systems.)

Whittaker, R. H. 1969. New concepts of kingdoms of organisms. Science 163:150–159. (Outlines the five-kingdom classification of biological organisms.)

Bibliography

Caullery, M. 1952. Parasitism and Symbiosis. Sidgwick and Jackson, London. 340 pp.

Cheng, T. C. 1970. Symbiosis: Organisms living together. Pegasus, New York. 250 pp.

Corliss, J. O. 1985. Concept, definition, prevalence, and host-interactions of xenosomes (cytoplasmic and nuclear endosymbionts). J. Protozool. 32:373–376.

Dudley, R. H. 1965. Partners in nature. Funk and Wagnalls, New York. 192 pp. (Popular book.)

Gotto, R. V. 1969. Marine animals: Partnerships and other associations. American Elsevier, New York. 96 pp. (Popular book.)

Henry, S. M. (ed.). Symbiosis, Vols. 1 and 2. Academic Press, New York. 478 pp., 443 pp.

Horsfall, J. G., and E. B. Cowling (eds.). 1979. Plant disease: An advanced treatise. Vol. 4, How pathogens induce disease. Academic Press, New York. 466 pp.

Horsfall, J. G., and E. B. Cowling (eds.). 1980. Plant disease: An advanced treatise. Vol. 5, How plants defend themselves. Academic Press, New York. 534 pp.

Horsfall, J. G., and A. E. Dimond (eds.). 1960. Plant pathology. Vol. 2, The Pathogen. Academic Press, New York. 715 pp.

Kafatos, M., and A. G. Michalitsianas. 1984. Symbiotic stars. Sci. Am. 251 (July): 84–94.

Linskens, H. F., and J. Heslop-Harrison (eds.). 1984. Cellular interactions. Encyclopedia of plant physiology, new series, Vol. 17. Springer-Verlag, Berlin. 743 pp. (Many chapters on intercellular interactions in plants, fungi, and algae.)

Margulis, L., and K. V. Schwartz. 1982. Five kingdoms: An illustrated guide to the phyla of life on earth. W. H. Freeman, San Francisco. 338 pp.

Read, C. P. 1970. Parasitology and symbiology: An introduction. John Wiley, New York. 316 pp.

Rosenblum, L. A., and H. Moltz (eds.). 1983. Symbiosis in parent-offspring interactions. Plenum Press, New York. 284 pp.

Scott, G. D. 1969. Plant Symbiosis. Institute of Biology, Studies in Biology, no. 16. Edward Arnold, London. 58 pp. (Compares similarities and differences among various symbiotic systems.)

Trager, W. 1970. Symbiosis. Van Nostrand Reinhold, New York. 100 pp. (Brief treatment of some mutualistic symbioses.)

Van Beneden, P.-J. 1876. Animal parasites and messmates. D. Appleton, New York. 274 pp.

Whittaker, R. H. 1959. On the broad classification of organisms. Q. Rev. Biol. 34:210–226.

CHAPTER 2

Viral Symbiotic Associations

A. Introduction

Virology is a twentieth-century field of specialization whose scope is interdisciplinary. We are familiar with many human viral diseases, such as poliomyelitis, the common cold, mumps, and chicken pox. Today, we know of the existence of about 1,000 different viruses but only a few of these have been studied intensively. New viral-host relationships are constantly being discovered. Viruses have been detected in all major categories of life forms, in bacteria, fungi, protoctists, plants, and animals. In spite of their diversity, viruses have many features in common. For example:

Viruses are distinguished by their small size, ranging from 15 to 200 *nanometers*[1] (nm) in diameter. They are not visible under a light microscope. Advances in electron microscopy have greatly facilitated their identification and description.

Viruses have only one type of nucleic acid (DNA or RNA), which is surrounded by a protein coat called a *capsid*.

The largest viruses have an additional covering, a lipoprotein membrane called the *envelope*, of which the lipid component is derived from the host cell.

Viruses are acellular and lack organelles such as nuclei, ribosomes, and mitochondria.

Many viruses can be crystallized.

Viruses are intracellular symbionts and will not multiply outside living cells.

In many cases, a virus infects a host cell through a direct interface between it and the cell surface. In other cases, viruses are transmitted through an *intermediate host* or *vector*. Once inside the cell, viruses follow biological processes of enzymatic synthesis, control, and replication. When viral particles are fully assembled within the host cell, the cell dies and a new generation of viral particles is released. It is also possible for a virus to infect host cells without killing them. In this chapter, we describe disease-producing viruses as well as viral-host associations that are asymptomatic or in concert with the host.

The nucleic acids of viruses may be single or double stranded (ss or ds) DNA or RNA. Most viruses possess 6 to 10 genes; some viruses, such as smallpox, have 200 to 300 genes. *Viroids*, the smallest viruses, are naked, infectious nucleic acid molecules whose genetic information does not code for a single protein.

1. VIRUS STRUCTURE. On the basis of structural features, viruses are grouped into three major types (Fig. 2.1).

1. *Icosahedral viruses* consist of a protein shell, the *capsid*, that is made up of subunits called *capsomeres*. The capsid is often a twenty-sided structure that encloses the nucleic acid. Some icosahedral viruses are naked, such as the poliomyelitus virus and mycoviruses, and some have an outer envelope, such as herpes and yellow fever viruses.

2. *Helical viruses* are rod-shaped or filamentous. They may be naked, such as tobacco mosaic virus and other plant viruses, or enveloped, such as influenza viruses and viruses that cause mumps and measles.

3. *Bacteriophages* are viruses that infect bacteria and often have a complex structure composed of a head, tail, end plate, and fibrils.

[1] one nanometer = one millionth of a millimeter

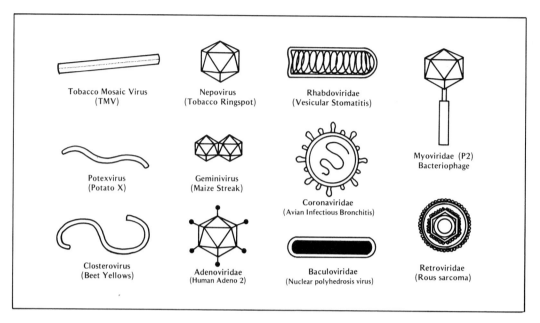

Figure 2-1. *Types of viral particles and their structural features.* (Adapted from Matthews, R. E. F. 1982. Classification and nomenclature of viruses. Intervirol. 17: 1–199.)

2. CLASSIFICATION OF VIRUSES. The classification of viruses is confusing to a beginning student because the system is a hybrid of several traditional approaches. Animal virologists generally have used a system modeled after Linnaeus's zoological classification, with families, genera, and species. Plant virologists have classified viruses according to their hosts.

In recent years, several attempts have been made to establish order in the comparative study of viruses. The International Committee on Taxonomy of Viruses (1982) has developed an outline of virus classification and nomenclature (Table 2.1).

Viruses may be investigated by examining infected host cells, by observing the effect of viral growth on a cell culture, or by measuring the antibody response that a virus evokes in a host animal. Much has been learned about viruses from their cytopathic effects on hosts and from electron microscopic observations of host cells, which reveal the structure of viral particles. Viruses occur in all life forms. In many naturally occuring viral-host symbioses, there are no recognizable cytopathic effects and such infections are called *persistent.*

3. VIRUS LIFE CYCLE. A typical virus life cycle includes phases of adsorption, infection, incubation, and maturation. *Adsorption* involves the attachment of viruses to a host cell. Specific regions on the host cell surface serve as *virus receptors.* Viral adsorption involves weak forces such as ionic or hydrogen bonds. In addition, there is an interplay at the receptor sites (Fig. 2.2). The *infection* stage occurs after the virus has penetrated the host membrane and entered the cytoplasm. Some viruses inject their nucleic acid into a host cell, leaving their protein coats behind as empty shells. Other viruses enter a cell with their protein coats intact. After a whole virus has entered a cell, the viral nucleic acid must be freed from its protein coat before the virus can replicate and initiate viral protein synthesis. Viral replication depends on the activity of a specific enzyme replicase. Viral messenger RNA directs the synthesis of protein. *Incubation* is a period of viral-directed metabolic activities that occur at the expense of the host cell. This phase is also known as the *latent period. Maturation* is the phase during which newly synthesized nucleic acid molecules and protein subunits assemble into *virions* (virus particles). Virions accumulate in either the cytoplasm or nucleus as inclusions. Mature virions may then be released following lysis of the host cell or may remain in a resting stage until the cell dies.

Table 2-1. *Outline of virus classification*

Classification	Group	Examples
ssRNA, enveloped	Togaviridae	Yellow fever virus
	Coronaviridae	Avian bronchitis virus
	Retroviridae	Cancer viruses
	Rhabdoviridae	Rabies virus
	Paramyxoviridae	Mumps and Measles viruses
	Orthomyxoviridae	Influenza and Flu viruses
ssRNA, nonenveloped	Phyto-picornaviridae	Tobacco mosaic viruses
		Tobacco necrosis virus
		Potato X virus
		Alfalfa mosaic virus
	Picornaviridae	Poliomyelitis virus
		Common cold viruses
ssDNA, nonenveloped	Parvoviridae	Papova viruses
		Hepatitis B virus
	Geminivirus	Corn streak viruses
dsRNA, nonenveloped	Reoviridae	Cytoplasmic polyhedral virus
		Wound tumor virus in plants
	Mycoviruses	Viruses in fungi
dsDNA, nonenveloped	Adenoviridae	Human respiratory viruses
	Iridoviridae	Iridescent viruses in insects
	Caulimovirus	Cauliflower mosaic virus
	Myoviridae	Polyhedral bacteriophages
dsDNA, enveloped	Baculoviridae	Nuclear polyhedral virus
	Herpesviridae	Herpes viruses
	Poxviridae	Smallpox virus
		Entomopoviruses

B. Viruses in Bacteria

Viruses in bacteria are commonly referred to as *bacteriophages* or *phages* and are distributed widely throughout the world of bacteria and cyanobacteria. Bacterial viruses offer an excellent model of virus-host interplay. The most widely studied phage system is that involving *T-even* (T2, T4, T6) phages and *Escherichia coli,* a common intestinal bacteria. Because of the ease of cultivating and the safety in handling, the bacteriophage-*E. coli* system has served as a model for numerous studies of basic viral mechanisms.

Bacteriophages have a tadpolelike shape with polyhedral heads attached to slender tails. T-even phages undergo several structural rearrangements when they attach to the wall of *E. coli.* At the tip of the base plate of each phage are fibrils, which establish initial contact at receptor sites on the host membrane. The tail fibrils fold in order to allow the base plate and its pins to become anchored to the bacterial cell surface. The viral tail sheath contracts and forces the tail tube into the cell wall. It is believed that lysozymic enzymes in the base plate are involved in the penetration of the bacterial cell wall. The final stage of penetration is the release of viral DNA into the bacterial cytoplasm (Fig. 2.3).

The *latent period* is the time between infection and the release of viral progeny from the host cell. Following infection of *E. coli* with T-even phage, the normal biochemical processes of the host stop and synthesis of bacterial nucleic acids and proteins no longer occurs. Phage DNA controls all of the cell's metabolic activity, causing the bacterial cell under attack to manufacture viral DNA. Messenger RNA is transcribed from the new viral DNA and viral specific proteins are produced inside the host cell using preexisting ribosomes and transfer RNA. The proteins form the building blocks for the phage head, contractile tail, and core, and for phage lysosomes. The sequence of protein synthesis is precisely controlled.

When an infected bacterial cell undergoes lysis, it releases 70 to 300 new phage particles. The entire infection process from initial contact

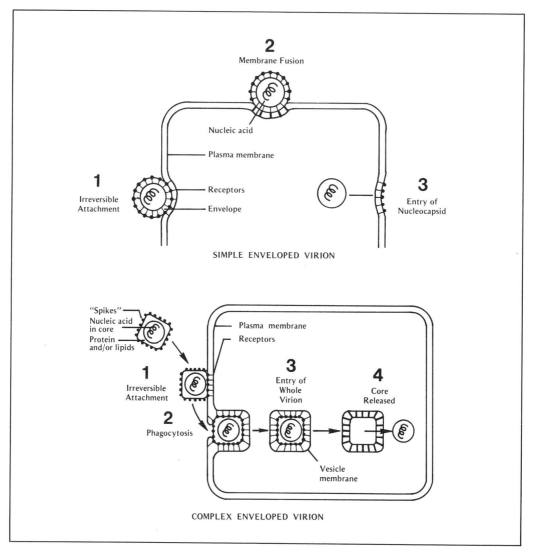

Figure 2-2. Interactions of viruses with the host cell membrane, including two modes of viral entry into cells. (Adapted from Simons, K., H. Garoff, and A. He-lenius. 1982. How an animal virus gets into and out of its host cell. Sci. Am. 246 (February): 58–66.)

to the release of new bacteriophages takes less than 30 minutes.

Some phages do not cause host cell lysis. The viral DNA of these temperate viruses becomes integrated into the circular chromosome of the host. This condition of the bacteriophage is called the *prophage state*. As a prophage, the virus DNA replicates along with the bacterial DNA and, through such an integration, transforms the genetic properties of the bacterium that determine virulence. This phenomenon is called *viral conversion*. The phage DNA as an inte-grated part of the bacterial chromosome serves as a *template* for replication of more viral DNA. Certain environmental stresses such as UV radiation, X-rays, or chemicals may activate the template phage and cause it to replicate. Lysis of the host cell results and a new generation of viral particles is ready to infect other host cells. A cell containing such a prophage is called *lysogenic* (Fig. 2.3). A virus that does not produce obvious signs of ill health in its host is called an orphan virus, since there is no disease name to highlight its existence.

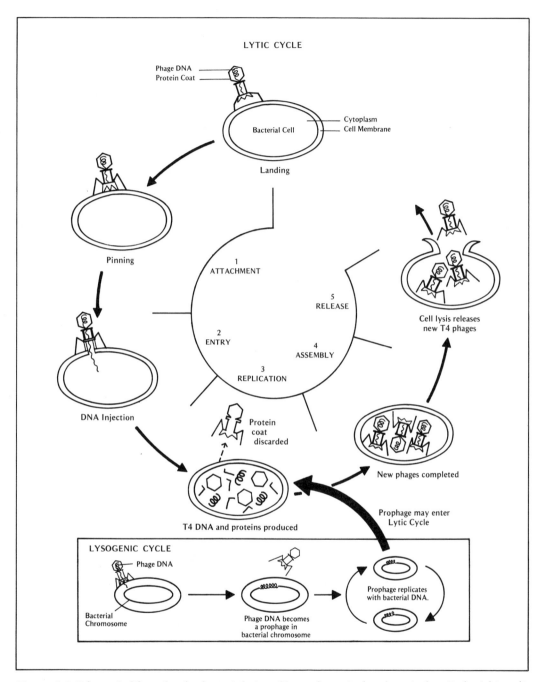

Figure 2-3. Schematic life cycle of a bacterial virus (T-even bacteriophage) on its host Escherichia coli. *Bacteriophage may enter lytic or lysogenic cycle.*

Viral conversions, which result in new genetic properties for the host, have been observed in *Clostridium botulinum, Corynebacterium diphtheriae, Salmonella* sp., and *Vibrio cholerae.* In *Salmonella,* bacteriophages called *epsilon phages* were the first viruses known to alter the *antigenic* properties of host bacteria. The cholera pathogen, *V. cholerae,* though rare in Western countries, threatens the rest of the world with sporadic outbreaks. These sudden outbreaks of the disease have been linked to the conversions of surface antigens induced by phages. Significant discoveries have been made on the nature of proteinaceous toxins of *C. diphtheriae* and *C. botulinum.* It has been confirmed that these toxins are the products of prophages residing on the host chromosome. Currently, scientists think that several toxins previously believed to be of bacterial origin may in fact be the products of phage genes.

Phages have also been known for their property of carrying along host DNA and introducing bacterial genes into new hosts upon infection. Such a phenomenon is known as *transduction.* In *Micrococcus aureus,* genes for antibiotic resistance have been transduced.

Phage lambda (λ) and other phages can establish stable lysogenic relationships. Phage λ inserts its DNA into a bacterial chromosome, and when the host cell divides, the prophage DNA replicates and then segregates. Phage λ represses the genes for its transcription, thereby preventing disruption of the host cell. *Lysogeny* is a stable relationship between bacteria and virus and is believed to be a common occurrence in nature. As a prophage, the virus does minimum damage to the host, and thus its presence may go undetected. For example, it took more than 20 years to detect the K-12 strain of *E. coli,* which carries phage λ. There is usually one viral genome per host chromosome and it is faithfully transmitted to all descendant bacterial cells. Isolates of almost all bacterial species have proven, on close examination, to be lysogenic to one or more phages. Genetic analysis of phage λ has shown it to have sophisticated mechanisms of biochemical regulation.

Phage P2 resembles phage λ, since it too, after its insertion into bacterial DNA, shuts off most of the viral genome by producing a repressor protein. Phage P1 is different, since it does not insert itself into the bacterial chromosome, but rather perpetuates itself as an independent plasmid. *P1 phage* also has repressor genes like phage λ and phage P2.

A lysogenic bacterium is immune to lysis by relatives of its prophage. This immunity is a new property of the bacterium. Some phages can alter host responses to other unrelated phages. Such a phenomenon is known as *prophage interference.* Immunity and modifying effects are examples of new properties conferred by the genome of a prophage onto its host bacterium.

Cyanobacteria contain viruses that are similar to *T-phages* in morphology and life cycle but are not genetically related. Such viruses are called *cyanophages* and they may inhibit growth of the infected cyanobacteria. In this respect, cyanophages may have a useful role, since "blooms" or population explosions of some cyanobacteria can be a nuisance in ponds, lakes, and reservoirs.

Viruses have been reported in several actinobacteria such as *Actinomyces bovis* and *Nocardia farcinia.* The viruses are called *actinophages* and they are abundant in soil. The virus resembles a T-phage but differs in the structure of its tail piece and in lacking a base plate and contractile fibers.

C. Viruses in Animals

1. INSECTS. Invertebrates harbor large numbers of a wide range of virus types. Most of our information on virus infection of this vast assemblage of animals comes from reports of viruses in insects.

Most viral infections of insects, under natural conditions, are subclinical or latent, and they are often difficult to detect. Insects may have evolved an efficient mechanism that prevents viral infections from reaching pathogenic levels. Under ecological stresses, however, these viral infections may be expressed, a situation that often results in host death. Below are some examples of well-established virus-insect symbioses.

a. Nuclear Polyhedrosis Viruses. Polyhedrosis viruses form polyhedral bodies in either the nucleus or the cytoplasm of infected insect cells. The spruce sawfly was first observed in 1930 infesting some 2,000 square miles of spruce forest in eastern Canada. By 1938, the insect had spread to an area of 12,000 square miles. It is believed that at that time the nuclear polyhedrosis virus infected and multiplied within these insects, and in four years the sawfly population decreased noticeably. Such viruses also infect caterpillars of butterflies and moths, and larvae of bees and flies. The rod-shaped virus, which contains DNA, multiplies within the host cell nucleus, which enlarges, and this ultimately causes the cell to burst. Infected larvae often cease feeding and move, *en masse,* to different areas of the plant. For example, infected nun moth larvae gather, in a characteristic manner, on tree tops where they change color, become fragile, and die when their epidermis ruptures and virus inclusion bodies are released. The polyhedral inclusion bodies contain DNA virus particles, which are enclosed either singly or in bundles within a crystalline protein. Because of their size, up to 15 μm in diameter, and their ability to refract light, inclusion bodies are easily seen under a light microscope. Upon liberation, the virus is disseminated to neighboring plants by rain and wind. When ingested by the larvae, the viruses migrate from the intestine to susceptible tissues in the hemocoel. Nuclear polyhedrosis viruses are also believed to be transmitted by means of the eggs of adult insects.

b. Cytoplasmic Polyhedrosis Viruses. Many cytoplasmic polyhedrosis viruses have been noted in butterfly and moth caterpillars. Two noteworthy examples of insects that have cytoplasmic polyhedroses are the tent caterpillar moth, *Malacosoma fragili,* and the gypsy moth, *Parthepria dispar.* Particles of these viruses are enclosed in polyhedral protein crystals, which appear in the cytoplasm of the infected cells and are clearly visible under the light microscope. The RNA-containing viruses are spherical (more accurately, they have a twenty-sided symmetry) and are restricted to the epithelial cells of the intestine. Infected larvae develop slowly and have disproportionately large heads and long bristles. In the advanced stage of infection, the infected cells disintegrate and release viral particles into the intestinal lumen, from which they are voided in feces. Viruses of this group can be transmitted by the insect's eggs.

c. Granulosis Viruses. These viruses are similar to nuclear polyhedral viruses but occur only in nuclei of cells of butterflies and moths. The infected tissues contain millions of granules, each containing one or two DNA-containing, rod-shaped, viral particles. The granulosis viruses can survive in soil for up to two years.

d. Pox Viruses. Pox viruses have a worldwide distribution and occur in several orders of insects such as Coleoptera, Diptera, Lepidoptera, and Orthoptera. The DNA virus has a complex lipoprotein outer envelope and develops in the host cell cytoplasm. Some pox viruses have large, spindle-shaped protein crystals.

e. Iridescent Viruses. One of the best known viral infections of insects involves the larvae of the crane fly. These larvae have a blue-green iridescence from the crystalline arrays of viral particles in their tissues. The virus also occurs in beetles, mosquitoes, and midges. Most insects acquire the virus orally during feeding. The virus, in addition to DNA, contains lipids and proteins. It resides in the host cytoplasm, and in the advanced stage of infection the virus may account for as much as one-quarter of the insect's weight. This is the highest proportion known for an animal viral infection. Iridescent viruses have been used as a biological control for mosquitoes.

f. Sigma Viruses. Sigma viruses are often present in some strains of *Drosophila* and can be detected by the sensitivity of infected flies to carbon dioxide. Infected flies become permanently paralyzed and die when they are exposed to concentrations of carbon dioxide that merely anesthetize uninfected flies. The virus is transmitted from one generation to the next. Some authorities question whether sigma is a virus at all.

2. OTHER INVERTEBRATES. During the past two decades, viruses have been discovered in shrimps, crabs, daphnids, isopods, oysters, octopods and squids, annelids, nematodes, flukes, hydra, and sponges.

Viruses in shrimps and crabs have been investigated in some detail. The hepatopancreatic epithelial cells of several species of shrimp harbor viral particles. Infections occur in the nuclei and can readily be detected by the presence of large intranuclear inclusion bodies. The shrimp hosts show little sign of the infection except for the loss of absorptive and secretory functions of the infected cells. Stress factors may play an important role in the development of advanced stages of viral-induced pathologies in shrimp.

A number of different viruses or viruslike particles have been reported in blue crabs. Herpeslike virus particles, *rhabdovirus,* and *reolike viruses* occur in blood cells and gill epithelia. Pathogenic effects, when pronounced, include the loss of hemolymph clotting ability, discoloration of the exoskeleton, and inhibition of feeding and molting.

Various oyster species have been examined for viral infections. A herpeslike virus was associated with increased mortality of the American oyster in Maine. The virus produces intranuclear inclusion bodies. Ecological stress, such as elevated water temperature caused by effluent from the cooling systems of power plants, is believed to be an important factor in increased viral-correlated mortality in the oysters.

Our knowledge of viral relationships in non-insect invertebrates is still in a state of infancy. From the examples discussed, it is clear that virus-animal interactions are widespread in nature. It is tempting to speculate on the role natural hosts play as reservoirs or vectors of viruses that are of economic and public health concern. The mechanism of cellular immunity and disease defense in nonvertebrate hosts requires attention from scientists in order to develop additional models of virus-host interactions.

3. VERTEBRATES. Viral disease in a vertebrate host results directly from viral replication cycles within the host. If many cells of an organ are infected by a virus, the biological malfunction of such an organ becomes noticeable and may ultimately lead to death of the host. Some viruses are limited to specific types of tissues; for example, influenza viruses replicate only in cells of the respiratory tissues. Other viruses such as smallpox can multiply in a wide range of tissues such as those of skin, lung, and other internal organs. A number of RNA viruses cause a total and rapid inhibition of host protein synthesis because the viral RNA interferes with the synthesis of host messenger RNA. The inhibitory effect of viruses on DNA synthesis is more gradual. During the infection process, membrane formation is stimulated along with production of dehydrogenase enzymes that are needed for ATP production. Most viruses in vertebrates produce noticeable cytopathic effects in host cells. These effects include:

Hyperplasia, or cellular proliferation; e.g., *herpes simplex virus*

Polykaryocytosis, or the formation of multinucleate structures known as syncytia; e.g., *measles virus*

Inclusion body formation in the cytoplasm and nucleus; e.g., *pox virus* and *herpes virus*

Malignant transformation, or the formation of tumors; e.g., *papova virus*

Vacuolation of either cytoplasm, e.g., SV 40; or the nucleus, e.g., *pig pox virus*

Necrosis, or the collapse of infected cells; e.g., *polio virus*

In *latent* viruses, such as *retroviruses,* the viral genome is in a suppressed or inactive form within the host cell.

Hyperplasia is the characteristic cytopathic effect of oncogenic viruses. Rapid cell proliferation produces tumorous masses of cells in the body, as in the case of *Rous sarcoma* in chickens. Latency is receiving considerable attention from researchers, especially in relation to tumor formation. The role of latent viruses in oncogenic lesions is currently being examined in a number of cases. *Reverse transcriptase,* or RNA-dependent DNA polymerase, facilitates the formation of DNA from the RNA genome of a virus, thus allowing the virus to become integrated in the host cell genome.

Some viruses become attached to the membranes of red blood cells and cause them to clump together, a condition known as *hemogglutination.* Scientists can distinguish some animal viruses from others by their ability to cause hemogglutination.

Death of a host cell from viral infection is believed to result from the release of degradative enzymes from the lysosome. At least 50 different kinds of lysosomic enzymes have been identified. Scientists believe that virus-induced changes in the lysosomic membrane, rather than any substances produced by the virus, are primarily responsible for lysis of the host cell.

a. Interferons. Interferons are glycoprotein molecules that are secreted by host cells in response to infection with DNA and RNA viruses. Most vertebrate species produce their own unique type of interferon. Even different tissues in the body produce different interferons. For example, interferon produced by human leukocytes is effective only against viruses that infect leukocytes. Interferons are detected within 24 hours of viral invasion and their concentration begins to decline a few days later. Interferon produced by an infected cell may constitute its primary defensive response until the host can produce antibodies. Although all types of nucleic acids injected into a host cell can stimulate interferon production, the most significant response is obtained with double-stranded (ds) RNA. Interferon does not affect viral metabolism in the host cell, but when neighboring healthy cells are treated with interferon, they become resistant to viral infection. The interferon molecules bind with the cell membrane and inhibit synthesis of viral nucleic acid and protein by inducing the host cell to produce two enzymes, phosphodiesterase and protein kinase. Interferons are effective against both viral and nonviral tumors. Host cells produce interferon molecules in exceedingly small quantities. At the present time, several biotechnology companies are attempting to clone human interferon genes in order to mass-produce this compound. Large amounts of this substance will allow studies to be made on its effectiveness against human cancers.

b. Somatic Cell Fusion. The measles virus, avian New Castle disease virus, and the Sendai virus cause cell fusion. The membranes of cells infected with these viruses become so altered that the cells fuse together. The mechanism of cell fusion is not clear, but a *fusion factor* consisting of a single glycosylated polypeptide has been identified. Virus-induced cell fusion has provided scientists with a powerful research tool for studies of the cell membrane, for introducing substances into cells, and for mapping genes. When cells of two different species fuse, the hybrid cell contains the chromosomes of both species in a single nucleus. As the hybrid cell multiplies, most of the chromosomes of one of the parent cells are lost. Hybrid cells containing only one chromosome of the target species are then examined for genetic expressions. Many human gene loci have been identified by human-mouse somatic cell hybridizations. Sendai virus is most frequently used in cell fusion studies.

c. Prions. Prions are infectious protein particles that do not appear to have nucleic acids. *Scrapie,* a nervous disorder of sheep and goats, and *kuru*

disease of the human nervous system are known to be caused by prions. Scientists suspect that *Alzheimer's disease,* the commonest form of human senile dementia, may also be caused by prions. Prions seem to represent the so-called *slow viruses* because they require long incubation periods during which the patient or host animal shows no disease symptoms. Once the disease manifests itself, it progressively worsens and eventually leads to death of the host. Prion molecules have a molecular weight of 50,000, which makes them about 100 times smaller than the smallest viruses. The discovery of prions by Stanley Prusiner in 1978 has extended the frontiers of virology and has raised questions on the evolutionary significance of nucleic acids and proteins and on the concept of infection.

4. CANCERS: CELLULAR TRANSFORMATION. In recent years, there have been major breakthroughs in our understanding of the basic mechanisms involved in the development of cancers and tumors. A tumor is a large mass of cancer cells, all of them having descended from a normal *founder cell.* Some inborn or external stimulus alters the cell, causing it to divide and eventually produce billions of cancer cells. Cancer genes, or *oncogenes,* have been discovered in the chromosomes of tumor cells. These genes become activated during the transformation of a normal founder cell into a cancerous cell. Thus, a cancer cell is one whose oncogenes have been activated by mutation or viral agents.

Cancer cells exhibit over 100 distinctive features in tissue culture. The most common features include:

loss of cellular territoriality, or the inability of cells to form a monolayer; instead, cancer cells grow over each other and form a tumor;
high rate of sugar import, mostly for use in cellular anaerobic metabolism; and
change in the immunological properties of the cell membrane.

By using different experimental approaches, researchers have confirmed that oncogenes are DNA segments. In addition, *proto-oncogenes,* which are slightly altered normal genes, have been discovered. The precursor of the *bladder-carcinoma oncogene* is a proto-oncogene that is part of the normal human DNA genome. Other human tumor oncogenes have also been traced to corresponding proto-oncogenes. Why these slightly altered genes are maintained in the human genome is not known. Does an organism carry the seeds of its own destruction?

Retroviruses contain RNA instead of DNA and some have a single proto-oncogene that can induce an oncogene. Recent studies have demonstrated that the proto-oncogenes of retroviruses originated from host cells. These genes are carried by retroviruses to new host cells. As of 1984, over 17 cellular proto-oncogenes have been described, each being associated with an oncogene and a particular retrovirus.

In summary, oncogenes can be activated by mutations or by retroviruses that carry proto-oncogenes. Bladder carcinoma is activated by mutations in humans and by retroviruses in rats. Proto-oncogene involvements have been confirmed for human carcinomas of the colon, lung, bladder, and pancreas, as well as for some sarcomas.

Acquired immune deficiency syndrome (AIDS) came to the attention of scientists in the early 1980s. The prevalence of this disease among homosexual men and users of intravenous drugs indicated its infectious nature. The disease is characterized by a progressive loss of helper-inducer T cells (T4 cells). In 1984, a retrovirus called human T-lymphotropic virus type III (HTLV-III) was identified as the causal agent of AIDS. The virus has been isolated from the patient's peripheral blood, semen, saliva, and tears. Retrovirus infections are known to persist throughout an animal's life. The AIDS virus has been isolated from patients months and years after a serological test has established the presence of viral antigens. The infected individuals may remain asymptomatic for periods up to four years. Approximately 13,000 cases of AIDS were identified in the United States by August 1985, and epidemiologists have projected that within a year 12,000 new cases will be diagnosed. The disease is increasingly being recognized as the leading cause of premature mortality when measured by years of potential life lost.

In recent years, modern experimental techniques have increased our understanding of animal-viral symbioses. In spite of all efforts, however, it has yet to be established how animal viruses enter a host cell, replicate, and undergo assembly. The polio virus, which is related to the foot-and-mouth disease virus and hepatitis A virus, is the best studied of the animal viral pathogens. The polio virus was the first of the animal viruses to (a) be grown in tissue-cultured cells; (b) have a vaccine developed against it; (c) undergo chemical and physical analyses of its RNA and protein subunit and their assembly; and (d) have its life cycle worked out within infected cells. Structural analysis of the polio virus, which was completed in 1985, has provided scientists with new opportunities to identify the antigenic sites on virus particles and to describe the relationship between capsid structure and host and tissue specificity. It is anticipated that within a few years new insights into the molecular mechanisms of viral pathogenesis will be discovered.

D. Mycoviruses: Viral Inhabitants of Fungi

Fungal virology is a relatively new interdisciplinary science. Mycoviruses share many common features, including the following. (a) Mycoviruses occur naturally and are endogenous to the host cell not only during replication but also during transmission. (b) Most mycoviruses have genomes that consist of double-stranded RNA (dsRNA). (c) Many cause asymptomatic infections and can be described as *latent* and *persistent.*

Mycoviruses occur in all of the major groups of filamentous fungi. Some examples of mycoviruses follow.

Agaricus bisporus viruses are a complex of six different types that cause degenerative diseases of cultivated mushrooms. These viruses are transmitted through hyphal fusions and are also carried by spores of the fungus.

Viruses of *Penicillium,* including species that produce the antibiotic penicillin, of *Gaeumannomyces graminis,* a causal agent of take-all disease in cereals, of *Helminthosporium maydis,* a causal agent of corn blight, and of *Helminthosporium victoriae,* which causes blight of oats.

Killer system viruses in the yeast *Saccharomyces cerevisiae* and in the corn smut fungus *Ustilago maydis.*

Agaricus bisporus viruses have been studied for their virulence and pathogenicity. Disease severity is correlated with the concentration of viral particles. Disease symptoms include deformities of the mushrooms because of degeneration of infected mycelium. The virus accumulates in the vegetative mycelium as well as in the fruiting body and spores. In spores, the virus particles occur in the cytoplasm; in the vegetative mycelium they accumulate in the vacuoles in crystalline arrays.

The yeast *Saccharomyces cerevisiae* consists of several strains. One of the strains (killer) produces a toxin that kills the cells of other strains (sensitive). The killer system in this yeast was earlier thought to have been a genetic phenomenon with three phenotypic expressions: killer, sensitive, and neutral. It is now believed that a virus is closely integrated with the yeast cells. The infected host cells contain segments of double-stranded RNA, which represent the viral genome and contain the determinants for toxins and immunity in the killer system. In the yeast, more than 20 nuclear genes may be required to regulate the dsRNA segment, which is maintained as a wild type allele. The viral genome is frequently lost through mutation and the daughter cells then become sensitive to the killer strains.

A killer system also exists in *Ustilago maydis,* where extra dsRNA genomes have been noted in killer and immune strains of the fungus.

Several viruses that infect plants are transmitted by fungi. It has been suggested that fungi may be alternate hosts for plant viruses. Tobacco mosaic viruslike particles have been found in fungi such as rusts and powdery mildews. Virulence of plant pathogenic fungi may be altered because of their mycoviral symbionts. For example, a mycoviral infection reduces the pathogenicity of *Gaeumannomyces graminis,* which causes take-all disease in cereals. Mycoviruses have been considered as a possible means of biological control of pathogenic fungi.

Mycoviruses offer an excellent model for understanding the molecular basis of the evolution of virus-host relationships. Mycoviruses produce stable associations with their host. Viral latency benefits host survival, whereas viral persistence benefits the virus. In this mutualistic symbiosis, the virus gains a more efficient means of transmission when the integrity of a host system is maintained. There is a close interaction between host genes and fungal viruses. Viral replication in fungi is controlled in such a way that its rate does not exceed the rate of host cell division. Host-mediated transmission of mycoviruses may sometimes lead to viral degeneration. For example, many viral particles in mushrooms lack nucleic acids.

E. Viruses in Algae

Viruses and viruslike particles have been reported in a number of algae. A virus capable of degrading the nuclei of a green alga, *Uronema gigas,* was first described in 1972. The virus consists of a large polyhedron capsid bounded by an outer membrane and double-stranded DNA, and some viral particles have a prominent tail. Algae in the 2 to 4 cell stage of division are the most susceptible to viral infection. How the virus penetrates the thick wall of the algal cell is not known. Viral DNA may be transmitted from one generation to the next without a breakdown of the host cells.

Viruslike particles have been observed in the nucleus of the green alga *Cylindrocapsa* sp. The particles are polyhedral and have a distinct membrane. The green alga *Chara corollina* was experimentally injected with a virus isolated from a diseased *Chara* and yellowing and necrosis developed on the inoculated alga. The virus was rod-shaped and appeared similar to tobacco mosaic virus particles.

F. Viruses in Plants

The "breaking of colors" in tulips, a condition that is produced by a virus, was beautifully illustrated in the still-life oil paintings by Dutch artists in the early seventeenth century. The streaking effect produced on infected flowers was much prized in Holland. In the heyday of 'tulipomania,' farmers traded away cattle, grain, and cheese for the much desired bulbs. One story tells of a girl who was given an infected bulb as part of her dowry. Infected tulips have been known for over 400 years. It was not until 1892, however, that a Russian scientist, Dimitri Ivanovski, showed in the tobacco plant the existence of a disease agent that could pass through a filter capable of holding back the smallest known bacteria. In 1898, a Dutch biologist, Martinus Beijerinck, confirmed Ivanovski's observation and described the infectious agent as *contagium vivum fluidum,* meaning "contagious living fluid." Since 1900, scientists have described over 600 plant disorders in which the causal agent is suspected to be a virus.

Viral infections in plants vary from clearing of leaf veins and chlorotic mottling to severe necrosis. Symptoms of a viral disease may be generalized throughout the plant or localized. A common sign of a virus infection in plants is retarded growth, which results in stunted plants, lowered productivity, and leaf deformation. Tumors are characteristic of several viral infections, notably, the *wound tumor of clover* and *swollen shoot of cacao*. The mechanism of tumor formation resulting from wound tumor virus infection is unknown. Viral plant infections vary widely. At one end of the spectrum are virulent viral infections that result in necrotic lesions—the cells die so rapidly that they are unable to pass on viral particles to neighboring cells. At the other end are *asymptomatic* or *latent infections,* in which the cells are not seriously damaged.

Most plant viruses are spread by fungi, nematodes, and arthropods such as leaf hoppers and aphids. Leaf hoppers, because they feed on phloem tissue, are vectors of viruses that become systemic throughout the plant. This type of viral infection is usually called *yellows.* Aphids feed on superficial parenchyma cells and are vectors of mosaic viruses. Most plant viruses produce inclusion bodies, which may be amorphous or crystalline and reside either in the cytoplasm or the nucleus. These inclusions, in most instances, consist of aggregates of viral particles. Plant inclusion bodies, which are similar to those of insect viruses, are large structures and easily seen under a light microscope. Plant cells infected with tobacco mosaic virus contain two kinds of inclusions, hexagonal crystals and irregular masses called X bodies.

Over 95% of all plant viruses possess single-stranded RNA (ssRNA). *Tobacco mosaic virus* (TMV) has single-stranded RNA and is one of the best studied of the plant viruses. In 1946 Wendell Stanley was awarded a Nobel Prize for crystallizing TMV particles, and in 1955 Heinz Fraenkel-Conrat showed that the RNA of TMV was infectious. When inoculated into plants, TMV-RNA initiates synthesis of more viral RNA and also of viral protein subunits, which, together with the RNA, assemble into typical rods of TMV (Fig. 2.4). An example of a double-stranded RNA (dsRNA) virus is the *wound tumor virus.* Recent discoveries have confirmed the existence of DNA plant viruses such as the cauliflower mosaic virus (dsDNA) and the bean mosaic virus *(ssDNA).*

Many plant viruses carry more than one RNA strand in each viral particle. There are 4 strands of RNA in alfalfa mosaic and cucumber mosaic viruses. Viral infection and the production of new viruses require the genetic information of 3 of the 4 RNA strands. Cowpea mosaic, tobacco ringspot, and tobacco mosaic viruses have 2 RNA strands and both are necessary for viral reproduction.

Gemini viruses and *cauliflower mosaic viruses* are the only plant viruses that contain DNA genomes. *Gemini viruses,* such as *bean mosaic virus* and *corn streak virus,* occur chiefly in the tropics and form viral aggregates in the nuclei of plant cells. The use of DNA viruses as vectors of beneficial genes into plants is being studied.

Associated with the tobacco necrosis virus (TNV) is a smaller unrelated virus, the *satellite virus.* The satellite virus is an obligate symbiont and requires the presence of TNV, the host virus, for its replication. The nature of the dependency is not known. The satellite virus is not needed by the tobacco necrosis virus, but when present it modifies the infectious nature of the TNV virus. Some virologists predict that similar satellite viruses will be discovered among bacteriophages and animal viruses.

Although virus-infected plants often recover and new growth may be completely free of symptoms, the virus still remains in the plant. Such a phenomenon is called *acquired immunity.* Some viruses can develop intracellular symbiotic associations in both plant and animal cells. These viruses have become well adapted to a "double life" of molecular biosynthesis in two distant life forms. Many plant viruses multiply in the tissues of their insect vectors. Leaf hoppers, for example, are well-known hosts of many systemic plant viruses. Development of the wound tumor virus in a leaf hopper's intestinal filter chamber and in its hemolymph has been well documented. The ability of insects to transmit a plant virus is genetically controlled.

G. Summary and Perspectives

Viruses are the smallest of all symbionts and they occur intracellularly in all forms of life. Some viruses kill their host cells, and others may live for years in concert with their hosts, often conferring on them new genetic properties. There are DNA viruses and RNA viruses, retroviruses, satellite viruses, viroids, and viruslike particles such as prions. Bacterial viruses, or bacteriophages, carry DNA from one host into other hosts. Insect viruses have been studied more closely than viruses of other animals. Some scientists believe that invertebrates may be natural reservoirs of viruses that infect hu-

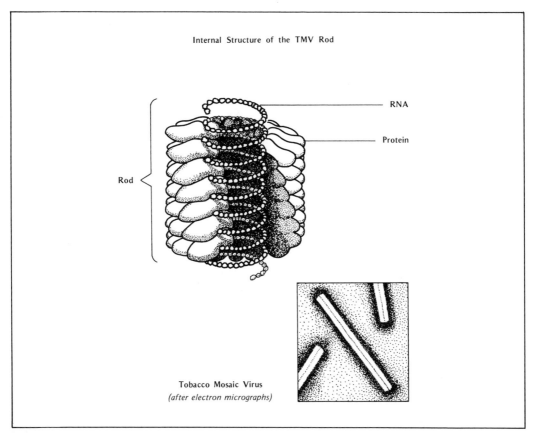

Internal Structure of the TMV Rod

RNA

Protein

Rod

Tobacco Mosaic Virus
(after electron micrographs)

Figure 2-4. *Morphology and detailed organization of the protein subunits and nucleic acid of tobacco mosaic virus.* (Adapted from Fraenkel-Conrat, H., and P. C. Kimball. 1982. Virology. Prentice-Hall, Englewood Cliffs, New Jersey.)

mans and their livestock. Viruses cause a variety of cytopathic effects in vertebrate hosts, including cancers. Vertebrate host cells respond to viral infections by producing defensive compounds such as interferons and antibodies. Mycoviruses usually form stable associations with their host, whereas many plant viruses cause virulent infections of host plants.

Viruses are significant disease agents of plants and animals and are among the most difficult parasites to control by external methods. Viruses may also have a useful role in human affairs. Some viruses are used as biological control agents for various types of pests; others such as lambda phages are commonly used as vectors in recombinant DNA research. Viruses in hosts that belong to different kingdoms have common patterns of behavior in their life cycle. Viruses represent the ultimate stage in the evolutionary development of host-parasite relationships. Their extreme reduction in structure and their ability to integrate themselves into the genome of the host cell have parallels with mitochondria and chloroplasts as well as bacterial symbionts, such as kappa, which also have developed close relationships with the host genome. Some viruses incorporate themselves into the chromosomes of their host cells and become thereafter a part of the genetic makeup of the host. Satellite viruses modify the infectious ability of the TNV virus, showing that even viruses are not immune from parasitic agents. In a similar fashion it is becoming increasingly recognized that the traits of some organisms are the result of viruses that infect these organisms. For example, pathogenic virulence of certain bacteria and protoctists has been correlated to the presence of viral genomes in these organisms. Advances in structural analysis of capsid protein and sequencing of the animal viral genome are providing new insights into the basic nature of viral symbioses, enabling scientists to discover how a virus enters, replicates, and assembles inside a host cell and to identify the molecular mechanisms of viral pathogenesis. The presence of viruses in many different forms of life suggests that viruses may have a much greater role in symbiotic associations than previously realized.

Review Questions

1. What features do viruses have in common?
2. What are viroids?
3. Describe the major types of viruses according to their structure.
4. Describe the stages of a typical virus life cycle.
5. Describe the structure of a T4 phage.
6. What is the prophage stage?
7. How has the polio virus furthered our understanding of animal-virus symbioses?
8. Explain the role of interferon in an infected cell.
9. Explain somatic cell fusion.
10. Name several diseases that may be caused by prions.
11. How do plant viruses differ from animal viruses?
12. Distinguish between oncogenes and proto-oncogenes.

Further Reading

Burke, D. C., and A. G. Morris (eds.). 1983. Interferons: From molecular biology to clinical applications. Thirty-fifth Symposium of the Soc. for General Microbiology. Cambridge Univ. Press, Cambridge. 337 pp.

Diener, T. O. 1981. Viroids. Sci. Am. 244 (January): 66–73.

Diener, T. O. 1982. Viroids and their interactions with host cells. Ann. Rev. Microbiol. 36:239–258. (A comprehensive review of "absolute parasites" of plants and possibly of animals.)

Lemke, P. A., and C. H. Nash. 1974. Fungal viruses. Bacteriological Reviews 38:29–56. (Comprehensive review of the structure, biochemistry, virulence, and transmission of fungal viruses.)

Mahy, B. W. J. 1985. Strategies of virus persistence. British Medical Bulletin 41:50–55. (Discusses how viruses persist in a population, individual, and cells. Antigenic variation is one of the principal devices by which some viruses exist successfully in host animals.)

Newton, A. A. 1982. Viruses: Exploiters or dependents of the host? Parasitology (UK) 85:189–216.

Payne, C. C. 1982. Insect viruses as control agents. Parasitology (UK) 84:35–77.

Prusiner, S. 1984. Prions. Sci. Am. 251 (October): 50–59. (An up-to-date review of prions by the scientist who discovered these infectious proteins.)

Reanney, D. C. 1982. The evolution of RNA viruses. Ann. Rev. Microbiol. 36:47–73. (Asks probing questions on the origin of RNA viruses, which occur in all forms of life.)

Sänger, H. L. 1984. Minimal infectious agents: The viroids. In: The Microbe, Part I, Viruses, 281–334, ed. B. W. J. Mahy and J. R. Pattison. Thirty-sixth Symposium of the Soc. for General Microbiology. Cambridge Univ. Press, Cambridge.

Simons, K., H. Garoff, and A. Helenius. 1982. How an animal virus gets into and out of its host cell. Sci. Am. 246 (February): 58–66.

Tinsley, T. 1984. Insect viruses as pesticides. In: Concepts in viral pathogenesis, 398–404, ed. A. Notkins and M. Oldstone. Springer-Verlag, New York.

Zinkernagel; R. M., H. Hengartner, and L. Stitz. 1985. On the role of viruses in evolution of immune responses. British Medical Bulletin 41:92–97. (Considers influences of viruses on the immune system and vice versa, often leading to biological balances that protect both the host and the virus. Role of HLA antigens and T cells are discussed.)

Bibliography

Bos, L. 1983. Introduction to plant virology. Longman, London and New York. 160 pp. (Provides a good historical account of discovery of viruses as infectious agents; also a good survey of important plant virus diseases.)

Diener, T. O. 1979. Viroids and viroid diseases. John Wiley, New York. 252 pp.

Fraenkel-Conrat, H., and C. Kimball. 1982. Virology. Prentice-Hall, Englewood Cliffs, New Jersey. 406 pp. (An excellent introductory text in virology; covers viruses of bacteria, plants, and animals.)

Fraenkel-Conrat, H., and R. R. Wanger (eds.). 1979. Comprehensive virology, Vol. 1–19. Plenum Press, New York.

Francki, R. I. B. 1985. Plant virus satellites. Ann. Rev. Microbiol. 39:151–174.

Joklik, W. K. 1980. Principles of animal virology. Appleton-Century-Crofts, New York. 373 pp.

Lemke, P. A. (ed.). 1979. Viruses and plasmids in fungi. Marcel Dekker, New York. 653 pp.

Luria, S. E., J. E. Darnell, D. Baltimore, and A.

Campbell. 1978. General virology. John Wiley, New York. 578 pp.

Mahy, B. W. J., and J. R. Pattison (eds.). 1984. The microbe, Part I, Viruses. Thirty-sixth Symposium of the Soc. for General Microbiology. Cambridge Univ. Press, Cambridge. 344 pp.

Matthews, R. E. F. 1985. Viral taxonomy for the non-virologist. Ann. Rev. Microbiol. 39:451–474. (An excellent review of recent developments in the taxonomy of viruses.)

Mims, C. A. (ed.). 1985. Virus immunity and pathogenesis. British Medical Bulletin 41:1–102.

Molitoris, H. P., M. Hollings, and H. A. Wood (eds.). 1979. Fungal viruses. Springer-Verlag, Berlin. 194 pp.

Notkins, A., and M. Oldstone (eds.). 1984. Concepts in viral pathogenesis. Springer-Verlag, New York. 409 pp.

Smith, K. M. 1976. Virus-insect relationships. Longman, London. 291 pp.

CHAPTER 3

Bacterial Associations

Bacteria as Symbionts of Other Bacteria, Protoctists, and Animals

A. General Characteristics of Prokaryotes

Bacteria are the dominant form of life on our planet in terms of numbers of individuals, and they are an indispensable part of human affairs. Many cause disease but some are also used to make vaccines and antibiotics, which combat disease. Bacteria play extremely important roles in the decomposition of waste material, in the production of medicines and food products such as cheese and yogurt, and in the new techniques of genetic engineering.

A major concept in biology is the distinction between organisms that are *noncellular* (viruses), *prokaryotic* (bacteria), and *eukaryotic* (fungi, protoctists, plants, and animals). Viruses have been discussed in chapter 2. In this chapter we focus on prokaryotes as symbionts. Prokaryotes are believed to be the simplest and most ancient *cellular* organisms.

Prokaryotes differ from eukaryotes in several ways (Fig. 3.1). The prokaryotic cell divides by binary fission or by budding, but never by mitosis. Each cell has only one chromosome and that lacks histone proteins. Sexuality occurs in some prokaryotes but it is distinctive in that the cytoplasm of the cells never fuse; only the chromosome, or part of it, passes from one cell to another. Eukaryotic cells divide by mitosis and meiosis, the latter being part of a sexual process that involves cell fusion and is well developed and regulated. The prokaryotic cell does not contain membrane-bound organelles such as

nuclei, mitochondria, and chloroplasts, and on the average is smaller than eukaryotic cells, usually less than 10 μm long. Ribosomes of prokaryotic cells have a sedimentation coefficient of 70S, compared to 80S for the ribosomes of eukaryotic cells. Motile prokaryotes have flagella that lack microtubules and consist of the protein flagellin. The cell walls of prokaryotes are made of peptidoglycans and never cellulose or chitin. Eukaryotic cells have membrane-bound nuclei, mitochondria, and , in the case of plants, chloroplasts. The cell walls of eukaryotes contain cellulose or chitin. Motile structures of eukaryotic cells consist of microtubules, which contain the protein tubulin, in a $9+2$ paired arrangement.[1]

B. Bacterial Symbionts of Bacteria

The phenomenon of bacteria parasitizing other bacteria has been studied for the past two decades. Most of the known bacterial symbionts of bacteria have been placed in the genus *Bdellovibrio* (*bdello* leech; *vibrio* comma-shaped). In 1962, Heinz Stolp, while searching for bacterial viruses from soil, observed *plaques* on cultures of the bacterium *Pseudomonas phaseolicola*. The plaques continued to grow in size, a feature not found in plaques caused by bacteriophage

---

[1] Margulis and Schwartz (1982) have revived an old word, *undulipodia*, for the motile structures of eukaryotes

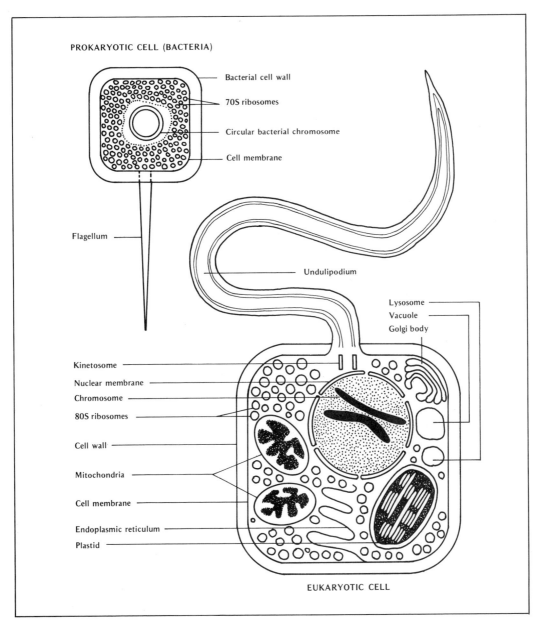

PROKARYOTIC CELL (BACTERIA)

Bacterial cell wall

70S ribosomes

Circular bacterial chromosome

Cell membrane

Flagellum

Undulipodium

Lysosome
Vacuole
Golgi body

Kinetosome

Nuclear membrane

Chromosome

80S ribosomes

Cell wall

Mitochondria

Cell membrane

Endoplasmic reticulum

Plastid

EUKARYOTIC CELL

Figure 3-1. Schematic comparison of prokaryotic and eukaryotic cells. (Adapted from Margulis, L., and K. V. Schwartz. 1982. Five kingdoms. W. H. Freeman, San Francisco.)

infection. Stolp examined the colonies with a phase contract microscope and observed small bacteria attacking the pseudomonads. Since this initial discovery, more than 300 research papers have been published on bdellovibrios.

Bdellovibrios are very small bacteria (1 to 2 μm long) whose natural habitat is the *periplasmic space* (i.e., space between the cell wall and plasma membrane) of other bacteria. Bdellovibrios attack sensitive bacteria, often 10 to 20 times larger than themselves, by penetrating their cell wall and growing and multiplying in the periplasmic space (Fig. 3.2). The symbiont breaks down the host's cellular constituents, and when the host cell dies, progeny of the symbiont are liberated. When bdellovibrios are first intro-

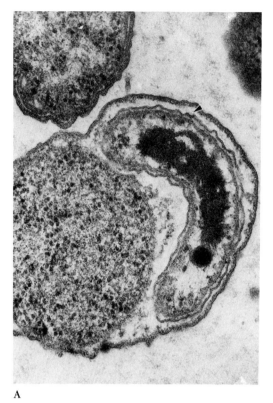

A

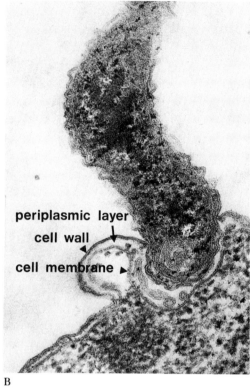

B

Figure 3-2. (A) Bdellovibrio *symbiont (arrow) in-side periplasmic space of host cell of* Escherichia coli. *(B) Middle stage of penetration of* Bdellovibrio bacteriovorus *into* E. coli. *Note invagination of host cell wall and membrane.* (Jeffrey C. Burnham, Medical College of Ohio.)

duced into a bacterial colony, they move quickly, about 100 cell lengths per second, and collide violently with a host bacterium. Just before colliding, the symbiont spins rapidly and within seconds after impact contact is made with the surface of the host cell. The end of the cell opposite the flagellum attaches to the host cell wall by means of tiny filaments. The symbiont penetrates the cell wall by making a hole, but exactly how it does this is not fully understood. Some scientists feel that the violent force of the collision or the spinning of the parasite, or both, punctures the host cell wall. Recent evidence, however, points to an enzymatic digestion of the outer membrane and the peptidoglycan layer of the host cell wall. Penetration is completed in 5 to 20 minutes. Young host cells are penetrated in less time than older or metabolically inactive cells. Following infection, the bdellovibrio establishes itself in the periplasmic space and grows in size. It then elongates into a large, he-

lical-shaped structure and divides to produce vibrio-shaped, motile individuals. The life cycle of a bdellovibrio is completed in one to three hours. The host cell, or part of it, swells into a globular body, the *bdelloplast,* because the cell wall is weakened by enzymes produced by the symbiont. Infected bacteria stop moving within seconds after a bdellovibrio attaches to their cells and their synthesis of nucleic acids and proteins halts following penetration. Permeability of the host cell membrane changes, allowing degraded products to leak into the periplasmic space. In the final stage of the infection, the host membrane breaks down and bdellovibrio progeny are liberated; the number of progeny per cell depends on the host species: for example, five for *Escherichia coli,* ten for *Pseudomonas flourescens,* and twenty for *Spirillum serpens.* The differences in the number of symbiont offspring per cell reflect the size and metabolic rate of the host and its ability to provide nutrients to the symbionts. Bdellovibrios are not true endosymbionts because they are outside the host cytoplasm.

All bdellovibrios are aerobic, Gram-negative bacteria. They possess *exoenzymes* that degrade nucleic acids, proteins, lipids, and peptidoglycan, and they are noted for their high rate of respiration, which is seven times that of *E. coli*.

In 1970, *bdellophages,* or viruses that attack *bdellovibrios,* were first noted. Bdellophages develop only in a sensitive bacterial host or in the presence of its extract. This is a triad, a unique compound symbiotic association involving a noncellular organism (the bacterial virus), a small prokaryote (the bdellovibrio), and a Gram-negative prokaryote (the host bacterium).

On the basis of nutritional characteristics, bdellovibrio strains are grouped into three categories.

Host-dependent strains require living or killed bacteria

Host-independent strains have lost the ability to parasitize bacteria but will grow heterotrophically on organic media.

Facultative-parasitic strains can multiply within living bacteria or grow heterotrophically on organic media without host bacteria.

Bdellovibrios attack only Gram-negative bacteria. Some strains attack a wide range of bacterial species, whereas others have a limited host range. Three species of *Bdellovibrio* are recognized on the basis of nucleic acid homology and DNA base composition analysis. The species are *B. bacteriovorus, B. starrii,* and *B. stolpii.* The predator-prey relationship between *Bdellovibrio* and its host cells is unique among the bacteria. *Bdellovibrio*-like bacteria have been observed in cyanobacteria and the green alga *Scenedesmus.* Other predatory bacteria include *Vampirococcus* and *Daptobacter.*

Recent studies have suggested that a mutualistic relationship exists between some nitrogen-fixing, blue-green bacteria and pseudomonads attached to the host filaments (Fig. 3.3). Nitrogen compounds excreted by the host are used by the attached bacteria, which in turn consume oxygen and liberate carbon dioxide when they respire organic compounds. Growth and nitrogen fixation of the host are stimulated by the reduced oxygen and increased CO_2 concentrations caused by the pseudomonads.

C. Bacterial Symbionts of Protoctists

1. *PARAMECIUM*. In 1865, Johannes Müller observed rod-shaped particles (bacteria) inside *Paramecium caudatum* and since then the literature on endosymbionts of ciliates has grown considerably. Bacteria have been observed in the micro- and macro-nuclei as well as the cytoplasm of protoctists. The bacterial symbionts of *Paramecium aurelia* have been studied the most thoroughly. The first symbiont to be discovered was kappa, a transformed bacterium that occurs in the cytoplasm of *P. aurelia.* Kappa particles were considered to be an example of cytoplasmic inheritance before their symbiotic nature was recognized.

All symbionts of *Paramecium* are rod-shaped, Gram-negative bacteria that cannot live outside their host. Thousands of symbiont particles may exist in one host cell. The bacteria have customarily been designated by Greek letters, but recently assigned to the genera *Caedibacter, Holospora, Lyticum, Pseudocaedibacter,* and *Tectibacter.* Some strains of the bacteria, called killers, are toxic to sensitive paramecia. Killer strains develop only when the bacterial symbionts contain phage or plasmids. Below is a list of some well-known symbionts of *P. aurelia.* All symbionts, except alpha, occur in the cytoplasm of the host.

Alpha *(Holospora caryophila)*. These appear in the macronucleus as short rods, crescents, or spirals, the shape depending on the growth rate of the host; a nonkiller.

Delta *(Tectibacter vulgaris)*. Generally nonmotile rods, which are often associated with other symbionts such as kappa and mu; a nonkiller.

Gamma *(Pseudocaedibacter minuta)*. Frequently appear as small doublets; a strong killer; death of sensitive paramecia results from vacuolization of the cell.

Kappa *(Caedibacter taeniospiralis)*. These bacteria vary in size. Some kappa particles have an inclusion body called "R," which is a tightly coiled proteinaceous ribbon that is refractile and shines brightly (Fig. 3.4). Kappas that have "R" bodies are called *bright* and are toxic, whereas *nonbright* kappas lack "R" bodies and are nontoxic. Bright forms of kappa lose their ability to reproduce. Nonbright kappas are infectious and can multiply in paramecia that have the genes to maintain

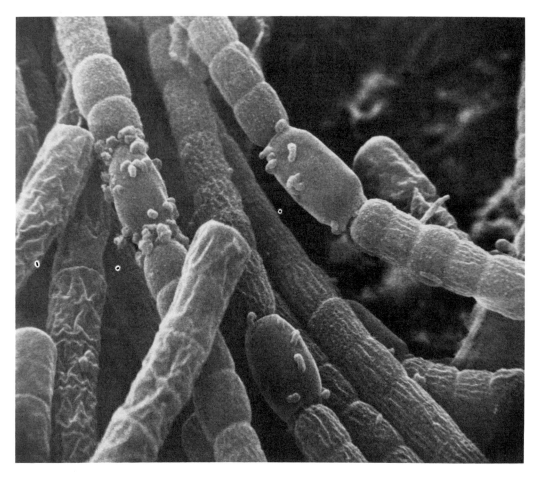

Figure 3-3. *Pseudomonad bacteria attached to heterocysts of* Aphanizomenon *sp.* (Hans W. Paerl, University of North Carolina.)

them. The formation of "R" bodies depends on the induction of defective bacteriophages or DNA plasmids that are present in non-bright kappa particles. The induction may occur spontaneously or result from stimuli such as ultraviolet light. Paramecia that contain bright kappa particles are called killer strains. Particles released by these strains may be ingested by sensitive paramecia, which are then killed by the toxic action of the particles. Symptoms of sensitive paramecia infected with killer particles, or even exposed to the medium in which a killer strain has grown, include spinning, vacuolization, paralysis, and rapid death. Most strains of paramecia are sensitive to toxic kappa particles. Killing is highly specific, each killer being resistant to

its own kappa symbiont but sensitive to other symbiont strains.

Lambda *(Lyticum flagellatum)*. Motile, large rods that are covered with flagella and cause rapid lysis of sensitive paramecia (Fig. 3.4).

Mu *(Pseudocaedibacter conjugatus)*. Mate killer; elongated rods whose killing action depends on the cell-to-cell contact that occurs during conjugation between mating paramecia. Particles of mu are transferred from killer to sensitive mate.

Sigma *(Lyticum sinuosum)*. The largest symbiont and a close relation of lambda; a curved rod and a rapid killer.

Another ciliate, *Euplotes aediculatus,* contains particles of the symbiont *Omikron.* The

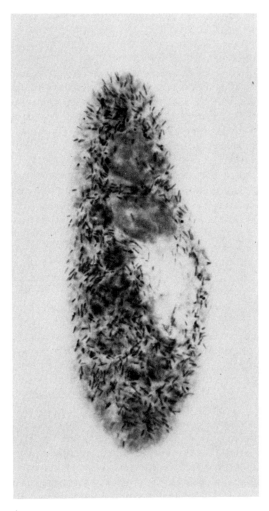

A

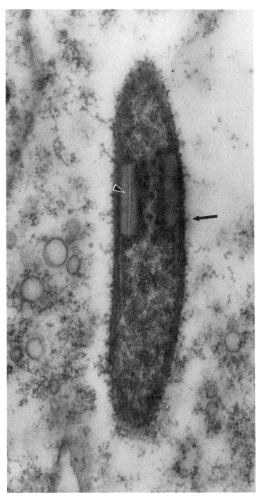

B

Figure 3-4. Bacterial symbionts of Paramecium. *(A)* Lyticum flagellatum *(lambda), endosymbiont of* Paramecium tetraurelia; *lambda appears as dark-staining rods in cytoplasm; osmium-lacto-orcein preparation. (B)* Caedibacter varicaedens *(kappa),* endosymbiont *of* Paramecium biaurelia. *Electron micrograph of a longitudinal section of a bright kappa particle (long arrow) with "R" body (short arrow) and spherical phages inside the coiled body.* (Louise B. Preer, Indiana University.)

symbiont, discovered in 1975, resembles a Gram-negative bacterium and is essential for the survival of the host.

2. AMOEBA. In 1966, Kwang W. Jeon noted a bacterial infection in a laboratory culture of *Amoeba proteus*. The bacteria initially killed the host amoebae but some amoebae became tolerant and survived. Within a few years, bacteria in the amoebae culture had become nonpatho-

genic. By 1971, the association had become so integrated that the amoebae could not survive without their symbionts. In 1979, Jeon estimated that each amoeba harbored an average of 42,000 bacteria. Furthermore, new amoebae could be induced to establish symbioses after about 200 cell generations. The bacteria appeared to be Gram-negative rods and they resembled *Escherichia coli*. In the amoebae, large numbers of symbionts become enclosed in membrane-bound vesicles of different sizes (Fig. 3.5). In order for any intracellular symbiont to become established, it must avoid destruction by the host cell. In addition, the division of the symbiont must be in balance with that of its host. Although free-living bacteria are commonly ingested and digested by *Amoeba proteus,* the symbiotic bacteria resist digestion. Jeon feels that the bacteria avoid lysosomal enzymes of the host because the vesicles that contain the symbionts do not fuse with lysosomes. How it is that this avoidance occurs is not known. Furthermore Jeon noted that within five years the number of bacteria per amoeba decreased from an average of 60,000 to 150,000 down to 42,000, suggesting that the host cell controlled the rate of symbiont multiplication. During this time there was also a decrease in adverse effects on the host amoebae. Both amoebae and bacterial symbionts undergo continuous changes in structure and physiology during the course of the symbiosis. A plasmid DNA in the bacterial symbiont may be responsible for the adaptive response.

The amoeba-bacteria association is an excellent model for studying the evolution of new symbiotic associations. Such an inducible symbiosis may be possible with other organisms.

D. Bacterial Symbionts of Animals

1. BACTERIAL LUMINESCENCE. Symbioses involving luminescent bacteria are very common in the marine environment. Light that is emitted by symbiotic bacteria, although of no direct benefit to the bacteria, is important for the host. Some marine fishes carry axenic cultures of luminescent bacteria in specially adapted light organs. Two genera, *Photobacterium* and *Vibrio (Beneckea),* contain all the known species of marine luminescent bacteria. Luminescent bacteria occur free-living in sea water, as plankton, and they also grow on the surface of dead marine animals. The light produced by the bacteria on dead animals attracts feeding organisms. Luminescent bacteria that are ingested by other organisms resist digestion and are expelled into the ocean and in this way are dispersed. Thus, bioluminescence may have an evolutionary selection value, since it enhances a symbiont's survival and propagation.

Biochemical pathways for bioluminescence are similar in most luminescent bacteria. In general, the enzyme luciferase mediates a reaction in which the pigment *luciferin* is oxidized to a product and light is emitted in the oxidation process. Another feature of the luminescent bacteria and fish symbiosis is·the regulation of the enzyme luciferase. Bacterial symbionts produce small amounts of compound called *autoinducers,* which must be present in high concentrations in order for the light-emitting system to function. Free-living bacteria produce low levels of autoinducers and therefore do not synthesize luciferase. When bacteria are concentrated in a host, however, autoinducers accumulate in sufficient quantity to generate the light-producing compounds.

Photobacterium and *Vibrio* are widely distributed in ocean waters. They live as ectosymbionts on teleost fish and as endosymbionts in squid and tunicates. They are heterotrophic on dead and decaying organisms, parasites of marine crustaceans, and commensals in the digestive tract or surface of marine fishes and crustaceans. *Vibrio* is the more cosmopolitan and nutritionally versatile genus. Luminescent bacteria are important symbionts in the alimentary canals of marine animals. The bacteria produce extracellular chitinase, which helps invertebrates such as mussels, scallops, and crabs digest chitin.

Some fish have light-emitting organs that contain large numbers of luminescent bacteria, all belonging to the genus *Photobacterium.* The bacteria receive nutrients from the host and are in a protected environment. Light produced by the bacteria frightens or diverts predators of the fish and also helps them to communicate with mates. Some teleost fish, such as the flashlight fish, harbor dense colonies of bacteria in specialized light organs. The bacteria produce light continuously but the fish controls light emissions by means of shutterlike structures on the organs. The angler fish has a modified dorsal-fin ray that contains a light organ at its tip. The organ contains luminescent bacteria and is dangled in front of the fish, like bait, to lure prey. The squid, *Loligo,* harbors bacteria in organs near its anus and when alarmed discharges luminescent secretions from these organs. The secretions distract or frighten other organisms and are effective substitutes, under dark conditions, for the black ink that the squid usually secretes. Recent studies have confirmed that tunicates house luminescent bacteria. The bacterial symbionts are believed to infect the tunicate eggs and in this way are passed on to new individuals. Luminescent bacteria are easily isolated from

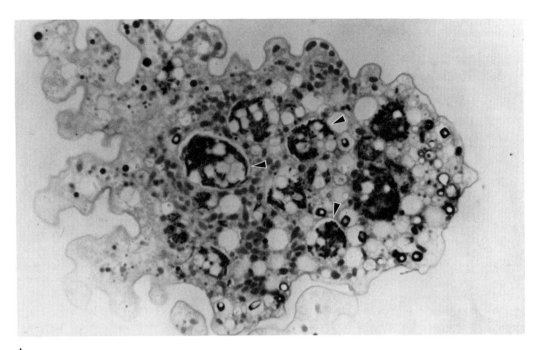

A

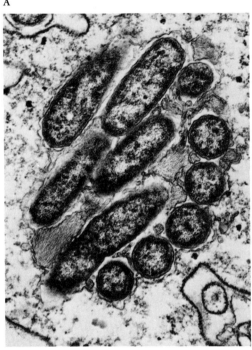

B

Figure 3-5. Symbiotic bacteria inside Amoeba proteus. *(A) An amoeba with symbiont-containing vesicles (arrows). (B) Electron micrograph of a medium-sized vesicle with bacteria.* (Kwang W. Jeon, University of Tennessee.)

the surface of dead marine animals. Meat of marine animals, when stored, will develop colonies of luminescent bacteria, which are harmless to man.

Freshwater and terrestrial animals such as mayflies, ants, millipedes, and caterpillars house luminescent bacteria. They also occur in the intestines of nematode larvae that attack caterpillars. After the larvae enter the host gut, they release bacteria into the caterpillar's hemolymph. The bacteria multiply and the caterpillar glows during the early stages of infection. The luminescence helps to disperse the nematodes by attracting organisms that feed on the caterpillars. The bacteria are pathogenic to the insect host but only in the presence of the nematodes. Similarly, the nematode cannot complete its life cycle without the bacteria.

2. BACTERIAL SYMBIONTS OF INSECTS. There are more species of insects than any other group of animals. Therefore, it is not surprising that there are numerous examples of symbiotic associations between insects and bacteria. The bacteria may be either extracellular or intracellular within the insect tissues.

Many insects, such as cockroaches, leaf hoppers, and aphids, have well-defined organs called *mycetomes.* A mycetome consists of *mycetocytes,* or highly specialized cells that contain symbionts. Mycetomes are usually in the body cavity mixed with fat tissues, although in some insects they occur in the intestines and Malpighian tubules.

Studies on insect symbioses during the past 15 years have used modern techniques of electron microscopy and biochemical analysis. Based on these studies, several generalizations can be made. (a) The primary intracellular symbionts belong to the order Eubacteriales and they vary in shape from rods to lobed vesicular bodies. (b) In most cases, the bacteria have two endogenous membranes and an outer, third membrane that is derived from the host. The cytoplasmic membrane (M1) encloses the bacterial cytoplasm and shows little specialization. A second membrane (M2) is believed to represent the lipopolysaccharide-lipoprotein layer, which is characteristic of Gram-negative bacteria. This structure is elastic and is involved in the formation of vesicles. Early in insect embryogenesis, the symbionts become surrounded by a third membrane (M3), which is of host origin. (c) All symbionts reproduce by binary fission. (d) The symbiont's ribosomes are significantly smaller than those found in the cytoplasm of the mycetocytes and about the same size as the ribosomes of free-living bacteria. (e) The endosymbionts possess DNA and highly ordered crystalline bodies. (f) Insect eggs become infected with bacteria, thus ensuring continued host association from one generation to another. (g) The symbionts excrete vitamins and nutrients that are used by the host.

Mycetocytes contain the usual organelles of a eukaryotic cell. The nucleus is generally enlarged, lobed, and polyploid. Unusual components of mycetocytes include transparent vacuoles and granular bodies. Transparent vacuoles are large, single-membrane structures that may contain glycogen granules. Granular bodies resemble peroxisomes in their fine structure. Associated with the mycetocytes are sheath cells, which in aphids form a small, thin layer of flattened cells around the mycetome. Sheath cells contain bacteria, called secondary symbionts, that are different from those in the mycetocytes. The bacteria are rod-shaped, more loosely bound by the host membrane (M3) than the primary symbionts of the mycetocytes, and have a cell wall similar to that of free-living bacteria.

Paul Buchner, an authority on insect symbioses, believed that the mycetocyte, or host cell, controls the rate of division of the primary symbionts. The number of dividing symbionts increases significantly just before embryogenesis and transovarial infection. Mycetocytes destroy some symbionts by lysosomal enzymes and in this way control the size of the bacterial population. Lysosomal breakdown of secondary symbionts is more pronounced and is a way of removing nonviable individuals or a means by which the host acquires nutrients from the symbionts, or both. The insect hemocoel is a hostile environment for primary mycetocytal symbionts, and they are rarely found in it except during a brief period of transovarial infection. Secondary symbionts, however, are often found in hemocytes, which may be an indication of their poor adaptation to the host.

In aphids, at a certain time during embryogenesis, mycetocytes release their bacteria, which then pass through the follicular membrane and enter the developing insect embryo. The symbionts lack the host-derived membrane (M3) during the infection process. In leaf hoppers, the process of symbiont release and infection of embryos involves two morphologically distinct forms of bacteria: (1) a nondividing form that has three membranes and is typical of those in mycetocytes, and (2) a dividing, extracellular form that has two membranes and lives only for a short time. Bacteria in the mycetocytes are transformed into infectious forms before they are released into the host hemolymph. The infectious forms multiply and enter a special host cell where they acquire their third membrane and assume the nondividing form. A few days later, the host cell disintegrates to liberate the symbionts, which are now ready to enter the developing insect embryo. From the above description, it is clear that the process of infecting embryo mycetocytes is complex and involves a number of intermediate steps, the significance of which has yet to be fully understood.

One of the more successful research efforts on this subject has been the production of *aposymbiotic* insects, that is, host individuals freed of mycetocytal symbionts. Such insects result when antibiotics that are added to insect feed inhibit division of the symbionts. Lysosomal enzyme treatments also have killed bacteria of cockroaches, aphids, and leaf hoppers and resulted in aposymbiotic strains of these insects. Mycetocytes have been isolated from insect hosts and grown in tissue culture medium. Mycetocytal bacteria contribute the following nutrients to the host: tryptophan and sulfur amino acids, lipids, and vitamin B complex.

Some scientists have suggested that the mycetocyte symbiosis may serve as a model for the study of the evolution of cell organelles. For example, the bacterial cell wall that is present in secondary symbionts indicates that they are not highly evolved. In contrast, the primary symbionts lack a cell wall and may represent an advanced evolutionary stage.

3. BACTERIAL SYMBIONTS OF RUMINANTS.
Herbivores live on a diet that is rich in cellulose. They can feed on plant tissues because they contain symbiotic bacteria and protozoa in specially adapted stomachs. Ruminants, such as deer, cattle, and sheep, have a four-chambered stomach (Fig. 3.6) and consume large quantities of vegetation at one time without much chewing. The ingested food moves to the first and second stomachs, the *rumen* and *reticulum,* respectively, and remains there for several hours. Watery saliva is added to soften the food. During most of the holding period, the food undergoes bacterial fermentation.

Ecologists often refer to a ruminant as a fermentation factory. The host animal, unable to enzymatically break down cellulose, is colonized by a mixture of microorganisms that live as symbionts in the rumen-reticulum part of the stomach. All ruminants, after they ingest food, relax and "chew their cud," a process, called *rumination,* that consists of regurgitating small quantities of food from the rumen into the mouth, where it is thoroughly chewed and mixed with saliva and bacteria. The well-chewed material is swallowed again and this time channeled into a third chamber of the stomach, the *omasum,* from which it is transferred to a fourth compartment, the *abomasum.* Posterior to the omasum is the digestive tract of a ruminant, which is similar to that of other vertebrates. During a twenty-four-hour period, a typical ruminant feeds for about eight hours, mostly in the morning and evening, and spends another eight hours in rumination.

The rumen, or fermentation chamber, has many unique characteristics. Its walls are semipermeable and selectively transport molecules to and from the blood of the animal. Volatile fatty acids, produced as a result of microbial fermentation, are absorbed into the blood. Bicarbonate ions from saliva are continuously being added to the rumen to maintain a pH of about

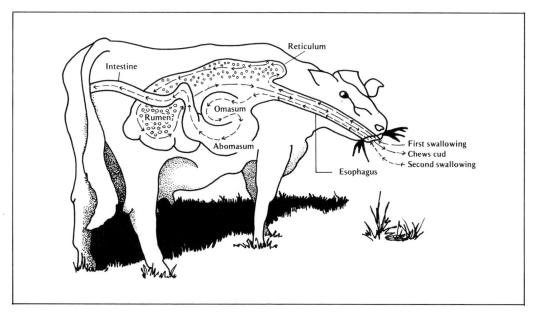

Figure 3-6. Diagram of the four-chambered stomach of a ruminant. Cellulolytic bacteria and pro- *tozoa live in the rumen-reticulum part of the stomach.*

BOX ESSAY

Life in the Alimentary Canal

The alimentary canals of most animals harbor large numbers of bacteria, several kinds of protozoa, and viruses. Many of these symbionts are neutral or beneficial, but some may be actually or potentially pathogenic. The large intestines of many vertebrates contain as high a bacterial concentration as that found anywhere in nature. The importance of intestinal bacterial symbionts can be studied experimentally by administering large doses of wide spectrum antibiotics to the host. This results in the partial sterilization of the alimentary canal. The antibiotic-treated animals can be compared with normal animals by measuring various biological parameters such as growth, maturity, and disease susceptibility. Sterile young may be obtained by Caesarean section from an animal treated with antibiotics and maintained in a germ-free environment.

Gnotobiology is the study of life in a germ-free state. In a gnotobiotic system, the sterile host can deliberately be inoculated with bacteria, or other symbionts, in order to study their effects. One of the major contributions from these studies has been the finding that bacteria of the vertebrate alimentary canal provide the host with vitamins. Rats treated with antibiotics develop symptoms of thiamine deficiency. Intestinal bacteria are the only source of vitamin K, which is necessary in host blood-clotting mechanisms. In addition, germ-free animals often have poorly developed antibody systems.

In humans, food and microbes enter the digestive system through the mouth. The microbes are dispersed throughout much of the intestine and many leave the body in the feces. Before birth, all humans are sterile. They become contaminated with microbes during the passage through the birth canal and through feeding. In breast-fed infants, the baby obtains pioneer microbes from the microbiota of the mother's skin. As the individual grows, a large number of symbionts, which include both aerobic and anaerobic organisms, become permanent residents in the intestine. *Escherichia coli* is the best-known resident symbiont of human intestines. Species of *Salmonella* and *Shigella* are also common. Anaerobic species of *Bacteroides* and *Clostridium* are the principal residents of the large intestine.

6.5. Peristalsis occurs at one to two minute intervals and mixes and moves the rumen contents. The rumen is almost totally free of oxygen and has a relatively constant temperature of about 39° C. Carbon dioxide and methane are the two principal gases of the rumen. Methane is a direct product of fermentation; carbon dioxide is produced from fermentation and by neutralization of bicarbonate ions.

Cellulolytic bacteria in the rumen degrade the cellulose of plant material into its constituent subunits, cellobiose and glucose. The sugar molecules are then fermented and compounds such as acetic acid, butyric acid, and propionic acid are produced and metabolized by the host. The ruminant converts propionic acid into carbohydrates such as lactose and glycogen. Rumen bacteria are strictly obligate anaerobes. Each milliliter of rumen contents contains between 10 to 100 billion bacteria. Some of the important cellulolytic bacteria include *Bacteroides succinogenes, Ruminococcus albus,* and *R. flavefaciens.* Important starch-digesting species include *Bacteroides amylophilus, Butyrivibrio fibrisolvens, Selenomonas ruminantium, Streptococcus bovis,* and *Succinomonas amylolytica.* Rumen symbionts are passed from generation to generation through salivary contamination of the young by nursing adults.

The waste products of some rumen bacteria serve as the raw material for other bacteria. For example, carbohydrate-fermenting bacteria produce hydrogen, carbon dioxide, and formate as waste products, which methanogenic bacteria use to produce methane. Similarly, plant proteins that enter the rumen are digested and fermented by bacteria and ammonia is liberated. Nearly all of the bacterial species in the rumen can use ammonia as a major source of nitrogen. Small amounts of amino acids are also produced and are used at once by the bacteria in protein metabolism.

Anaerobic ciliates are also an important component of the rumen. There are two major groups of ciliates in the rumen, the *holotrichs* and the *entodiniomorphs*. The holotrichs have cilia all over their cell surface and include the genera *Dasytricha* and *Isotricha* (Fig. 8.3). Holotrichs convert sugar into a type of starch that is chemically similar to plant starch. Entodiniomorphs have a firm pellicle that is often drawn out posteriorly into spines and cilia, and are represented by the genera *Diplodinium,* and *Entodinium.* Studies have shown that cellulose digestion by rumen protozoa is insignificant compared to that by bacteria.

Bacteria and protozoa in the rumen are beneficial to each other. Digestion of starch and proteins by the protozoa releases sugars and amino acids, which the bacteria use to grow and reproduce. The protozoa, in turn, ingest bacteria and use their protein as a source of amino acids. The protozoa in the rumen stimulate bacterial growth, which in turn increases bacterial ingestion by the protozoa. As a result, there is a greater turnover of bacterial carbon and nitrogen. The ruminant participates in this symbiosis by providing nutrients for the rumen symbionts and by regulating the physical and chemical conditions of its fermentation chamber. The ruminant meets its protein needs and also obtains many vitamins by digesting some of its rumen residents.

The rumen system is an excellent model for studying symbiotic associations under stable environmental conditions. On a practical basis, the rumen fermentation is of special significance to man. Cattle and other domesticated ruminants are efficient converters of the bioenergy of green plants into a source of animal protein, woolen fiber, and leather products.

In summary, ruminants benefit from their association with microbes by having the cellulose of the plant material that they eat broken down by the symbionts into usable compounds. Bacterial and protozoan symbionts interact in a mutually beneficial manner that results in increased rates of carbon and nitrogen turnover in the rumen ecosystem. Finally, ruminants obtain their dietary protein and vitamins by digesting some of the bacterial-protozoan cell mass. The strict environmental constraints of the rumen fermentation chamber allow only a few well-adapted species of bacteria and protozoa to survive.

4. BACTERIAL SYMBIONTS OF TUBE WORMS.

One of the most unusual habitats of a symbiotic association is the recently discovered deep-sea vents of the Pacific Ocean. Hot water rises from these vents as a result of water seeping through cracks in the ocean floor and being boiled by the underlying molten rock. The water has a high concentration of hydrogen sulfide. These vents are identical to the hot springs found on land, the only difference being that the marine vents are about two miles below the surface of the ocean. As surprising as it may seem, these remote, and seemingly inhospitable, habitats contain a rich and diverse group of organisms. One of the strangest members of this group is the red tube worm, *Riftia pachyptila* (phylum Pogonophora). Large colonies of this giant worm are attached to rocks along the path of the water that flows from the hydrothermal vents. The temperature of the water around the worms is about 10° C.

Pogonophorans live inside stiff, chitinous shells, which they secrete around themselves. Their body consists of a long trunk, a short anterior piece, and a short posterior piece, which anchors the animal to the tube and to the ground. Numerous long, cilia-bearing tentacles project from the anterior end of the animal. Pogonophorans that live in regions such as the Galápagos rift may be up to 3 meters long. They glide partially in and out of their tubes and move with the currents of water that flow from the vents. The animals lack a mouth and a digestive tract, which raises the obvious question of how they obtain nutrients.

Part of the body cavity of a tube worm, the *trophosome,* is highly vascularized and contains dense colonies of symbiotic Gram-negative bacteria, which are the source of nutrients for the worms. The bacteria are *chemoautotrophs,* which means that they can use chemical energy from inorganic compounds instead of light energy to make organic carbon compounds. The chemoautotrophic bacteria fix carbon dioxide using energy (ATP) and hydrogen ions that they obtain from the oxidation of hydrogen sulfide. Two important enzymes of the Calvin-Benson cycle, in which CO_2 is fixed in photosynthetic organisms, have been found in high concentrations in the trophosome. The enzymes are ribulose bisphosphate carboxylase and ribulose 5-phosphate kinase. The worms use the carbon products synthesized by the bacteria and they also use the bacteria directly for food. The well-developed circulatory system in the trophosome and the high oxygen-binding capacity of the blood hemoglobin are adaptations that ensure an adequate supply of O_2 and CO_2 to the bacteria.

Chemoautotrophic bacteria that oxidize sulfur compounds are common in the vent water as well as in the tube worms. These bacteria may also be the source of nutrients for other marine animals, such as clams and polychaetes, that live in these specialized habitats. The vent communities differ from other communities in that they do not depend on photosynthesis as the primary step in the food chain.

E. Summary and Perspectives

Bacteria are prokaryotes whose cells are fundamentally different in organization and chemistry from those of eukaryotes. Bacteria are symbionts in diverse organisms, from one-celled paramecia to humans. Bdellovibrios are parasitic bacteria that forceably enter other bacteria. Their mode of entry is being used as a model of how ancestral prokaryotes may have become colonized by symbiotic organisms.

Paramecium aurelia is host to different types of rod-shaped, Gram-negative bacteria such as kappa, strains of which are killers. Why organisms such as *P. aurelia* are more predisposed to form symbiotic unions than others is not clear. It may be the result of a deficiency in the defensive barriers of these organisms. In a similar fashion, most lichen associations are formed by fungi that belong to the class Ascomycetes and most of these fungi associate with only a few types of green algae belonging to the genus *Trebouxia*. Further, one dinoflagellate genus, *Symbiodinium,* is the photosynthetic symbiont of a vast array of different marine organisms.

Laboratory studies have shown that *Amoeba proteus* can establish artificial symbioses with bacteria within a few years. The artificially induced symbiosis between the bacteria and *A. proteus* illustrates how symbiotic associations may have evolved and suggests that even today symbiotic unions are being formed and dissolved continually in natural situations. Within five years, bacteria that began as pathogens of *A. proteus* assumed a role of organelles and became indispensable to their host. It is hard to distinguish between a symbiont that spends its entire life in a host cell and an organelle. Inducible symbioses are also possible with other organisms such as the fungi and algae of lichens, nitrogen-fixing bacteria and the roots of plants, and mycorrhizal fungi and tree roots.

Luminescent bacteria occur in different marine organisms and terrestrial insects and the light they produce may lure prey, attract feeding organisms, or frighten predators. The diversity of organisms that house luminescent bacteria suggests that bioluminescence has a selective value in terms of the evolution of organisms.

Bacteria are common in insects and are housed in specialized cells called mycetocytes, which form unique organs called mycetomes. The life cycles of bacterial symbionts inside insects is complex and not clearly understood. Aposymbiotic insects are ones that have been freed of mycetocytal symbionts, usually by means of antibiotics fed to the insects.

Ruminant animals contain large populations of cellulolytic bacteria in their stomachs along with various types of protozoa, which also digest cellulose. Rumen bacteria, in addition to their relationship with their host, also interact mutually with rumen protozoa.

Sulfur-oxidizing bacteria are symbionts of tube worms, which live near deep-sea vents in the Pacific Ocean. The bacteria synthesize carbon compounds, which are used by the host.

Review Questions

1. Distinguish between prokaryotic and eukaryotic cells.
2. Explain how bdellovibrio symbionts attack other bacteria.
3. Why are there so many different types of bacterial symbionts in *Paramecium aurelia?*
4. Why is the symbiosis of *Amoeba proteus* and bacteria a good model for studying the origin of new symbiotic systems?
5. How are luminescent bacteria useful to marine animals?
6. Distinguish between mycetomes and mycetocytes.
7. How does the mycetocyte symbiosis serve as

a model for studying the evolution of cell organelles?

8. Describe the types of organisms present in the rumen-reticulum part of the ruminant stomach and their symbiotic interactions.

9. Describe the role of the chemoautotrophic bacteria in marine tube worms.

Further Reading

Cavanaugh, C. M., S. L. Gardiner, M. L. Jones, H. W. Jannasch, and J. B. Waterbury. 1981. Prokaryotic cells in the hydrothermal vent tube worm *Riftia pachyptila* Jones: Possible chemoautotrophic symbionts. Science 213:340–341.

Coleman, G. S. 1975. The role of bacteria in the metabolism of rumen entodiniomorphid protozoa. In: Symbiosis: Symposia of the Soc. for Experimental Biology, no. 29, 533–558, ed. D. H. Jennings and D. L. Lee, Cambridge Univ Press, Cambridge.

Gibson, I. 1974. The endosymbionts of *Paramecium*. Crit. Rev. Microbiol. 3:243–273.

Görtz, H.-D. 1983. Endonuclear symbionts in ciliates. Int. Rev. Cytology, supplement 14:145–176.

Grassle, J. F. 1985. Hydrothermal vent animals: Distribution and biology. Science 229:713–717.

Hastings, J. W., and K. H. Nealson. 1977. Bacterial bioluminescence. Ann. Rev. Microbiol. 31:549–595.

Hastings, J. W., and K. H. Nealson. 1981. The symbiotic luminous bacteria. In: The prokaryotes, Vol. 2, 1332–1345, ed. M. P. Starr, H. Stolp, H. G. Trüper, A. Balows, and H. G. Schlegel. Springer-Verlag, Berlin.

Hobson, P. N., and R. J. Wallace. 1982. Microbial ecology and activities in the rumen. Crit. Rev. Microbiol. 12:165–225, 253–320.

Houk, E. J., and G. W. Griffiths. 1980. Intracellular symbiotes of the homoptera. Ann. Rev. Entomology 25:165–187.

Hungate, R. E. 1975. The rumen microbial ecosystem. Ann. Rev. Ecol. Syst. 6:39–66.

Jeon, K. W. 1983. Integration of bacterial endosymbionts in amoebae. Int. Rev. Cytology, supplement 14:29–47.

Mackowiak, P. A. 1983. Our microbial associates. Is the human body's normal flora a liability or an asset? Nat. Hist. 92 (April): 80–87. (Author examines the symbiosis of normal human microflora and fauna in terms of mutualism or parasitism.)

Moulder, J. W. 1985. Comparative biology of intracellular parasitism. Microbiol. Reviews 49:298–337. (A good review article on intracellular parasites.)

Preer, J. R., Jr., and L. B. Preer. 1984. Endosymbionts of protozoa. In: Bergey's manual of systematic bacteriology, Vol. 1, 795–811, ed. N. R. Krieg. Williams and Wilkins, Baltimore.

Reisser, W., R. Meier, H.-D. Görtz, and K. W. Jeon. 1985. Establishment, maintenance, and integration of mechanisms of endosymbionts in protozoa. J. Protozool. 32:383–390.

Starr, M. P. 1975. *Bdellovibrio* as symbionts: The associations of *Bdellovibrios* with other bacteria interpreted in terms of a generalized scheme for classifying organismic associations. In: Symbiosis. Symposia of the Soc. for Experimental Biology, no. 29, 93–124, ed. D. H. Jennings and D. L. Lee. Cambridge Univ. Press, Cambridge.

Stolp, H. 1979. Interaction between *Bdellovibrio* and its host cell. Proc. R. Soc. Lond. B. 204:211–217.

Troyer, K. 1984. Microbes, herbivory, and the evolution of social behavior. J. Theor. Biol. 106:157–169.

Varon, M. 1974. The bdellophage three-membered parasitic system. Crit. Rev. Microbiol. 3:221–241.

Wolin, M. 1979. The rumen fermentation: A model for microbial interactions in anaerobic ecosystems. In: Advances in microbial ecology, Vol. 3, 49–77, ed. M. Alexander. Plenum, New York.

Wolin, M. J. 1981. Fermentation in the rumen and human large intestine. Science 213: 1463–1468.

Bibliography

Agrios, G. N. 1978. Plant pathology. Academic Press, New York. 703 pp.

Buchner, P. 1965. Endosymbiosis of animals with plant microorganisms. Interscience, New York. 901 pp.

Guerrero, R. C. Pedrós-Alió, I. Esteve, J. Mas, D. Chase, and L. Margulis. 1986. Predatory prokaryotes: Predation and primary consumption evolved in bacteria. Proc. Nat. Acad. Sci. 83:2138–2142.

Jannasch, H. W. 1985. The chemosynthetic support of life and the microbial diversity at deep-sea hydrothermal vents. Proc. R. Soc. Lond. B. 225:277–297.

Jeon, K. W. (ed.). 1983. Intracellular symbiosis. Int. Rev. Cytology, supplement 14. 379 pp.

Kelley, D. P., and N. G. Carr (eds.). 1984. The microbe Part II, Prokaryotes and Eukaryotes. Thirty-sixth Symposium of the Soc. for General Microbiology. Cambridge Univ. Press, Cambridge. 349 pp.

Krieg, N. R. (ed.). 1984. Bergey's manual of systematic bacteriology, Vol. 1. William and Wilkins, Baltimore. 964 pp.

Lysenko, O. 1985. Non-spore forming bacteria

pathogenic to insects: Incidence and mechanism. Ann. Rev. Microbiol. 39:637–695. (A good review of bacterial associations with insect pests.)

Margulis, L., and K.V. Schwartz. 1982. Five kingdoms: An illustrated guide to the phyla of life on earth. W. H. Freeman, San Francisco. 338 pp.

Nealson, K., D. Cohn, G. Leisman, and B. Tebo. 1981. Coevolution of luminous bacteria and their eukaryotic hosts. In: Origins and evolution of eukaryotic intracellular organelles, 76–91, Ed. J. F. Fredrick. New York Academy of Sciences, New York.

Preer, L. B. 1981. Prokaryotic symbionts of *Paramecium*. In: The prokaryotes, Vol. 2, 2127–2136, ed. M. P. Starr, H. Stolp, H. G. Trüper, A. Balows, and H. G. Schelegel. Springer-Verlag, Berlin.

Preer, L. B., and J. R. Preer, Jr. 1977. Inheritance of infectious elements. In: Cell biology: A comprehensive treatise, Vol. 1, 319–373, ed. L. Goldstein and D. M. Prescott. Academic Press, New York.

Starr, M. P., H. Stolp, H. G. Trüper, A. Balows, and H. G. Schlegel (eds.). 1981. The prokaryotes: A handbook of habitats, isolation, and identification of bacteria. Vol's. 1 and 2. Springer-Verlag, Berlin. 2,271 pp.

CHAPTER 4

Bacterial Associations

Bacteria as Symbionts of Plants

A. Introduction

Much of life on earth depends on green plants. Their photosynthesis transforms the radiant energy of the sun into the chemical energy of organic molecules that other organisms use, directly or indirectly. Plants in nature do not exist in biological isolation, but rather are associated with different types of organisms. Bacteria are a major resident community on and around plants, and obtain energy and nutrients either through decomposition of dead plant material or from the exudates of living plants. Some bacteria form close mutualistic associations with plants, such as legumes, and in some tropical plant families bacteria are regular inhabitants of nodules on the margins of leaves. Over 200 bacterial species cause plant diseases such as soft rots, leaf spots, yellowing, blights, vascular wilts, and galls. In this chapter, we describe some well-known examples of bacterial symbionts of plants.

B. Agrobacterium *and* Crown Gall Disease

Crown gall disease affects approximately 100 families of plants and causes millions of dollars worth of damage each year to agricultural and ornamental plants. The disease is caused by *Agrobacterium tumefaciens,* a rod-shaped, Gram-negative bacterium that is closely related to the *Rhizobium* bacteria that form nodules on the roots of legumes. *Agrobacterium tumefaciens* lives in the soil and infects plants through wounds that the plants suffer either during seed germination or from insects or nematodes. In-

fection is usually near the crown of a plant, that is, the junction between the stem and the roots. The bacteria stimulate the plant cells around the wound to multiply and form a tumorlike growth, or gall (Fig. 4.1). The disease generally affects only wounded plants because it is the new cells developing around the wound that are most susceptible to bacterial infection. Galls generally appear several weeks after a plant is infected and may be small, a millimeter or more in diameter, or large, weighing close to 100 lbs., depending on the type of plant host. Plants rarely die from crown gall disease, but they become weakened, grow more slowly, and produce fewer flowers and fruits. Infected plants are susceptible to other disease-forming agents, such as fungi.

The infectious process of crown gall disease is being studied intensively. Each cell of *A. tumefaciens* contains one or more large *plasmids.* A plasmid is a circular piece of DNA that lies in the cytoplasm and replicates independently of the main chromosome of the bacterium. The plasmid responsible for tumorigenesis contains about 100 genes and is called the *Ti (tumor-inducing) plasmid.* After the bacterium attaches to a plant cell, the plasmid passes through the host cell wall and part of it, called *T-DNA* or *transferred DNA,* becomes integrated into the host chromosome. The genes on T-DNA are expressed along with other genes on the plant chromosome. The integrated bacterial genes cause the plant cell to produce chemical compounds that it normally would not produce and to multiply abnormally. In effect, the plant cell is partly controlled by genes from the bacterium. Crown gall disease is an example of *ge-*

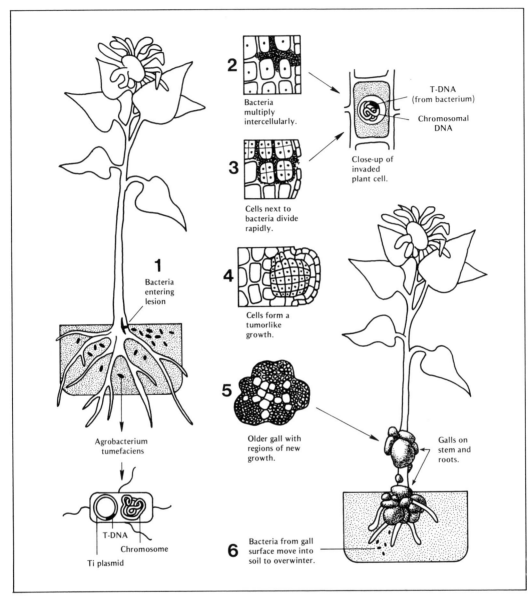

Figure 4-1. Genetic colonization of a flowering plant by Agrobacterium tumefaciens.

netic parasitism in which one organism, *A. tumefaciens,* uses its genes to parasitize the cells of another organism. Once the plasmid is incorporated into the host chromosome, it replicates along with the plant's chromosomes. Thus, all the progeny of a cell with T-DNA will have copies of the T-DNA and the presence of the bacterium no longer is necessary for the disease to progress.

The rapid and uncontrolled division of the plant cells, caused by T-DNA, leads to the formation of tumorlike growths that are characteristic of the disease. Genes on the T-DNA also are responsible for the production of plant growth hormones, such as *auxin* and *cytokinins,* which further stimulate the growth of the tumor. Other genes on the T-DNA segment code for the synthesis of an unusual group of chemical substances called *opines*. Opines are derived from amino acids and are produced only by plant cells infected with T-DNA. Opines do not have a role in the formation of tumors and serve only as sources of carbon, nitrogen, and energy for *Agrobacterium* containing the Ti plasmid. Opines cause a fourfold increase in the rate of transfer of Ti plasmids to avirulent strains of *A. tumefaciens* that are in the soil near the crown galls. Some scientists have referred to this phenomenon of plasmid transfer as *molecular altruism.* The altruistic behavior of an individual reduces its own fitness but increases the survival of the population to which it belongs. Opines also stimulate plasmid exchange between strains of *Agrobacterium* in the crown gall tissue.

Six different strains of *A. tumefaciens* are recognized, based on their ability to metabolize specific types of opines. For example, the *octopine strain* uses opines such as *octopine,* derived from the amino acid arginine, *lysopine,* derived from lysine, *histopine,* derived from histidine, and *octopinic acid,* derived from ornithine. The *nopaline strain* uses *nopaline,* derived from arginine, and *nopalinic acid,* derived from ornithine. Most strains have a wide host range but a few are limited in the types of plants they can infect.

That a segment of a plasmid of *A. tumefaciens,* a prokaryotic cell, can fuse with a chromosome of a eukaryotic plant cell is surprising. It is equally intriguing that genes of a prokaryotic chromosome are expressed within a eukaryotic cell. This is understandable in part because the T-DNA includes genetic sequences that can be "read" by enzymes of the eukaryotic cell.

The ability of *A. tumefaciens* to insert part of its plasmid into a plant chromosome has considerable implications for the improvement of agricultural crops. Through the techniques of genetic engineering, *A. tumefaciens* could be used to introduce desired traits into plant cells. Scientists hope to replace part of the T-DNA segment of the Ti plasmid with a segment that carries beneficial genes such as those for herbicide resistance. This hybrid segment could then be incorporated into the chromosome of a plant cell, which could then be regenerated to form a new, and possibly improved, plant.

An important event in the *A. tumefaciens* infection of plants is the binding of the bacterium to the host cell wall, a process that may be genetically controlled by the *A. tumefaciens* chromosome. A lipopolysaccharide constituent of the bacterial cell wall binds to a specific component of the plant cell wall. Only young, actively dividing plant cells, such as those that form around a wound, seem to have this component.

One way to control *A. tumefaciens* biologically is to interfere with its binding mechanism. In this way, *agrocin 84,* a compound obtained from *Agrobacterium radiobacter,* prevents infections caused by nopaline strains of *A. tumefaciens.* It appears that the genes for agrocin 84 synthesis are carried by a small plasmid.

Agrobacterium tumefaciens infects a wide variety of plants. Angiosperms and gymnosperms are susceptible to infection but monocotyledonous plants, as well as mosses and ferns, generally are resistant. The nature of this resistance is not understood. It may result from the inability of the bacterium to bind to the cell walls of these plants or from a failure of the T-DNA to become integrated in the host DNA.

Josef Schell and his colleagues from Belgium and Germany have proposed the concept of *genetic colonization* to explain how agrobacteria introduce genetic information into the host genome. By this process, the symbiont manages to secure some of the host's photosynthetic products in a manner that gives it a selective advantage over other competing organisms.

A recent study reported on the base sequence homology of the Ti plasmid and sections of DNA isolated from an uninfected plant. A similar situation exists between the nucleic acids of retroviruses and those of noninfected mammalian hosts. It is possible that a mammalian tumor virus and the Ti plasmid of *Agrobacterium* have a common mechanism for the incorporation of their DNA into a host chromosome. Crown gall is the only known example of a disease that persists in the absence of the inciting agent. The Ti plasmid, in part at least, may represent ancient plant genes that have become incorporated into *A. tumefaciens.* A similar situation may exist for the nitrogen-fixing bacterium *Rhizobium.*

In summary, the three principal features of the *Agrobacterium* symbiosis are *tumor production, opine production,* and *Ti plasmids.* Tumors occur when a plant cell with a T-DNA segment derived from a bacterial Ti plasmid divides in an uncontrolled fashion and produces opines. Synthesis of these molecules is controlled by T-DNA. Opines permit the symbiont to exploit the host's carbon and nitrogen energy sources. Ti plasmids have two major features: they contain genes that allow the bacteria to use opines as energy molecules; and they possess DNA that is transferred to the host genome.

C. Nitrogen-fixing Symbioses

Plant growth is limited by how much nitrogen is in the soil. Nitrogen is an important element for living organisms and is used to make amino acids, proteins, and nucleic acids. Nitrogen is common in our atmosphere, but plants cannot use it in its elemental form and must absorb it from the soil in the form of nitrates. The supply of nitrogenous compounds in the soil is continually replenished by bacteria that fix atmospheric nitrogen into ammonia. Other soil bacteria then convert ammonia into nitrites and nitrates. Nitrates absorbed by plant roots are converted back to ammonia, which is used to form the complex molecules needed by plants.

Nitrogen-fixing bacteria commonly form mutualistic relationships with plants. The bacteria contain the enzyme *nitrogenase,* which can catalyze complex reactions, involving N_2, hydrogen ions, and free electrons, that lead to the formation of ammonia. In order for these reactions to occur, the bacteria need energy-rich compounds, such as ATP, and also electrons, both of which are obtained by respiring sugars supplied by the plant. Because plants manufacture sugar by means of photosynthesis, it is clear that there is a close relationship between photosynthesis and nitrogen fixation. An increase in the rate of photosynthesis stimulates an increase in nitrogen fixation. Bacteria that fix nitrogen are a vital part of different ecosystems, since they are the primary suppliers of nitrogen to the soil.

Bacteria need several elements in order to fix nitrogen.

1. Access to atmospheric nitrogen
2. A nitrogenase enzyme complex
3. Large amounts of ATP (12 moles ATP per mole N_2 reduced)
4. An anaerobic environment
5. A supply of Fe and Mo, which are required for nitrogenase activity
6. Temperatures below 30° C.
7. A means of regulating the amounts of ATP available for the system and for controlling fluctuating ammonia production

Because oxygen readily denatures nitrogenase, nitrogen fixation must take place under low concentrations of oxygen. Symbiotic bacteria fix nitrogen in specialized structures, such as nodules, heterocysts, and vesicles in which low levels of oxygen can be maintained. Another important constraint on bacterial nitrogen fixation is the high-energy cost of the process. A symbiont's capacity for nitrogen fixation is controlled largely by the host because it regulates how much energy, in the form of sugar, the symbiont receives. In some cases as much as 30% of the photosynthetic products of a host plant may be used to support the processes of nitrogen fixation and assimilation.

As the bacterial symbionts fix nitrogen, the activity of their ammonia-assimilating enzymes declines because of host inhibition, and they begin to excrete much of the nitrogen they fix as ammonia. The host cells contain enzymes to convert the excreted ammonia into useful compounds. Such adaptations exist in the *Rhizobium*-legume and *Anabaena-Azolla* symbioses.

1. *RHIZOBIUM*-LEGUME SYMBIOSIS. *Rhizobium* is a genus of rod-shaped, motile bacteria that live in the soil, usually in areas where legumes grow. These bacteria form symbiotic associations with the roots of legumes such as alfalfa, clover, pea, and soybean. The bacteria stimulate the roots to form unique structures called *nodules,* within which nitrogen fixation occurs (Fig. 4.2). In soil that is deficient in nitrogen, legumes with their built-in nitrogen suppliers, the nodules, have a great advantage over other plants. Root nodules containing bacteria can supply a plant with all the nitrogen it needs. Plants without nodules must obtain their nitrogen from the soil, where it is often in short supply.

a. Infection Process. Infection of a legume begins when bacteria attach, by one end of their cell, to newly formed root hairs of plant seedlings (Fig. 4.2). It is not clear how this attachment occurs, but there is evidence that complementary chemical bonding may be involved. Proteins called *lectins* on the surface of root hairs bind to polysaccharides on the bacterial cell wall. In some instances, the binding is specific; that is, only certain lectins will bind to certain polysaccharides, which means that some strains of rhizobia infect only certain types of

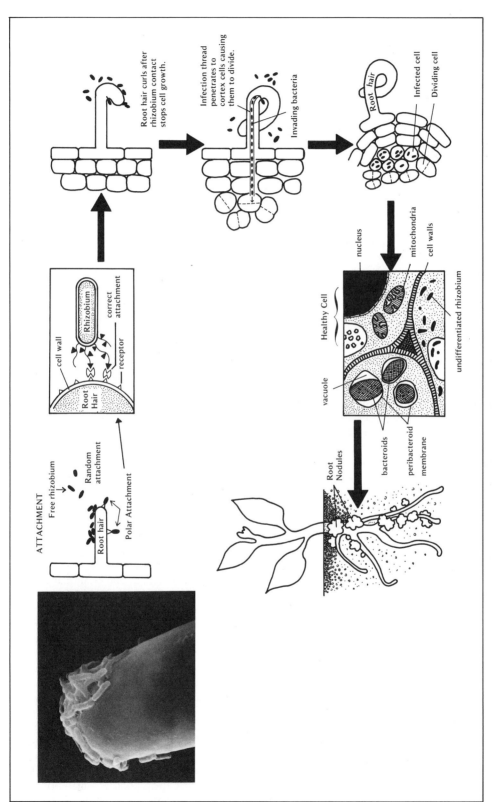

Figure 4-2. *Stages in the infection of legume roots by Rhizobium species. Insert photograph: A scanning electron micrograph of the attachment of cells of Rhizobium trifolii to a clover root hair tip.* (Insert: Frank B. Dazzo, Michigan State University.)

legumes. For example, *Rhizobium japonicum* commonly infects soybeans; *Rhizobium trifolii* infects only white clover. Legumes stimulate nodule-forming bacteria to grow around their roots, possibly because of substances secreted from the roots.

When bacteria attach to a root hair, the hair curls in response. The exact mechanism of hair curling is unknown. The bacteria then penetrate the cell wall and come in contact with the plasma membrane. Cell wall synthesis by the root hair is redirected toward the site of infection. Instead of depositing new wall material along the growing tip of the root hair, the cell now forms walls at the point of bacterial infection. As a result of this redirected cell wall synthesis, a tubular *infection thread* forms and grows inwardly toward the root cortex. The nucleus of the root hair doubles in size and directs the growth of the infection thread. The infection thread contains rhizobial cells, which are surrounded by a slimy substance in which they multiply. The infection thread is surrounded by cell wall except at its tip, where there is only naked plasma membrane.

b. Rhizobium *Transformation: Bacteroids.* The infection thread grows through the root cells and as it does, cells of the cortex, some distance away from the thread, are stimulated to divide. What induces this initial division is now known. Many of the cells that divide are thought to be tetraploid; that is, they have double the usual number of chromosomes. These dividing cortical cells form the nodules. The nodule cells become colonized by bacteria from the infection threads that eventually reach and penetrate the host cells. It is not known how the thread penetrates a cortical cell, but after it does so, the bacteria are released into the cytoplasm. The bacteria occur either singly or in groups within the cell and they are surrounded by a *peribacteroid membrane,* which originates from the host plasma membrane.

After the bacteria are released from the infection thread, they are called *bacteroids.* The cells of many strains of *Rhizobium* undergo radical changes in morphology and physiology when they become bacteroids. Some bacteroids are 40 times larger than the small rods they develop from and their numbers can almost fill a plant cell. Each legume host determines the size and shape of its enclosed bacteroids and also the number of bacteroids contained within a peri-

bacteroid membrane. One strain of *Rhizobium* may form different types of bacteroids in different species of legumes. Bacteroids are nonmotile and assume a wide variety of shapes. They have thin walls that allow for easy passage of nutrients from the plant to the bacteroids and the export of nitrogen from the bacteroids. The transformation of rod-shaped rhizobia into bacteroids inside the legume cells initiates the process that leads to nitrogen fixation. The bacteroids are similar to nitrogen-fixing organelles. They cannot use the nitrogen they fix, but rather depend on the plant for their nitrogen and carbon compounds.

There are two groups of *Rhizobium,* fast-growing and slow-growing species. The peanut plant is an example of a legume that forms nodules with slow-growing rhizobia. Some scientists place the slow-growers in a separate genus called *Bradyrhizobium.*

c. *Role of Oxygen-binding Proteins.* Legume root nodules contain large amounts of a proteinaceous red pigment called *leghemoglobin.* This protein is a direct product of the symbiosis, since its globin chains are synthesized by the plant and its heme groups are produced by the bacteroids. The pigment forms only after the symbiotic association has taken place, which suggests that the plant genes that code for the globin part of leghemoglobin are expressed as a result of the infection. The pigment is located in the cytoplasm of the plant cell. The function of the pigment is to bind and store oxygen, which is required by the bacteroids. The binding process also helps to maintain a low concentration of free oxygen in the nodule. Nitrogenase is inactivated at high oxygen concentrations. Leghemoglobin is similar in structure and function to hemoglobin found in the red blood cells of mammals.

d. *Nodule Biology.* Legume root nodules may be spherical or club-shaped. They do not develop near the root tips but occur in abundance in older parts of the root system. Nodules are outgrowths of plant tissue and they consist of uninfected tissue as well as tissue that contains bacteroids. The vascular system of the plant extends into the nodules and is the means by which nutrients pass in and out of the nodule.

Some strains of *Rhizobium* that nodulate legumes can also form nodules with a nonleguminous plant, *Parasponia,* a member of the elm family. This suggests that there is nothing unique about the genetic makeup of legumes that enables them to develop nitrogen-fixing symbioses with *Rhizobium*. It is possible that many other flowering plants have the necessary genes to allow for "controlled rhizobial infections" and subsequent development of nodules. As indicated earlier, the fixation of atmospheric nitrogen uses significant amounts of energy-rich compounds, which are supplied by the plant. Such an energy-dependent process may explain why many plants cannot afford to form symbiotic associations with nitrogen-fixing bacteria.

Nodules have a limited life span. Only those that function at maximum capacity are maintained on the plant. When nodules decompose, bacteria that have remained in the infection threads and the bacteroids present in the nodules are released into the soil. Whether these bacteria and bacteroids can reinfect other root hairs is not known.

e. Genetics of Nitrogen Fixation.

Legume mutualism is determined by many genes, among them genes called *nod* and *fix,* which represent nodulation and nitrogen fixation, respectively. Most important among the *fix* genes are the *nif* genes, which code for the synthesis of the enzyme nitrogenase. In most of the fast-growing *Rhizobium* species that have been examined, these genes are located on a large plasmid called *pSym* (symbiotic). Scientists have demonstrated a close linkage between the nitrogenase (*nif*) and nodulation (*nod*) genes on the *pSym* plasmid. *Nif* genes from *Klebsiella pneumoniae* have been cloned and their sequences have been found to be similar to those of *Rhizobium*. Currently, attempts are being made to sequence *nod* genes as part of an intensive research effort on the genetics of *Rhizobium* nodulation. The formation of a nodule in the host tissue is a complex, multi-step process involving recognition, binding receptors, infection, host cell multiplication, and cell enlargement. Mutants have been discovered for a number of nodular stages. These mutants are being used to discover how the process of nodulation is governed and controlled. Both bacterial and plant genes are involved in the formation of nodules.

Many aspects of the *Rhizobium*-legume symbiosis are not fully understood. Research on this symbiosis continues because of its agricultural significance and fundamental importance to natural plant communities. Efforts to increase food production in order to feed a growing world population require new supplies of fertilizers, of which nitrogen is a key element. It is more economical, as well as environmentally sound, to increase nitrogen production through biological systems, such as that of *Rhizobium* and legumes, than to produce nitrogen fertilizers by expensive chemical processes.

2. ACTINORHIZAL SYMBIOSIS. Nitrogen-fixing nodules that are different from those formed by *Rhizobium* occur on plants that are not legumes. The organisms responsible for these nodules are actinobacteria, filamentous bacteria that grow in the soil. Actinobacteria that form nitrogen-fixing symbioses belong to the genus *Frankia*. The associations are called *actinorhizae* and they develop on the roots of woody dicots. Symbiotic *Frankia* form numerous terminal swellings, called *vesicles,* within which nitrogen fixation occurs. Nodules of actinorhizae also contain hemoglobin, which is similar to leghemoglobin in *Rhizobium*-legume root nodules. The stages of infection that lead to actonorhizal nodules are similar to those of *Rhizobium*. The sequence of events is this: Root hair curling, infection through root hairs, growth of the bacterial filaments through the plant cell, rapid division of cortical cells to form nodules, invasion of the cortical cells by bacteria, and formation of hyphae, vesicles, and spores in the plant cells (Fig. 4.3). Depending on the type of host plant, one bacterial strain may form club-shaped or spherical vesicles. In some hosts the bacteria do not form vesicles but nitrogen fixation still occurs. The earliest stage of the association may be a mutual recognition that involves lectins. Strains of *Frankia* have been isolated from over 160 species of woody dicotyledonous plants, including species of *Alnus* (alders), *Casuarina* (beefwood), *Comptonia* (sweet fern), and *Myrica* (bayberry). Plants with actinorhizae are important sources of nitrogen in woods, fields, bogs, and roadsides, both from their leaf litter and decomposition of dead plants. *Frankia* strains have been cultured on artificial media. Their growth is much slower than that of most free-living bacteria and they assume a variety of shapes. The molecular genetics of *Frankia* is being studied in laboratories worldwide to determine if the nitrogen-fixing and nodulation genes of this actinobacterium are similar to those of *Rhizobium*.

Figure 4-3. Branches of the actinobacterium Frankia *sp. inside host cell of* Eleagnus umbellatus. (William Newcomb, Queen's University, Canada.)

3. CYANOBACTERIA AND PLANTS. Cyanobacteria are another group of nitrogen-fixing bacteria. They are photosynthetic bacteria that form associations with fungi, a few algae, some bryophytes, the fern *Azolla,* some cycads, and the angiosperm genus *Gunnera.* Considering how important nitrogen is to plants, it is surprising that there are so few symbiotic associations between plants and cyanobacteria. Cyanobacteria fix nitrogen within *heterocysts,* which contain nitrogenase. Recent studies of the photosynthetic cyanobacterium *Anabaena* by Robert Haselkorn and his colleagues of the University of Chicago have revealed that gene rearrangement takes place when vegetative cells of the cyanobacterium differentiate into heterocysts. Such genetic shuffling is common in the differentiating cells of advanced organisms, but its occurrence in prokaryotic forms was unexpected.

a. Azolla. *Azolla* is a genus of small, aquatic ferns that float on the surface of freshwater ponds and marshes throughout the world. The plant has tiny roots and a short, branched stem that is covered with small overlapping leaves (Fig. 4.4). The ferns multiply rapidly by vegetative reproduction. They can double their biomass in two days if growth conditions are optimal and quickly cover large bodies of water with a dense and sometimes impenetrable growth. Each leaf of *Azolla* is divided into a dorsal and ventral lobe. The ventral lobe floats on the water and contains filaments of the cyanobacterium *Anabaena azollae,* which lives in symbiosis with the fern. The bacteria are inside mucilage-filled cavities of the leaves (Fig. 4.4). The cavities are sealed so that cyanobacterial filaments cannot escape and other organisms cannot enter. Sexual structures that the fern uses to overwinter contain *Anabaena* spores, so that when germination occurs in the spring the new individuals of *Azolla* are infected by the bacteria. Thus, no stage in the life cycle of the fern is free of cyanobacteria.

Anabaena fixes nitrogen in the leaf cavities of *Azolla* and supplies the fern with all the nitrogen it needs. With its own built-in fertilizer plant, the fern can grow in waters that are deficient in nitrogen. The fixed nitrogen, or ammonia, that is released by *Anabaena* is absorbed by specialized hairs in the leaf cavity. Each cavity has about twenty randomly located simple hairs that transport sucrose from the fern to the cyanobacteria, where it is used in the nitrogen-fixation process. Each cavity also has two branched hairs that are always near a vein. These hairs are thought to assimilate the ammonia produced by the cyanobacteria, form amino acids, and transport them to the fern.

The leaf cavity that contains cyanobacteria acts as a greenhouse. Inside this space, the bacterial symbiont has a moist, protected environment and its nutrient needs are provided by the fern. The extent to which *Anabaena azollae* contributes to the total photosynthesis of the symbiosis is not clear. The isolated *Anabaena* symbiont fixes carbon dioxide and undergoes normal photosynthesis, but this process is inhibited when the bacterium is in the leaf cavity. The fern inhibits glutamine synthetase activity of the cyanobacterium. *Glutamine synthetase* is an enzyme that is involved in the assimilation of ammonia. When this enzyme is inhibited, the *Anabaena* symbiont excretes rather than assimilates the ammonia it produces by nitrogen fixation.

A

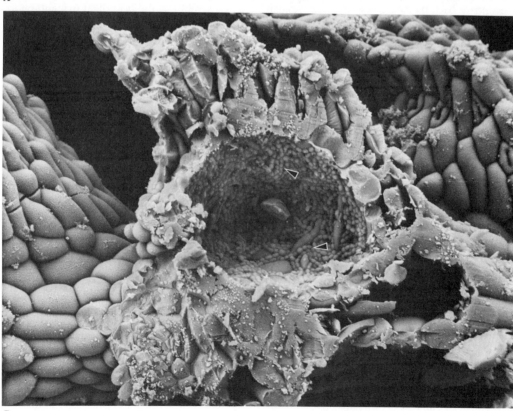

B

Figure 4-4. Azolla caroliniana. *(A) Fern showing branching pattern and roots. (B) Scanning electron micrograph of a section through a leaf cavity, show-* ing filaments of Anabaena azollae (arrows). (Part (*A*): Steve B. Dunbar, C. F. Kettering Research Lab. Part (*B*): Sandra K. Perkins, C. F. Kettering Research Lab.)

Anabaena filaments are attached to the growing tip, or meristem, of the *Azolla* plant. From this region the bacteria become incorporated into each new leaf as it develops. If the *Anabaena* colony on the meristem is sparse or absent, then bacteria-free leaves of *Azolla* will form. The bacteria on the meristem do not fix nitrogen because they lack heterocysts. Such bacteria receive nitrogenous compounds from the fern.

The development of the fern and *Anabaena* is synchronous. As each new fern leaf develops, the bacteria that become trapped inside the leaf cavity multiply, form many heterocysts (25% to 30% of total cell number) and begin to fix nitrogen. Cells of the *Anabaena* symbiont enlarge and divide slowly when they are in the leaf cavity.

Coryneform bacteria also grow in *Azolla* leaf cavities, along with *Anabaena*. Whether these secondary bacteria have a role in the symbiosis is not known. They may produce some of the slime that is found in the leaf cavities, which is important for the growth of the *Anabaena* filaments.

Experimental studies on the *Azolla-Anabaena* symbiosis have been hindered by the inability of bacteria-free (aposymbiotic) ferns to re-establish the symbiosis with isolated *Anabaena* symbionts. The specific recognition factors that are necessary to bring about the symbiosis are not known.

There are six species of *Azolla* but whether they all contain *Anabaena azollae* as a symbiont has not been determined. Detailed studies to identify the cyanobacteria in the different species of *Azolla* have not been made. Morphologically, the species of fern differ according to their reproductive structures. Physiologically, the species differ only in their optimum growth temperature.

Much research is being conducted on the *Azolla-Anabaena* association. *Azolla* is used as a green manure in Asian countries such as China and Vietnam, because it grows well in the warm, stagnant waters of rice paddies and provides nitrogen for the rice plants. Farmers grow *Azolla* alongside rice plants or mix the fern with the soil of the paddies. The use of natural fertilizers, such as *Azolla,* has a long history in Asian countries. Natural fertilizers are an important and inexpensive alternative to commercial fertilizers.

b. **Bryophytes.** Hornworts, such as *Anthoceros punctatus,* and thallose liverworts, such as *Blasia pusilla,* form symbiotic associations with the cyanobacterium *Nostoc.* Filaments of *Nostoc* live in mucilage-filled cavities on the undersurface of the bryophyte gametophytes. The bacteria fix nitrogen into ammonia, which they excrete into the cavities. The bryophytes absorb the ammonia and use it to synthesize proteins and nucleic acids. The bryophytes provide the bacteria with carbon compounds such as sucrose and a protected place in which to live.

Both symbionts are altered after they enter into the association. The *Nostoc* symbionts virtually stop reproduction and do not photosynthesize. Their cells have less pigment and fewer storage granules than free-living cyanobacteria and they appear to be nitrogen-starved. This suggests that the symbiotic cyanobacteria excrete most of the nitrogen they fix. As in the *Azolla* symbiosis, the host plant may inhibit the cyanobacterial enzymes that assimilate ammonia. The heterocyst frequency of *Nostoc* in the gametophyte cavities is much higher than that of the same symbiont growing outside the cavities (30% to 43% versus 3% to 6%, respectively). When *Blasia* becomes infected with *Nostoc,* its cavities form filamentous extensions of tissue that closely associate with the cyanobacteria. These extensions increase the area of contact between the symbionts and facilitate the exchange of nutrients.

c. **Cycads.** Cycads are early seed plants that were common and widespread during the Mesozoic era. Only about 10 genera and 100 species of cycads still grow in tropical and subtropical regions of the earth. About one-third of the known cycads contain symbiotic cyanobacteria in their roots. The bacteria are located in specialized branches of the root that, unlike normal roots, are negatively geotrophic. Instead of growing down into the ground, these roots grow on and above the soil surface. The roots develop nodules that contain cyanobacteria. Because of their warty appearance, these specialized structures are called *coralloid roots.* The roots are loosely organized and gases circulate easily in and out of them.

The endosymbionts of cycads are strains of *Anabaena* or *Nostoc.* The cyanobacteria penetrate the coralloid roots through cracks or openings and move between the middle lamellae of cells until they reach the middle of the root cortex. Small host cells in this area degenerate and form mucilage-filled spaces within which the cyanobacteria divide. Other host cells in the cyanobacterial zone elongate, possibly because of growth factors excreted by the cyanobacteria, and develop fingerlike extensions that may facilitate the exchange of nutrients between the symbionts.

Cyanobacteria of cycad roots fix nitrogen, which they supply to the cycad in the form of ammonia. Cycads that grow in the forests of the Southern Hemisphere are important for the nitrogen economy of those forests.

d. **Gunnera.** *Gunnera* is an angiosperm genus with about 50 species, all of which are perennial herbs. The plants grow commonly in wet areas of the tropics and Southern Hemisphere. Near the base of the leaf petioles, there are secretory glands that contain the cyanobacterium *Nostoc punctiforme.* Filaments of *Nostoc* multiply in the mucilage that fills the glands and then penetrate cells at the base of the glands. This association is unusual because the cyanobacterial symbionts are intracellular; that is, they are surrounded by the plasmalemma of the host cell. Each gland cell forms fingerlike extensions that increase the area of contact with the endosymbiont. The transformed glands are called nodules by some scientists and are compared with the nodules formed in the *Rhizobium*-legume symbiosis. Like other cyanobacterial symbionts, the *Nostoc* endosymbiont of *Gunnera* does not photosynthesize, has many heterocysts, and fixes nitrogen at a high rate. The *Nostoc* symbiont supplies *Gunnera* with all the nitrogen it needs and receives carbohydrates from the plant. Thus, in symbiosis the *Nostoc* endosymbiont behaves as a heterotroph.

Symbiotic associations between cyanobacteria and plants have a number of common features. Symbiotic cyanobacteria usually move in a slow, gliding manner and can penetrate cavities and spaces of plant organs and form colonies. After a population of cyanobacteria is established in a plant, growth of the endosymbiont either stops or slows down. The host plant may regulate the growth and population size of the endosymbiont, but exactly how this is done is not known. Symbiotic cyanobacteria have more heterocysts and fix nitrogen at higher rates than free-living cyanobacteria. The symbiotic forms have larger vegetative cells, fewer nutrient reserves, do not photosynthesize, and cannot assimilate the nitrogen they fix. The host plant inhibits the nitrogen-assimilating enzymes of the endosymbiont. As a result, the ammonia produced by the symbiont is excreted and passed to the host plant, which then converts it to other compounds. Symbiosis alters host plants. They produce hairs and extensions of cells that increase the surface contact between host and cyanobacteria and facilitate the exchange of nutrients.

D. Mycoplasmas: The Smallest of Bacteria

Mycoplasmas are a new class of bacteria and are the smallest cellular organisms known that reproduce. They were discovered by two groups of Japanese scientists in 1967. These prokaryotes lack the characteristic cell wall of other bacteria and were first seen in the phloem cells of plants with *yellows disease.* Yellows disease was long suspected to be caused by a virus because it was transmitted by insects and the infectious agent passed through bacterial filters. Antibiotic treatment of sick plants, however, restored them to good health. Mycoplasmas are resistant to penicillin and other antibiotics that inhibit cell wall formation but are sensitive to antibiotics such as tetracycline. Today, we know of several different kinds of mycoplasmas and mycoplasma like organisms (MLOs).

Plants infected with phloem-specific mycoplasmas are often stunted and have small flowers and fruits. *Corn stunt* and *citrus stubborn* are two diseases that are caused by the mycoplasma *Spiroplasma. Aster yellows, lethal yellowing of coconut palms, elm phloem necrosis, peach X disease,* and *pear decline disease* are other plant diseases suspected of being caused by mycoplasmas. These obligate symbionts occur intracellularly in sieve tube elements of the phloem. Sieve tube cells are a highly specialized environment. A large volume of nutrients passes through these cells to areas of the plant where active growth is taking place or to storage areas of the roots. A unique feature of these cells is their high sugar concentration and hence high turgor pressure gradient. Additionally, sieve cells contain significant concentrations of K^+, Mg^+, Cl^- and PO_4^- ions, amino acids, proteins, and ATP molecules. It is this uniquely special-ized cellular environment to which the sym-bionts have become adapted.

Leaf hoppers and other insects feed on phloem sap, and the chemical constituents of their hemolymph are strikingly similar to those of the sieve tube cell. Insect hemolymph has high concentrations of organic and inorganic nutrients, which result in a high osmotic pres-sure. Leaf hoppers feeding on phloem cells ac-quire mycoplasmas that multiply rapidly in the alimentary canal, hemolymph, salivary glands, and ovaries of the insects. Mycoplasmas are eventually transported to the salivary glands from which they are transferred to new phloem cells during a leaf hopper's feeding. The same mycoplasmas can reproduce in different hosts, such as a flowering plant and an insect, because the intracellular environments of the hosts are remarkably similar. The lack of cell walls in my-coplasmas may be viewed as an evolutionary consequence of intracellular symbiosis.

In recent studies, mycoplasmas have been ob-served in the leachates of flowers, suggesting that flowers may also be habitats for these pro-karyotes. Possibly, they survive in nectar and are transmitted by pollinating insects. Indeed, in-sects could be the principal host of these pro-karyotes, and the flowers may represent only transient habitats.

Animal-associated mycoplasmas first gained the attention of researchers when tissue culture cells obtained from respiratory and urogenital tracts developed large populations of these or-ganisms. Scientists are beginning to realize that mycoplasmas may be a normal constituent of human and animal mucous membranes. Myco-plasmas have been reported from cells of the oral and nasal cavities, urogenital tracts and lungs of cattle and humans. Some authors esti-mate that 20% of all human pneumonia may be caused by *Mycoplasma pneumoniae*. The signif-icance of mycoplasmas in human and animal health is poorly understood because they can-not be grown in an isolated culture.

E. Summary and Perspectives

Many species of bacteria cause plant diseases. *Agrobacterium tumefaciens* causes crown gall disease in many plants. The infection process of this disease is a subject of much study because the bacterium has the potential of becoming an

important vector for carrying genes into plant cells using techniques of genetic engineering. The *A. tumefaciens* symbiosis may serve as a model for studying whether genetic information passes between symbionts. In other associations as well, including those of rhizobia and le-gumes, lichens, and mycorrhizas, there are mor-phological transformations of the symbionts and new chemical compounds produced, which fu-ture studies may show are caused by genetic col-onization of a host by its symbiotic partner. At present, little is known about gene flow be-tween the partners of a symbiosis, but consid-ering that genetic exchange occurs between *A. tumefaciens* and plant cells as well as between the organelles of a eukaryotic cell, it appears likely that a similar situation exists between part-ners of other symbioses, especially highly inte-grated ones. A recent finding of gene transfer from the pony fish to its symbiotic luminescent bacteria supports this view.

Some bacteria form mutualistic, nitrogen-fix-ing symbioses with plants. The nitrogen fixed by these organisms is an important part of different ecosystems throughout the world. The best-known nitrogen-fixing symbiosis is that between *Rhizobium* and legumes. Intensive research on this symbiosis is being conducted in order to understand the genetics of nitrogen fixation and nodulation. The long-term goal of such research is to be able to introduce nodulating bacteria into crop plants and thus eliminate the need for artificial nitrogen fertilizers. Unfortunately, ni-trogen fixation involves complex interactions between the host plant and its microbial sym-bionts, and therefore the symbiosis must be studied as an integrated whole in order to un-derstand it fully. *Frankia* is an actinobacterium that forms nitrogen-fixing nodules on the roots of many trees and shrubs and fixes nitrogen in vesicles. The infection process as well as the for-mation of hemoglobin in nodules is similar to that in the *Rhizobium*-legume symbiosis.

Cyanobacteria also fix nitrogen and form sym-bioses with plants such as *Azolla,* bryophytes, and cycads as well as with lichen-forming fungi. Nitrogen is fixed in specialized structures called heterocysts, which are similar in function to the vesicles of *Frankia.* In all nitrogen-fixing sym-bioses, the host inhibits the nitrogen-assimilat-ing enzymes of the bacteria and thereby causes ammonia to be excreted from the symbiont, to be used by the host.

Mycoplasmas are the smallest-known cellular

organisms. They cause many plant diseases and also have been found in different animal tissues.

Review Questions

1. Explain what is meant by genetic colonization with regard to crown gall disease.
2. What are the two major features of Ti plasmids?
3. List some of the conditions that are necessary for bacteria to fix nitrogen.
4. Describe the stages of infection in the *Rhizobium*-legume symbiosis.
5. Would you expect the nitrogen-fixing genes of *Frankia* to be similar to those of *Rhizobium?*
6. Why is the *Azolla-Anabaena* symbiosis of economic importance?
7. Where are the endosymbionts of cycads located?
8. Why is the *Gunnera-Nostoc* symbiosis unusual?
9. Name some diseases caused by mycoplasmas.

Further Reading

Brill W. J. 1977. Biological nitrogen fixation. Sci. Am. 236 (March): 68–81.

Brill, W. J. 1981. Agricultural microbiology. Sci. Am. 245 (September): 199–215. (Describes the *Rhizobium*-plant symbiosis and how *Agrobacterium tumefaciens* may be used in gene transfer among plants.)

Chilton, M. D. 1983. A vector for introducing new genes into plants. Sci. Am. 248 (June): 50–59. (Considers *Agrobacterium tumefaciens* as a carrier of desirable genes to various cultivated plants.)

Fincher, J. 1984. Tailored genes. Horticulture 62 (April): 50–57. (Use of *Agrobacterium tumefaciens* to create custom-made plants.)

Haselkorn, R., J. W. Golden, P. J. Lammers, and M. E. Mulligan. 1985. Organization of the genes of nitrogen fixation in the cyanobacterium *Anabaena.* In: Nitrogen fixation research progress, 485–490. ed. H. J. Evans, P. J. Bottomley, and W. E. Newton. Martinus Nijhoff, Dordrecht, Netherlands.

Lumpkin, T. A., and D. L. Plucknett. 1980. *Azolla:* Botany, physiology, and use as a green manure. Economic Botany 34:111–153.

Marx, J. L. 1985. How rhizobia and legumes get together. Science 230:157–158.

Millbank, J. W. 1974. Associations with blue-green algae. In: The biology of nitrogen fixation, 238–

264, ed. A. Quispel. American Elsevier, New York.

Nester, E. W., and T. Kosuge. 1981. Plasmids specifying plant hyperplasias. Ann. Rev. Microbiol. 35:531–565.

Peters, G. A. 1978. Blue-green algae and algal associations. BioScience 28:580–585.

Peters, G. A., and H. E. Calvert. 1983. The *Azolla-Anabaena azollae* symbiosis. In: Algal symbiosis, 109–145, ed. L. J. Goff. Cambridge Univ. Press, Cambridge.

Reisser, W. 1984. Endosymbiotic cyanobacteria and cyanellae. In: Cellular interactions: Encyclopedia of plant physiology, new series, Vol. 17, 91–112, ed. H. F. Linskens and J. Heslop-Harrison. Springer-Verlag, Berlin.

Roberts, G. P. 1981. Genetics and regulation of nitrogen-fixation. Ann. Rev. Microbiol. 35:207–235. (Discusses the genetics of *nif* genes, gene mapping, and gene expression, as well as how symbiosis is regulated.)

Sequeira, L. 1984. Plant-bacterial interactions. In: Cellular interactions: Encyclopedia of plant physiology, new series, Vol. 17, 187–211, ed. H. F. Linskens and J. Heslop-Harrison. Springer-Verlag, Berlin.

Stewart, W. D. P. 1978. Nitrogen-fixing cyanobacteria and their associations with eukaryotic plants. Endeavour, new series 2:170–179.

Stewart, W. D. P., P. Rowell, and A. N. Rai. 1980. Symbiotic nitrogen-fixing cyanobacteria. In: Nitrogen Fixation, 239–277, ed. W. D. P. Stewart and J. R. Gallo. Academic Press, New York.

Torrey, J. G. 1978. Nitrogen fixation by actinomycete-nodulated angiosperms. BioScience 28: 586–592.

Vance, C. P. 1983. *Rhizobium* infection and nodulation: A beneficial plant disease? Ann. Rev. Microbiol. 37:399–424. (An excellent comprehensive review.)

Verma, D. P. S., and S. Long. 1983. The molecular biology of *Rhizobium*-legume symbiosis. Int. Rev. Cytology, supplement 14:211–245. (Provides an excellent account of developments in the biology of *Rhizobium* and considers its evolutionary perspective from a pathogen to an organelle.)

Bibliography

Alexander, M. (ed.). 1984. Biological nitrogen fixation: Ecology, technology, and physiology. Plenum Press, New York. 247 pp.

Fay. P. 1983. The blue-greens (Cyanophyta-Cyanobacteria). Institute of Biology, Studies in Biology, no. 160. Edward Arnold, London. 88 pp.

Giles, K. L., and A. G. Atherly (eds.). 1984. Biology

of the Rhizobiaceae. Int. Rev. Cytology, supplement 13. 336 pp.

Kosuge, T., and E. W. Nester (eds.). 1984. Plant-microbe interactions: Molecular and genetic perspectives. Vol. I. Macmillan, New York. 444 pp.

Maramorosch, K. (ed.). 1974. Mycoplasma and mycoplasmalike agents of human, animal, and plant diseases. Ann. N.Y. Acad. Sci. 225:5–532.

Postgate, J. R. 1982. The fundamentals of nitrogen fixation. Cambridge Univ. Press, Cambridge. 252 pp.

Pühler, A. (ed.). 1983. Molecular genetics of the bacteria-plant interactions. Springer-Verlag, Berlin. 393 pp.

Schell, J., M. V. Montagu, M. DeBeuckeleer, M. DeBlock, A. Depicker, M. DeWilde, G. Engler, C. Genetello, J. P. Hernalsteens, M. Holsters, J. Seurinck, B. Silva, F. Van Vliet, and R. Villarroel. 1979. Interactions and DNA transfer between *Agrobacterium tumefaciens,* the Ti-plasmid, and the plant host. Proc. R. Soc. Lond. B. 204:251–266.

Sharifi, E. 1984. Parasitic origins of nitrogen-fixing *Rhizobium*-legume symbioses: A review of the evidence. BioSystems 16:269–289.

Silver, W. S., and E. C. Schroder (eds.). 1984. Practical applications of *Azolla* for rice production. Martinus Nijhoff, The Hague, Netherlands. 227 pp.

Verma, D. P. S., and Th. Holm (eds.). 1984. Genes involved in microbe-plant interactions. Springer-Verlag, Vienna. 393 pp.

CHAPTER 5

Symbiosis and the Origin of the Eukaryotic Cell

A. Introduction

The concept of the cell as the basic unit of life is the core of modern biology. During the past thirty years cell biologists have vastly expanded our knowledge of the internal organization of cells and the biochemical and evolutionary processes of many different organisms. We study cells to learn how life began on earth and to understand the dynamic processes that led to the evolution of present-day eukaryotic cells. Symbiosis is increasingly being recognized as the cornerstone in the origin and organization of eukaryotic cells.

As we have learned in chapter 3, the fundamental distinction in cell biology is between prokaryotic and eukaryotic cells. Bacteria have prokaryotic cells, whereas all other living organisms, such as plants, animals, fungi, and protoctists, have eukaryotic cells. The prokaryotic cell has only one protein-synthesizing unit, whereas the eukaryotic cell has three: the nucleocytoplasm, mitochondria, and chloroplasts. All eukaryotic cells share the same fundamental organization, which suggests that this type of cell arose early during cell evolution. Many scientists believe that the eukaryotic cell arose from symbiotic associations between prokaryotic cells. As stated by F. J. R. Taylor (1974), "The eukaryotic 'cell' is a multiple of the prokaryotic 'cell'"

B. Theories on the Origin of Eukaryotic Cells

Since the mid-1960s, several theories have been proposed to explain the origin of organelles in eukaryotic cells. There are three main theories.

1. SERIAL ENDOSYMBIOSIS THEORY. The *serial endosymbiosis theory* (S.E.T.), also known as the *exogenous theory,* best explains the origin of organelles such as mitochondria, chloroplasts, and microtubular complexes. These organelles may have arisen by successive symbiotic events about three billion years ago. The first step in this process was the invasion of anaerobic prokaryotic host cells by aerobic prokaryotes, which in time became transformed into mitochondria. At a later stage, eukaryotic heterotrophs acquired cyanobacteria and gained photosynthetic capabilities. These symbiotic cyanobacteria eventually developed into chloroplasts. The S.E.T. also suggests that symbiotic associations between host cells and spirochetes resulted in microtubular organelles of motility such as undulipodia (eukaryotic flagella) (Fig. 5.1).

2. AUTOGENOUS THEORY. The central assumption of the *autogenous theory* is that organelles evolved within the cell by progressive compartmentalization. Cell structures such as the nuclear membrane, endoplasmic reticulum, Golgi bodies (dictyosomes), vacuoles, and lysosomes are believed to have evolved from elaborations of the host cell membrane.

According to the autogenous theory, the independent nature of mitochondria and their ability to metabolize nutrients are evolutionarily advanced traits. The symbiotic theory, however, considers mitochondrial autonomy as a primitive trait and integration with the nucleus and cytoplasm as evolutionarily advanced. Some evidence supports the autogenous theory. For example, the mitochondrial genome of yeasts has

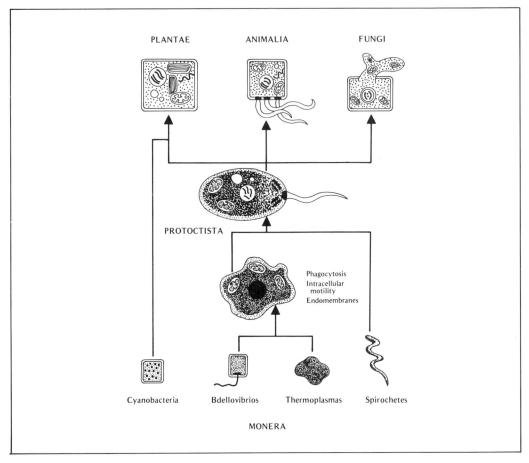

Figure 5-1. *Model of the symbiotic origin of eukaryotic cells.* (Adapted from Margulis, L. 1981. Symbiosis in cell evolution. W. H. Freeman, San Francisco.)

a eukaryotic type of organization in having introns and repetitive sequences. In the autogenous theory, the basic separation of living organisms is between plants and animals and not between prokaryotic and eukaryotic organisms. The theory further states that algae evolved from photosynthetic bacteria and certain other algae gave rise to plants while others lost their chloroplasts and gave rise to fungi and animals. Thus, all organisms developed from prokaryotic ancestors by a series of mutations and along a single line of descent. This theory assumes the existence of organisms that provide a bridge between cyanobacteria and green algae, the so-called uralgae. Despite many studies and much speculation, however, uralgae have not been discovered. The continuing belief by some sci-

entists that such forms exist has been called the *botanical myth* by Lynn Margulis (1981).

3. FUNGAL THEORY. Cavalier-Smith (1981) has argued that the first eukaryote was a fungus that contained mitochondria. As support for this theory, he cites the primitive features of fungi such as the small size of their cells, nuclei, and genomes, lack of repetitive DNA, Golgi with unstacked cisternae, and the lack of 9 + 2 cilia.

The theories on how the eukaryotic cell arose have sparked imaginative experimental approaches in molecular biology, electron microscopy, and biochemistry. Increased research interest in the symbiotic origin of cells has created a new biological subdiscipline, *endocytobiol-*

ogy, and has led to the appearance of international journals on symbiosis, which publish new advances from cell biologists and symbiologists. This new interdisciplinary perspective on symbiosis has stimulated healthy debate in this vigorously expanding field of study.

C. Origin and Evolution of the Eukaryotic Cell

1. ANCESTRAL HOST CELL. What kinds of present-day organisms exist that might be similar to early host cells whose symbiotic associations evolved into the modern eukaryotic cell? Scientists agree that the primitive host cell was a proto-eukaryotic cell with the following features: (a) it was large in size and had membrane systems; (b) it could carry out fermentation of sugars; and (c) it had some type of ATP-generating system and an actomyosin system. Present-day prokaryotic organisms are unsuitable models for the ancestral host cell since their cell wall is a barrier to phagocytosis. Early bacteria may have entered host cells in the same way *Bdellovibrio,* a present-day parasitic bacterium, enters its hosts, forcibly penetrating the wall and membrane of the host cell. The life history of *Bdellovibrio* symbionts is described in chapter 3. Two other organisms have been considered as possible ancestral host cells.

a. **Pelomyxa palustris.** This multinucleated, giant amoeba is considered to be a living example of the ancestral proto-eukaryote. The amoeba has a highly developed vacuolated cytoplasm, lacks mitochondria and Golgi bodies, and contains several types of endosymbiotic bacteria, which are transmitted to successive generations. The endosymbionts are tightly enclosed in vacuoles within membranes provided by the host. When the amoeba begins to divide, these vacuoles migrate to the nuclei and become bound to the nuclear membrane. The newly formed amoebae contain one to three nuclei along with associated bacteria.

b. **Thermoplasma acidophilum.** This mycoplasmalike, free-living prokaryote lacks the characteristic bacterial cell wall. *Thermoplasma* is now classified as an archaebacterium and has features of the hypothetical proto-eukaryotic host cell. *Thermoplasma acidophilum* has the smallest genome of any known free-living organism

and its genome resembles that of eukaryotic cells in having histonelike proteins associated with its DNA. In addition, actinlike filaments have been reported in this organism.

2. ENDOMEMBRANOUS ORIGIN OF CELL ORGANELLES. Cytoplasmic membranes of the eukaryotic cell continuously change their form and structure but always maintain a continuum with the nuclear envelope and the outer plasma membrane. Organelles that are derived from cytoplasmic membranes contain salt solutions, organic molecules, proteins, carbohydrates, and digestive enzymes. In some fungi, Golgi bodies consist of single cisternae, whereas in other eukaryotes the Golgi bodies have stacks of cisternae. Golgi bodies in animals, algae, and fungi are closely associated with the endoplasmic reticulum or nuclear envelope, whereas in plants such close associations are rare. Microtubules and actin filaments are often found close to Golgi bodies. Golgi bodies, endoplasmic reticulum, and vacuoles are morphologically similar. Evidence from electron microscopy shows that vacuoles are derived from the enlarged vesicles of the endoplasmic reticulum or Golgi bodies. Vacuoles and lysosomes are single membranous structures containing substances for osmoregulation, excretion, or digestion.

3. ENDOSYMBIOTIC ORIGIN OF MITOCHONDRIA, CHLOROPLASTS, AND MICROTUBULES. Many proteins and nucleic acids can be viewed as "living fossils" because they have been conserved from the dawn of life. They have been found in prokaryotic cells as well as eukaryotic cells and are believed to have a common ancestry.

The serial endosymbiosis theory states that mitochondria, chloroplasts, and flagella are transformed bacterial symbionts that live within the host cell. This idea is not new, since early scientists, such as Ivan Wallin, were convinced that mitochondria were bacteria that lived inside other cells. In 1927, Wallin used the term *symbionticism* to describe microsymbiotic complexes such as those of mitochondria and cells. Wallin felt that "symbionticism" played an important role in the origin of species. As early as 1883, A. F. W. Schimper proposed that the chloroplasts of plants were cyanobacteria (formerly blue-green algae) that lived symbiotically inside plant cells. These insightful ideas were not accepted by the general scientific community because they were impossible to prove experimentally. As with many scientific hypotheses, the ideas were eventually forgotten. The revitalization and expansion of these hypotheses is the work of Lynn Margulis, who has published books and articles promoting the symbiotic origin of eukaryotic cells. Her lucid and forceful arguments have been instrumental in the general acceptance of the symbiotic theory.

The S.E.T. is strongly supported by findings from structural, molecular, and biochemical studies of eukaryotic cell organelles. The DNA of chloroplasts and mitochondria is different from that of the host nucleus, in that it is circular and histone-free, like that of prokaryotes. Ribosomes of these organelles are also similar to those of prokaryotes both in size and in the nucleotide sequence of their RNA. Chloroplasts and mitochondria divide by fission, mutate, and synthesize proteins. The mitochondria and chloroplasts of eukaryotic cells are only semi-independent. For example, most of the proteins needed for the development and function of chloroplasts are coded for by genes in the nucleus and synthesized on ribosomes in the cytoplasm. The proteins are then transported into the chloroplasts. Some proteins are a joint product of the nucleus and chloroplast, for example, ribulose bisphosphate carboxylase, which is an important enzyme in photosynthesis. Such cooperative interaction between the nucleus and chloroplast suggests that the symbiosis between chloroplasts and plant cells is an ancient one. Similar relationships exist between mitochondria and the host cell nucleus.

There is evidence to suggest that during evolution much of the genome of the bacterial ancestors of mitochondria and chloroplasts was transferred to the host nucleus. These organelles contain much less genetic material than independent bacteria, and segments of mitochondria and chloroplast DNA are identical to segments of nuclear DNA. The same type of DNA that occurs in more than one organelle is called *promiscuous DNA.*

a. Mitochondria. The origin of mitochondria differs from that of chloroplasts. It is not clear whether mitochondria have had a single or multiple origin, since there is evidence to support both possibilities. Determining the type of living bacterium that could represent the ancestor of mitochondria is difficult. Mitochondria cannot be cultured outside a cell and, therefore, cannot be compared directly with other bacteria. Comparisons can be made only through molecular and biochemical characteristics. Mitochondria show greater integration with the host eukaryotic cell than do chloroplasts and have double membranes. The outer membrane is believed to represent the host cell vacuole membrane, and the inner membrane is thought to be that of the symbiont. Mitochondria not only have smaller ribosomes than those of the eukaryotic cell cytoplasm but also have an antibiotic sensitivity similar to that of prokaryotic cells. Mitochondrial DNA is varied in size; there is two to three times more genome in protoctist and plant mitochondria than in the mitochondria of animals.

Fungal mitochondria have a unique genetic code. For example, the codon UGA, which is a stop codon in prokaryotes and eukaryotes, is a secondary codon for tryptophan in mitochondria. Again, the sequences of tRNA in mitochondria resemble neither those of prokaryotes nor those of eukaryotes. Mitochondrial mRNA genes have intervening sequences, or introns, along with post-transcriptional processing (removal of introns and splicing of exons). These features were previously thought to occur only in the eukaryotic nuclear genome.

The strongest evidence for the endosymbiotic origin of mitochondria comes from the sequencing of proteins and nucleic acids. For example, the sequences of cytochromes, ferredoxin, and 5S rRNA are more similar to those of photosynthetic prokaryotes such as *Rhodopseudomonas* than to those of the nucleocytoplasm of the eukaryotic cell. It has been demonstrated from sequencing rRNA from mitochondria of fungi, mice, and humans, from nuclei of animals, yeast, and corn chloroplasts, and from *E. coli* that the greatest similarity in conserved sequences was between mitochondria and prokaryotes.

Earlier it was thought that the ancestors of mitochondria were aerobic, nonphotosynthetic bacteria. New evidence suggests a strong parallel in the electron transport system between the denitrifying bacterium *Paracoccus* and mitochondria. *Paracoccus denitrificans,* a free-living aerobic bacterium, has a respiratory system similar to that of the mitochondria in yeast and animals. The existence of this bacterium suggests that similar types of bacteria lived millions of years ago, when the original endosymbiotic event occurred. It is not known what type of

BOX ESSAY

Symbiotic Jumping Genes

One of the interesting revelations of the twentieth century has been the discovery that an animal or plant cell contains two to three different genomes. In animal cells there are separate genomes in the nucleus and mitochondria, whereas plant cells have an additional genome in the chloroplasts. The genomes in the organelles represent the genes of ancient and transformed symbiotic bacteria.

There is growing evidence that the separate genomes of cells are not static but rather move, in part, between each other. *Promiscuous DNA* is that DNA which moves from mitochondria and chloroplasts to the nucleus and between chloroplasts and mitochondria. The movement of DNA sequences between chloroplasts and mitochondria in plants such as corn, peas, beans, and spinach appears to be very common. There may also be a reciprocal transfer of DNA from the nucleus to the organelles and from the mitochondria to chloroplasts, but these routes have not been

determined as clearly. The transfer of DNA appears to be a continuous process but the exact mechanism by which the transfer occurs is not clear. A vector such as a plasmid may be involved, or the transfer may occur after fusion of the organelles.

Genes move not only between organelles but also between organisms. *Agrobacterium tumefaciens* transfers part of its plasmids to plant cells, thereby causing crown gall disease. This is an example of gene transfer between a prokaryote and a eukaryote. A reverse transfer has been reported, between a eukaryote and a prokaryote. Transfer of the gene for the enzyme copper-zinc superoxide dismutase has been thought to have occurred between the pony fish and its symbiotic, light-producing bacterium *Photobacterium leiognathi*. The bacterium lives in the luminescent organ of the pony fish and is closely associated with its host. The structure of the enzyme in the bacterium is very similar to its structure in eukaryotes.

---

host these bacteria associated with in earlier times.

Recent evidence further suggests that *Paracoccus* is closely related to the Rhodospirillaceae (purple nonsulfur bacteria). It is thought that in an anaerobic environment the symbiont had photosynthetic capability, but as the environment became aerobic the photosynthesis became suppressed as the symbiont acquired greater efficiency in respiration. Studies have demonstrated a striking parallel in the DNA homologies of bacteria such as *Agrobacterium, Rhizobium,* and rickettsia, which are now placed in the same subgroup of purple bacteria. It is interesting to note that some strains of these bacteria form obligate symbioses with eukaryotic organisms.

b. Chloroplasts. The chloroplast more closely resembles a prokaryotic cell than does the mi-

tochondrion. There are many different types of plastids and they probably originated from multiple symbioses that involved cyanobacteria and eukaryotic algae. Most chloroplasts, like mitochondria, are surrounded by two membranes. The inner membrane represents the plasma membrane of the original prokaryotic symbiont, and the outer membrane represents the vacuolar membrane formed by the host around the symbiont. The chloroplasts of green algae, plants, and red algae have two enveloping membranes. The chloroplasts of *Euglena* have three outer membranes, and those of brown algae and diatoms are surrounded by four membranes. It is clear that these different types of plastids could not have evolved from a common ancestor. Cyanobacteria are the most likely ancestors of the plastids of red and green algae, the latter group giving rise to green plants.

The chloroplasts of *Euglena* are thought to

have arisen when a eukaryotic flagellate ingested unicellular green algae that were not digested and subsequently became symbiotic. The first, or outer, membrane around the *Euglena* chloroplast is the vacuolar membrane that the host cell formed around the endosymbiont. The second and third membranes represent the plasma membrane and chloroplast membrane, respectively, of the original green algae. In the brown algae and diatoms, the outer two of the four membranes around the chloroplast originate from the endoplasmic reticulum of the host cell.

Several unusual algae were once thought to be "missing links," that is, intermediate stages in the cyanobacteria-chloroplast evolution. *Cyanidium caldarium, Cyanophora paradoxa,* and *Glaucocystis nostochinearum* are eukaryotic, unicellular algae that contain bluish green structures called *cyanelles,* which are modified cyanobacteria. How the host cell acquired the cyanobacteria is not clear. One possibility is that some eukaryotic algae lost their chloroplasts and ingested cyanobacteria, which were eventually transformed to cyanelles and assumed the role of chloroplasts. Having once had chloroplasts, such colorless eukaryotic hosts would be better equipped to maintain externally acquired cyanobacteria. Most cyanelles will not grow separately in culture and the DNA content of some, such as *C. paradoxa,* is reduced, like that of true chloroplasts. It is not known whether there are genetic interactions between cyanelles and the host nucleus as there are in the case of true chloroplasts. Thus, although cyanelles have a superficial resemblance to chloroplasts and have reduced DNA, the present information is too limited to support the view that they are primitive chloroplasts.

Evidence for the endosymbiotic origin of chloroplasts has been collected from molecular data, such as DNA-RNA hybridization of chloroplasts, nucleocytoplasm, and cyanobacteria and sequencing of plastocyanin, ferredoxin, and cytochrome c as well as tRNA, and 16S and 5S rRNA. All of these studies have indicated a close relationship between chloroplasts and blue-green prokaryotes. In particular, the 5S sequences of chloroplast rRNA show a greater similarity to those from cyanobacteria than to those from eukaryotic cytoplasm.

One criticism of the endosymbiotic origin of chloroplasts and mitochondria is that it fails to explain how genes were transferred from the symbiont to the host nucleus. Recently, it was demonstrated that when foreign nucleic acids were injected into the cytoplasm of a frog oocyte, some DNA fragments migrated and accumulated in the nucleus of the oocyte. Similarly, plant nuclei are known to incorporate foreign DNA as shown in the case of *Agrobacterium tumefaciens,* which was described in chapter 4.

c. Microtubules and Nuclei. A third aspect of the S.E.T. involves organelles that are used for motility. Margulis (1981) restricts the term *flagella* to motile structures or organelles found in prokaryotes. She calls the organelles of motility found in eukaryotes *undulipodia.* Flagella, as they are defined by Margulis, lack microtubules and contain the protein flagellin. In contrast, undulipodia are made up of microtubules, which consist of tubulin and are arranged in characteristic pairs, a $9 + 2$ arrangement. Margulis suggests that undulipodia developed from tubulin-containing spirochetes, which are helically shaped, motile bacteria. The spirochetes formed symbiotic associations by attaching themselves to the outer surface of nonmotile, heterotrophic protoctists and eventually evolved into undulipodia.

Symbiotic associations in the hindgut of termites and wood-eating roaches show that protoctist symbionts such as *Mixotricha paradoxa* are covered with a complex arrangement of spirochetes that are responsible for the movement of the protoctist. The symbiotic theory for the origin of undulipodia is supported by the following evidence: (a) presence of protein similar to alpha tubulin in free-living spirochetes and those associated with symbiotic protoctists, (b) tubulin-containing spirochetes found in environments devoid of eukaryotes.

Margulis also proposes that symbiotic spirochetes gave rise, via mutations, to centrioles, chromosomal kinetochores, and the spindle of the mitotic apparatus. With the development of mitosis came an explosion of many new species of eukaryotic microorganisms, some of which eventually gave rise to fungi and animals.

The origin and evolution of the eukaryotic nucleus is clouded in uncertainty and the following hypotheses have been proposed as explanations.

1. The nucleus arose autogenously when prokaryotic DNA became enclosed within an invagination of the plasma membrane, thus cre-

ating two compartments, the DNA-containing nucleus and non-DNA cytoplasm. Each unit then evolved separately.

2. The nucleus originated endosymbiotically from a prokaryotic cell that contained DNA and RNA.

3. The nucleus evolved from the microtubule-containing spirochetes that were ingested by a *Thermoplasma*-like host cell.

D. Circadian Rhythms

Circadian rhythms are biochemical oscillations with four main characteristics: (1) low frequency, (2) temperature insensitivity, (3) persistence in enucleated cells, and (4) resistance to metabolic toxins and protein synthesis inhibitors. Glycolytic oscillations are well known in many cells and there are regular pulses of cyclic AMP secretion by certain cells during the aggregation stage of cellular slime molds. Circadian rhythms have been studied in detail in protoctists such as *Acetabularia, Chlamydomonas,* and *Tetrahymena*. Scientists are attracted to the endosymbiotic theory of the eukaryotic cell for its ability to explain the occurrence of two primary oscillations as properties of distinct partners in the original symbiosis of host and symbiont. Scientific evidence in support of this hypothesis has yet to emerge, but the level of curiosity has increased the number of publications on this topic.

E. Host Cell: An Intracellular Ecosystem

Ecosystems are understood as ecological units of the biosphere, with well-defined boundaries and characteristic species of organisms. Each organism in an ecosystem generally forms colonies. From such a perspective, intracellular space represents an extreme in microhabitat. F. J. R. Taylor (1983) defines a cell as a *cytocosm* and identifies four features of this unique habitat: (1) a stable environment for the cell following symbiont intrusion (such a constant environment inhibits sexuality of the symbiont and encourages it to stay in only one phase of its life cycle); (2) extremely slow rate of succession; (3) exchange of nutrients and materials between the symbiont and host (the exchange is controlled by the selective permeability of the host

membrane); and (4) colonization of cell cytoplasm by new symbionts, bringing forth processes for evolutionary change.

The eukaryotic cell, then, is one of the oldest, smallest, and most intensive ecosystems. Individual organisms in the intracellular ecosystem should be genetically independent in order to reproduce successfully. Interdependent relationships between the individuals of the ecosystem involved division of labor in cellular functions. Modern-day eukaryotic cells have a variety of regulatory mechanisms, some in the form of negative feedback systems, that maintain and perpetuate the ecosystem.

The level of integration achieved by symbionts in the host cell is expressed in terms of evolutionary history. Symbioses discussed in this book show a wide range of adaptations between the symbionts and their ecosystems.

Intracellular parasitism is a nutritional relationship in which the parasitic symbiont obtains its food from the host cell. Intracellular parasites may become lifelong residents in the host cell, but some can also live in the extracellular environment. An intracellular symbiont must enter the host cell, overcome host defense mechanisms, reproduce, avoid killing the host cell, and finally exit from the old cell to a new one. Some specific examples of intracellular parasitism discussed in detail in this text are *Bdellovibrio, Plasmodium, Trypanosoma, Leishmania,* and *Toxoplasma*.

F. Symbiosis and Parasexuality

According to Margulis (1981), symbiosis is a form of *parasexuality,* a special type of reproduction found in some fungi that does not involve the usual stages of the sexual process. Two fungi may fuse and one may transfer its nuclei to the other. The nuclei remain together in the hyphae. Each hyphal cell, therefore, may contain two genetically different nuclei, a condition known as *heterokaryosis*. The different nuclei may fuse occasionally and undergo a mitotic reduction division that produces nuclei with new genetic traits. In a symbiotic association, the genes of both partners are close together but never actually fuse. Natural selection acts on both genomes as a unit. Many symbioses are more fit than their individual partners and some have cycles during which the individual and symbiotic states alternate.

G. A New Terminology

The discovery of the composite, symbiotic nature of the eukaryotic cells has raised questions concerning terminology. The word *cell* has been used traditionally to designate the basic unit of living organisms and it has been assumed that each cell has a single genome. Some scientists believe that the new concept of the polygenomic nature of eukaryotic cells necessitates the introduction of new terms and a revision of the cell theory, which states that all living organisms consist of cells. F. J. R. Taylor (1974) has introduced terms such as *procell* and *eucell* to distinguish between prokaryotic and eukaryotic cells and also words such as *monad, dyad, triad,* and *quadrad* to refer to cells according to the number of genomes they contain. For example, bacteria would be monads, and plants would be triads. Taylor also suggests replacing the word *multicellular,* which refers to organisms that consist of many cells, with terms such as *polymonad, polydyad,* and *polytriad.*

H. Summary and Perspectives

Studies on the origin of the eukaryotic cell have resulted in some of the most exciting findings in the field on symbiosis. Several competing theories have stimulated much research on the molecular biology of cell organelles such as mitochondria and chloroplasts, and prompted a search for the modern-day prototype of the ancestral prokaryote host that evolved into a eukaryote. The serial endosymbiosis theory (S.E.T.) states that eukaryotes evolved from prokaryotic symbioses; the autogenous theory holds that organelles in the eukaryotic cell arose from compartmentalization of primitive ancestral cells. The fungal theory states that the first eukaryote was a fungus that arose from an aerobic bacterium. The nature of the primitive host cell in the S.E.T. is not clear, but two living examples of what it may have been like are the giant amoeba *Pelomyxa palustris* and the mycoplasmalike prokaryote *Thermoplasma acidophilum.* The serial endosymbiosis theory provides the strongest evidence that mitochondria and chloroplasts and also undulipodia may have evolved from the endosymbioses of ancestral prokaryotic cells. The most convincing evidence comes from the sequence homology between nucleic acids of chloroplasts and mitochondria and those of free-living bacteria. Chloroplasts of algae and plants probably evolved from different symbiotic events between cyanobacteria and eukaryotic algae. Cyanelles are transformed cyanobacteria that live inside certain unicellular algae and have been considered to be an intermediate stage, or "missing link," in the evolution of chloroplasts. Undulipodia (eukaryotic flagella) and other microtubule-containing structures such as centrioles, kinetochores, and spindle apparatus may have arisen from symbiotic spirochetes. The origin of other organelles such as the nucleus, Golgi bodies, and endoplasmic reticulum may be explained better by the autogenous theory.

Circadian rhythms are biochemical oscillations that may be explained by the endosymbiotic theory as properties of two genetically distinct partners, but firm evidence is lacking.

The host cell is a unique ecosystem (cytocosm) or microhabitat. Some intracellular parasites have adapted to this habitat and live their entire lives in the host cells. Symbiosis has been considered to be a form of parasexuality, a special type of sexual process found in some fungi. Because of the composite nature of the eukaryotic cell, new terms such as *monad* (an organism with one genome, such as a prokaryote), *dyad* (an organism with two genomes, such as an animal, fungus, or protoctist), and *triad* (a plant, which has three genomes) have been proposed to replace the word *cell.*

Review Questions

1. Name the theories proposed to explain the origin of the eukaryotic cell. Which is the most accepted theory and why?
2. Which living organisms are similar to the ancestral host cell that evolved into the modern eukaryotic cell? Describe the primitive features of these organisms.
3. Which cell organelles may have had an endomembranous origin?
4. Explain the term *symbionticism.*
5. Describe the characteristics of mitochondria and chloroplasts that are similar to those of prokaryotes.
6. Explain why *Paracoccus denitrificans* is thought to be similar to ancestral bacteria that gave rise to mitochondria.
7. Distinguish between the chloroplasts of plants, diatoms, and *Euglena.* How did these differences arise?

8. What are cyanelles?

9. Symbiotic spirochetes are thought to have given rise to which organelles?

10. List several ways by which the nucleus of a eukaryotic cell may have arisen.

11. How can the endosymbiotic theory explain the phenomenon of circadian rhythms?

12. Why is a host cell considered to be an ecosystem and what are the features of this microhabitat?

13. How is symbiosis similar to parasexuality?

14. Define the terms *monad, dyad,* and *triad.*

Further Reading

Cavalier-Smith, T. 1980. Cell compartmentation and the origin of eukaryotic membranous organelles. In: Endocytobiology, Vol. 1, 893–942, ed. W. Schwemmler and H. E. A. Schenk. Walter de Gruyter, Berlin.

Gray, M. W., and W. F. Doolittle. 1982. Has the endosymbiont hypothesis been proven? Microbiol. Reviews 46:1–42.

Kunicki-Goldfinger, W. J. H. 1980. Evolution and endosymbiosis. In: Endocytobiology, Vol. 1, 969–984, ed. W. Schwemmler and H. E. A. Schenk. Walter de Gruyter, Berlin.

Margulis, L. 1971. Symbiosis and evolution. Sci. Am. 225 (August):48–57.

Margulis, L. 1981. Symbiosis in cell evolution. W. H. Freeman, San Francisco. 419 pp. (A detailed treatment of the serial endosymbiosis theory.)

Margulis, L., L. P. To, and D. Chase. 1981. Microtubules, undulipodia, and *Pillotina* spirochetes. Ann. N.Y. Acad. Sci. 361:356–368. (A review of two concepts of the origin of microtubule systems of eukaryotic cells, the endogenous concept and the symbiont concept.)

Moulder, J. W. 1985. Comparative biology of intracellular parasitism. Microbiol. Rev. 49:298–337. (A review of the interactions of intracellular parasites with vertebrate hosts.)

Schwemmler, W. 1980. Endocytobiology: A modern field between symbiosis and cell research. In: Endocytobiology, Vol. 1, 943–967, ed. W. Schwemmler and H. E. A. Schenk. Walter de Gruyter, Berlin.

Stern, D. B., and J. D. Palmer. 1984. Extensive and widespread homologies between mitochondrial DNA and chloroplast DNA in plants. Proc. Natl. Acad. Sci. 81:1946–1950. (Presents evidence that DNA transfer occurs between chloroplast and mitochondria.)

Taylor, F. J. R. 1974. Implications and extensions of the serial endosymbiosis theory of the origin of eukaryotes. Taxon. 23:229–258.

Taylor, F. J. R. 1983. Some eco-evolutionary aspects of intracellular symbioses. Int. Rev. Cytology,
supplement 14:1–25. (Considers the cellular and environmental consequences of marine intracellular symbioses.)

Taylor, F. J. R. 1980. The stimulation of cell research by endosymbiotic hypotheses for the origin of eukaryotes. In: Endocytobiology, Vol. 1, 917–942, ed. W. Schwemmler and H. E. A. Schenk. Walter de Gruyter, Berlin.

Taylor, F. J. R., and P. J. Harrison. 1983. Ecological aspects of intracellular symbiosis. In: Endocytobiology, Vol. 2, 827–842, ed. H. E. A. Schenk and W. Schwemmler. Walter de Gruyter, Berlin.

Weisburg, W. G., and C. R. Woese, M. E. Dobson, and E. Weiss. 1985. A common origin of rickettsiae and certain plant pathogens. Science 230:556–558. (Ribosomal RNA sequence comparisons have revealed evolutionary relationships of rickettsiae to agrobacteria and rhizobacteria as well as to mitochondria.)

Whatley, J. M., and F. R. Whatley. 1984. Evolutionary aspects of the eukaryotic cell and its organelles. In: Cellular interactions. Encyclopedia of plant physiology, new series, Vol. 17, 18–58, ed. H. F. Linskens and J. Heslop-Harrison. Springer-Verlag, Berlin.

Bibliography

Carlile, M. J., J. F. Collins, and B. E. B. Moseley (eds.). 1981. Molecular and cellular aspects of microbial evolution. Thirty-second Symposium of the Soc. for General Microbiology. Cambridge Univ. Press, Cambridge. 368 pp. (Articles on the evolution of the major groups of microorganisms.)

Cavalier-Smith, T. 1981. The origin and early evolution of the eukaryotic cell. Thirty-second Symposium of the Soc. for General Microbiology. Cambridge Univ. Press, Cambridge, 368 pp. (Presents the fungal theory for the origin of the eukaryote cell.)

Fredrick, J. F. (ed.). 1981. Origins and evolution of eukaryotic intracellular organelles. Ann. N.Y. Acad. Sci. 361:1–512.

Margulis, L. 1982. Symbiosis and the evolution of the cell. In: Yearbook of science and the future, 104–121. Encyclopedia Britannica Yearbooks, Chicago.

Margulis, L. 1985. Undulipodiated cells. BioScience 35:333. (A plea to substitute the term *undulipodia* for *eukaryotic flagella* and *cilia.*)

Margulis, L., and K. V. Schwartz. 1982. Five kingdoms: An illustrated guide to the phyla of life on earth. W. H. Freeman, San Francisco. 338 pp. (A catalog of the world's living organisms.)

Martin, J. P., and I. Fridovich. 1981. Evidence for a natural gene transfer from the pony fish to its bioluminescent bacterial symbiont *Photobacter leiognathi.* J. Biological Chemistry 256:6080–6089.

Reanney, D. C., and P. Chambon. (eds.). 1985. Genome evolution in prokaryotes and eukaryotes. Int. Rev. Cytology 93:1–368. (Includes a comprehensive treatment of mitochondrial and chloroplast genomes.)

Schenk, H. E. A., and W. Schwemmler (eds.). 1983. Endocytobiology, Vol. 2, Intracellular space as oligogenetic ecosystem. Walter de Gruyter, Berlin. 1,071 pp.

Schwemmler, W., and H. E. A. Schenk (eds.). 1980. Endocytobiology, Vol. 1, Endosymbiosis and cell biology: A synthesis of recent research. Walter de Gruyter, Berlin. 1,060 pp.

Taylor, D. L. 1981. Evolutionary impact of intracellular symbiosis. Ber. Deutsch. Bot. Ges. 94:583–590.

Tribe, M. P., and P. A. Whittaker. 1981. Chloroplasts and mitochondria. 2d ed. Institute of Biology, Studies in Biology, no. 31. Edward Arnold, London. 82 pp.

Tribe, M. P., P. A. Whittaker, and A. Morgan. 1981. The evolution of eukaryotic cells. Studies in Biology, no. 131. Edward Arnold, London. 60 pp.

Vidal, G. 1984. The oldest eukaryotic cells. Sci Am. 250 (February):48–57.

Wallin, I. E. 1927. Symbionticism and the origin of species. Williams and Wilkins, Baltimore. 171 pp. (The first published opinion that mitochondria were bacteria that formed microsymbiotic complexes with cells.)

CHAPTER 6

Fungal Associations

Fungi as Symbionts of Protoctists and Animals

A. Introduction

Fungi are familiar to most people as mushrooms in fields and woods or as mold on stale bread and decaying fruit. There are over 100,000 species of fungi, most of them microscopic. Fungi grow practically everywhere and produce large numbers of spores that are in the air, in soil or water, and on plants and animals. Under favorable conditions, the spores of many fungi will germinate and produce colonies within a few days.

Fungi are an important part of human affairs. Some species are of benefit to humans but others are extremely destructive. Penicillin, an antibiotic obtained from a fungus, has saved countless lives; yeasts are fungi that are used by people throughout the world to make bread and alcoholic beverages. Some fungi have caused destructive plant diseases. The great Bengal famine of 1945–1946 began with an infection of rice crops by the leaf blight fungus *Helminthosporium oryzae.* The infection reduced rice yields by 75% to 90% and resulted in the deaths of an estimated two million people. Between 1870 and 1890, the British people switched from drinking coffee to tea because a rust infection had destroyed the coffee trees in Sri Lanka. In the United States large populations of two common trees have been destroyed by fungal infections. The American chestnut has been virtually eliminated by the chestnut blight fungus, *Endothia parasitica,* and the American elm has been decimated by the fungus *Ceratocystis ulmi.* Other destructive pathogens are the rust and smut fungi that infect cereals. Fungi cause many chronic illnesses of humans as well as other animals. Ringworm and athlete's foot are familiar fungal infections of skin tissue. Many types of allergies and food poisoning are caused by common soil and airborne fungi.

B. Characteristics of Fungi

Fungi, like animals, are *heterotrophic,* eukaryotic organisms. For their nutrition they depend on organic molecules produced by other organisms. Fungi may be *biotrophic, necrotrophic,* or *saprophytic.* Biotrophic fungi obtain nutrients only from living cells of host organisms. Necrotrophic fungi obtain food from the dead cells of organisms they kill. Saprophytic fungi live off dead organisms. Fungi were long thought to be nonphotosynthetic plants mainly because they have cell walls. During the past three decades, *mycologists* (scientists who study fungi) have recognized the unique features of these organisms. Most biologists today recognize fungi as constituting a separate kingdom.

Fungi secrete extracellular enzymes that break down organic molecules, which are then absorbed and assimilated by the fungus. This mode of nutrition differs from that of animals, which ingest food, and from plants, most of which can manufacture food through photosynthesis. Many fungi form microscopic branching filaments called *hyphae,* which grow and usually produce a circular colony called a *mycelium.* A fungal hypha grows only at its tip and may be partitioned by cross walls to form a row of cells. Some fungi lack cross walls and a few, such as yeasts, are unicellular. Fungi usually reproduce asexually by producing spores either at the tips of hyphae or in sporangia that are produced on specialized hyphae. Many species produce reproductive structures, called *fruiting bodies,* that contain spores that are products of sexual reproduction. Fungal spores have characteristic shapes, sizes, and markings and are often used for species identification.

A fungal cell has the usual eukaryotic organelles. Food is stored as glycogen or lipid. The cell wall is composed of fibrillar and matrix components; the fibrillar portion contains a network of *chitin* microfibrils, which make the wall rigid, and the matrix contains protein and mannose. Cellulose, which is common in plant cell walls, is absent in fungi.

The process of mitosis and meiosis is unique for most fungi. When the nucleus divides, its membrane does not break down, as in other organisms, but rather pinches in and forms two daughter nuclei. Further, fungal chromosomes do not align along the equatorial plate of the cell during metaphase.

Many fungi reproduce sexually. Their gametes are not motile and the male gametes are carried to the female gametes by wind or water. In some fungi the male and female sex organs grow toward each other and fuse. Nuclei from these sex organs form pairs but do not fuse. Each pair of nuclei (N + N) is called a *dikaryon.* Fungi commonly have a dikaryotic phase, which may be a major portion of their life cycle. Eventually, the cell nuclei fuse to produce a diploid (2N) nucleus, which undergoes meiosis.

Fungi have exploited a large number of ecological niches in the biosphere. As saprobes, they constitute a major link in the decomposition of dead plants and animals and thus in the recycling of carbon. Some saprobic fungi will invade an injured living host and produce disease. These fungi are called *opportunistic pathogens.* Biotrophic fungi have developed mutualistic associations with algae, plants, protoctists, and animals.

C. Classification

Fungal taxonomy has undergone many revisions during the past twenty years. The following is a brief synopsis of the major groups of fungi as they are recognized today. It is intended to familiarize the reader with the groups from which we selected examples of fungal symbioses.

KINGDOM FUNGI (MYCOTA)

Phylum: Zygomycota (zygomycetes)
  Class: Zygomycetes; predatory fungi of insects; vesicular-arbuscular mycorrhizal fungi
  Class: Trichomycetes; obligate symbionts of arthropods
Phylum: Ascomycota (sac fungi)
  Class: Hemiascomycetes; yeasts; fungi that cause leaf curl and witches-broom disease
  Class: Plectomycetes; saprophytes and plant pathogens such as powdery mildews
  Class: Laboulbeniomycetes; obligate symbionts of arthropods
  Class: Pyrenomycetes; lichen-forming fungi; plant pathogens that cause Dutch elm disease, ergot of cereals, and plant wilts and rots
  Class: Discomycetes; lichen-forming fungi; saprophytes and plant pathogens that cause brown rot of fruits and gray molds of vegetables
Phylum: Basidiomycota (club fungi)
  Class: Teliomycetes; rust fungi; smut fungi
  Class: Hymenomycetes; toadstools; jelly fungi; ectomycorrhizal fungi
  Class: Gasteromycetes; puff balls, stink horns; ectomycorrhizal fungi
Phylum: Deuteromycota (Fungi Imperfecti)
  Class: Blastomycetes; yeastlike fungi
  Class: Hyphomycetes; condial fungi; nematode-trapping fungi
  Class: Coelomycetes; plant pathogens causing anthracnose of cotton, beans, and flax

D. Symbionts of Protoctists and Animals

1. **PROTOCTISTS.** Soil amoebae and testaceous (with shells) rhizopods are common hosts of specialized groups of zygomycetes. The fungi use adhesive hyphae to capture their hosts. Some fungal species do not penetrate their host, whereas others are internal symbionts. Each species of fungus produces a characteristic assimilative organ that digests the host cytoplasm (Fig. 6.1). *Stylopage rhabdospora* traps soil amoebae by means of thin, sticky strands of mycelia. At the point of contact with the amoeba the fungus produces penetration pegs that enter the protoctist and branch to form assimilative organs. Species of *Cochlonema* and *Endocochlus*

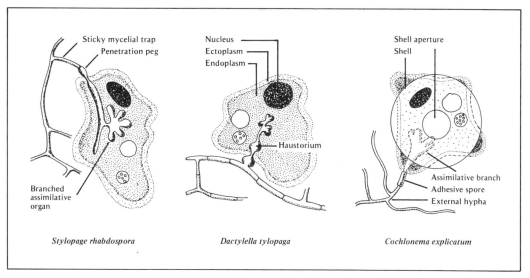

Figure 6-1. Examples of fungal parasitism of soil amoebae. (Adapted from Cooke, R. 1977. The biology of symbiotic fungi. John Wiley, London.)

are obligate endosymbionts of soil amoebae. *Cochlonema explicatum* produces adhesive spores that attach to amoebae, germinate, and then produce germ tubes that enter the host cytoplasm. Once inside the host, the tip of the germ tube swells and elongates. The amoeba gradually becomes sluggish and after it dies fungal spores are released from the host.

A few hyphomycetes, such as *Dactylella tylopaga,* also capture rhizopods. The morphological adaptations of these fungi, such as adhesive hyphae and assimilative organs, are similar to those of predaceous zygomycetes.

Fungal attacks on soil amoebae are widespread in nature. These fungi are obligate symbionts whose axenic culture has not been successful because their spores will germinate only in the presence of a host. It is not known what specific nutrients the fungi obtain from the amoebae.

2. NEMATODE-TRAPPING FUNGI. An American mycologist, Charles Drechsler, pioneered the study of fungi that trap microscopic organisms. From 1933 to 1975, Drechsler published papers on fungi that attacked nematodes and amoebae. In England, Charles Duddington's book *The Friendly Fungi: A New Approach to the Eelworm Problem* described the curious behavior of these soil fungi and captured the imagi-

nation of scientists and amateurs. *Nematologists* (scientists who study nematodes) explored the possibility of using carnivorous fungi to control nematode pests of plants.

Nematodes are roundworms that are common inhabitants of soil. In addition to feeding on bacteria and fungi, nematodes parasitize plants and animals. Most fungi that trap nematodes are hyphomycetes but some are zygomycetes. These fungi are common in decaying organic matter. When a wandering nematode touches a hypha of the zygomycete *Stylopage hadra,* the fungal cytoplasm moves rapidly to the point of contact, where a drop of viscous fluid is released. The nematode sticks to the fluid and, after a brief struggle, becomes immobilized. At this time, the fungus penetrates the nematode's body by means of a slender peg, which grows and branches and fills the host body cavity with hyphae. The hyphae of *S. hadra* become sticky only after they are touched by a moving organism.

It is not fully understood what stimulates a nematode-trapping fungus to form traps or how it attracts and kills its prey. Some *nematophagous fungi* grown in axenic culture form traps only in the presence of nematodes. In one study, a culture broth in which nematodes had grown contained substances, collectively called *nemin,* which induced trap formation in *Arthrobotrys* spp. Recent studies have indicated that amino

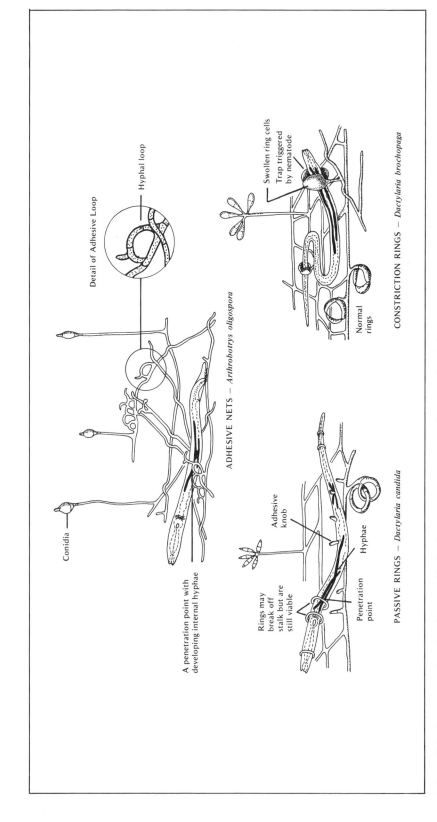

Detail of Adhesive Loop

Hyphal loop

Conidia

A penetration point with developing internal hyphae

ADHESIVE NETS − *Arthrobotrys oligospora*

Swollen ring cells
Trap triggered by nematode

Normal rings

CONSTRICTION RINGS − *Dactylaria brochopaga*

Rings may break off stalk but are still viable

Adhesive knob

Hyphae

Penetration point

PASSIVE RINGS − *Dactylaria candida*

Figure 6-2. Three types of mechanisms used by hyphomycetes to capture soil nematodes. (Adapted from Barron, G. L. 1977. The nematode-destroying fungi. Canadian Biological Publications, Ontario, Canada.)

acids such as valine and leucine, carbohydrates such as arabinose and ribose, and glycerol stimulated trap formation in some fungi. Some scientists feel that the fungus produces chemicals that attract wandering nematodes to the traps. It is also possible that fungal spores may appear as food to nematodes. Toxins produced by the fungus may be the cause of death of captured nematodes. Fungi use nematodes to supplement nutrients they obtain from dead organic matter.

Hyphomycetes that capture nematodes produce specialized hyphal traps, which may or may not be adhesive depending on the fungal species. Following are examples of common types of traps.

a. Adhesive Nets and Branches. These structures are characteristic of many nematode-trapping fungi. *Arthrobotrys oligospora* produces a complex of three-dimensional nets and loops. The fungus produces short lateral branches, which grow, curl, and then fuse with the parent hypha. A branch may arise from the loop hypha and produce another loop. Repeated loop production results in a network of adhesive hyphae. When a roundworm contacts a trap, it becomes caught and as it struggles it entangles itself in other traps and eventually is unable to move. A penetration peg may develop from any point of the fungal network. In the body cavity of the nematode, the peg produces a bulbous structure from which the assimilative hyphae arise. Most of the cell contents of the host are digested and absorbed by the fungus (Fig. 6.2).

b. Passive Rings. These traps are typical of *Dactylaria candida*. Each ring consists of three or four cells and is supported on a stalk. The diameter of a ring is barely wide enough to accommodate the head of a nematode. If a nematode tries to pass through the ring, it gets caught and is unable to retract. The ring often becomes detached from the mycelium during a nematode's struggle to free itself, but the ring remains viable and ultimately a peg from one of its cells enters the host. The killing action of this type of fungus is slow and frequently a nematode will have several rings around its body as it moves through additional traps. In addition to passive rings, *D. candida* also captures nematodes by means of short, adhesive knobs (Fig. 6.2).

c. Constriction Rings. These structures are the most sophisticated trapping mechanisms produced by fungi, such as *Arthrobotrys anchonia* and *Dactylaria brochopaga* (Fig 6.2). A typical ring consists of three cells supported on a one to two cell hyphal stalk. As the nematode's head moves through the ring, it contacts the ring cells and causes them to swell instantly, thereby trapping the nematode by constriction. Once a nematode is captured, a penetration peg from one of the ring cells grows into the body cavity and forms assimilative hyphae. Only the inner surface of the ring swells and the process is completed in 0.1 seconds. The mechanism of ring constriction has been a subject of several studies but is still poorly understood. Hot water, dry heat, and glass microneedles can trigger the constriction response.

3. ENDOSYMBIONTS OF NEMATODES. Endosymbiotic fungi of nematodes have several common features. These include (a) almost total absence of any hyphal structure outside the nematode, (b) exit tubes for the release of spores to the outside of the host, (c) resting spores that allow the fungus to survive in the absence of a suitable host, and (d) host infection resulting from ingestion of a spore. Following are examples of fungi that are internal symbionts of nematodes.

a. **Meria coniospora (*hyphomycete*).** The fungus produces tear-shaped spores that are sticky at their pointed end. The spores attach to the head of a nematode and produce penetration pegs from which infection hyphae develop. Young hyphae with a characteristic wavy appearance grow in the host body cavity. When mature, some hyphae break through the body wall of the host and produce spores (Fig. 6.3).

b. **Harposporium anguillulae (*hyphomycete*).** Spores of this fungus are crescent-shaped and have a pointed end. When the spores are ingested by a nematode, they become embedded in the muscular esophagus and the nematode is unable to feed. The spores germinate and produce hyphae, which grow throughout the body cavity of the host. At maturity the fungal colony develops erect hyphae, which produce numerous spores or conidia. The fungus also produces thick-walled *chlamydospores* that can withstand unfavorable environmental conditions (Fig. 6.3).

c. **Nematoctonus (*basidiomycete*).** Some species of this fungus capture nematodes by means

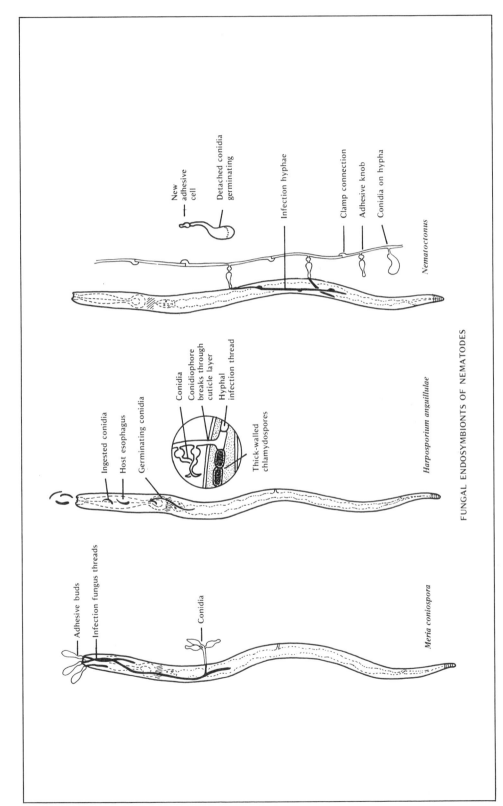

Figure 6-3. *Fungal endosymbionts of nematodes.* (Adapted from Barron, G. L. 1977. The nematode-destroying fungi. Canadian Biological Publications, Ontario, Canada.)

of adhesive knobs or aerial hyphae; other species are endosymbiotic. In the trapping species, each trap consists of a secretory cell that is swollen at both ends, constricted in the middle, and surrounded by a viscous, sticky fluid. When a nematode touches the fluid, a bond forms that is so strong that it rips open the cuticle of a struggling host. The fungus penetrates the captured nematode and fills its body cavity with hyphae. Endosymbiotic species produce sausage-shaped conidia on small, stalked hyphae. After dispersal, a conidium germinates and produces an adhesive cell at the tip of the germ tube. The infection process begins when the cell becomes attached to the cuticle of a wandering nematode (Fig. 6.3). A toxin produced by the germinating conidium immobilizes the nematode, which dies soon after it contacts the conidium.

Many nematodes parasitize cultivated plants, causing reduced yields and predisposing the host plants to secondary invasions by bacteria and fungi. Heavy nematode infestations often result in the premature death of plant seedlings. Scientists have developed a variety of control measures against soil nematodes such as fumigation with toxic chemicals and resistant varieties of plants. The use of symbiotic fungi as a way to control pathogenic nematodes has been a focus of many scientific studies. The results, in general, have not been successful when compared to those of soil fumigations. Nevertheless, many chemicals that are successful soil fumigants have now been banned for health reasons and farmers are left with fewer alternative measures to control nematode pests.

4. ARTHROPODS. The phylum Arthropoda includes over 70% of all the known species of animals. Insects, with an evolutionary history of 300 million years and more than 800,000 species, are the largest class of animals and represent one of the great success stories in evolution. Insects have occupied almost every conceivable ecological niche in the terrestrial and aquatic habitat and have developed associations with fungi that range from casual to intimate. Fungal symbionts and insect hosts show a variety of morphological, physiological, and ecological coadaptations, and also have some common characteristics. Both contain chitin as a structural component, in the walls of fungi and in the exoskeleton of insects. Many fungi and insects are small and produce large numbers of offspring, and both groups play an important role in the recycling of organic compounds.

a. Symbionts of Insects: Pathogenic Associations. Species of *Beauveria, Cordyceps, Coelomomyces, Entomophthora,* and *Metarrhizium* are well-known examples of fungi that are pathogens of insects. Some of these fungi have been used successfully in the biological control of insect pests. The symbionts infect the insect hosts, assimilate their hemolymph, digest soft tissue, and produce *mycotoxins,* which kill the host.

Many characteristics of insect-fungus symbioses can be generalized. Fungal spores usually attach to the insect cuticle in a specific manner, possibly through physical and chemical compatibilities between the fungus and its host. A high humidity (greater than 95% in some species) is necessary for spore germination. The spore produces a germ tube whose tip swells and attaches to the cuticle. Penetration pegs then develop and enter the host. On agar medium, a germinating spore produces protease, lipase, and chitinases, enzymes believed to be responsible for the digestion of the host integument. The symbiont may take several days to reach the insect *hemocoel,* a blood-filled space in the tissues. If the insect molts before the fungus reaches the hemocoel, the symbiont is discarded along with the old cuticle. The fungus multiplies in the hemocoel by small, budding bodies called *blastospores.* The developing fungus produces a toxin that kills the host within 48 hours. The host suffers from tremors and loss of coordination prior to death. The fungus then enters the mycelial phase and develops hyphae, which consume all the organs of the host. The insect body becomes mummified and hyphae grow out of the body. These hyphae produce *conidia* that are dispersed by wind or water. Following are some well-known examples of *entomogenous* (insect-inhabiting) *fungi.*

ENTOMOPHTHORA (ZYGOMYCETE). Species of this fungus have been used to control West African black flies, which are carriers of a nematode that causes human blindness. The fungus is an obligate parasite of the insect. Aphids, muscoid flies, caterpillars of butterflies and moths, and grasshoppers are common hosts of other species of *Entomophthora.* In the Soviet Union, pea aphids on legumes have been suppressed by *Entomophthora* infections. One major limitation in the commercial use of this fungus as a

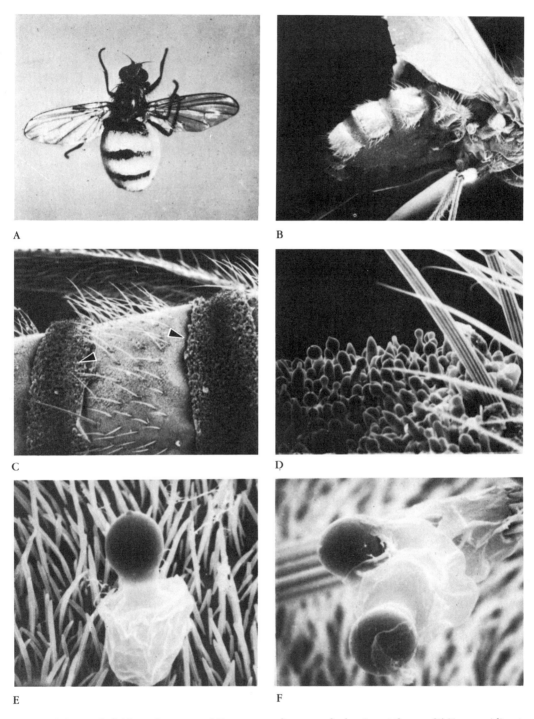

A

B

C

D

E

F

Figure 6-4. Conidial life cycle stages of Entomophthora muscae, *parasite of the onion fly. (A) Infected female fly. (B, C) Lateral view of fly, showing fungal sporulation (arrows). (D) Conidiophores emerging from insect abdomen. (E) Single coni-* *dium attached to insect thorax. (F) Two conidia attached to an abdominal seta of the host fly. Note mucilage attached to the conidia.* (Raymond I. Carruthers, USDA, Boyce Thompson Institute.)

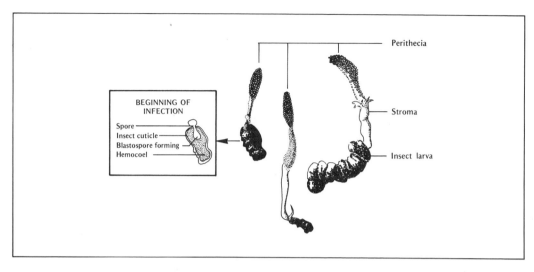

Figure 6-5. Cordyceps *infection of an insect pupa.*

biological control agent is the relatively short life of its fungal spores, about two weeks. One curious aspect of *Entomophthora* is that it produces single-spored sporangia that are "shot" from the tips of specialized hyphae. An infected insect becomes surrounded by a "halo" of white sporangia that are discharged from hyphae that grow out of the insect's body (Fig. 6.4).

CORDYCEPS (ASCOMYCETE). *Cordyceps* attacks many different types of insects and spiders (Fig. 6.5). The infected hosts shrivel and dry after death but resist decay because of an antibiotic compound, cordycepin, produced by the fungus. After the host dies, the fungus produces a cylindrical structure (*stroma*) that grows out from the host and produces many fruiting bodies (*perithecia*), which contain spores. The ancient Chinese were familiar with the mummified silkworms and cicadas produced by *Cordyceps* infections. A Chinese custom was to place precious stone carvings of the infected insects in the mouths of their dead relatives in the hope of immortalizing them. Caterpillars infected with *Cordyceps* have been used in Chinese medicine for curing many illnesses ranging from drug addiction to the effects of overeating.

BEAUVERIA BASSIANA (HYPHOMYCETE). This fungus is commonly isolated from dead, overwintering insects. In the Soviet Union, large quantities of the fungus are prepared commercially and used to control the Colorado potato beetle. The fungus reduces the longevity of the adult insects and also causes a high larval mortality.

Beauveria conidiospores remain viable in soil for up to two years. Efforts are being made in several other countries, including the United States, to use *Beauveria* to supress Colorado potato beetle infestations.

METARRHIZIUM ANISOPLIAE (HYPHOMYCETE). This fungus shows promise for controlling the rhinoceros beetles that have destroyed many coconut trees in the South Pacific. The beetle was accidentally introduced into the region in the 1930s. Beetle-infested trees do not produce fruit and new plantings are destroyed in the seedling stage. The application of fungus around the base of an infected tree dramatically improves its health, since the fungus kills the beetles. Similar efforts to control rhinoceros beetle larvae with *M. anisopliae* in southern India have been successful. The fungus is frequently isolated from soil-inhabiting insects and is also being used to control mosquito larvae and the sugar cane spittlebug. In Brazil, spittlebug control by *M. anisopliae* application is successfully being extended to many acres of sugar cane and pasture land.

b. Symbionts of Insects: Nonpathogenic Associations. There are many fungus-insect associations in which the host is not harmed. The associations are obligatory for one or both partners. The following are examples of nonpathogenic symbioses.

LABOULBENIOMYCETES. The Laboulbeniales is a group of microscopic ascomycetes that are ec-

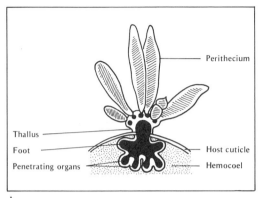

A

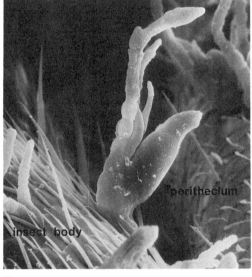

B

Figure 6-6. Laboulbenia *parasitism of insects. (A) Schematic drawing of the fungus with its penetrating organs and perithecia. (B) Scanning electron micrograph of fungus perithecia emerging from the insect cuticle.* (Part (*B*): Gerald Van Dyke, North Carolina State University.)

tosymbionts of insects and other arthropods. The fungi appear as black or darkish bristles to the naked eye and do not harm the insects. The fungi are obligate symbionts, and although some species have a wide host range, most live only on certain insect species.

Incredibly, a fungus may be restricted to specific locations on the insect such as mouth parts, legs, or wings. Species of *Laboulbenia* and *Stigmatomyces* are common on flies and beetles (fig. 6.6). *Stigmatomyces ceratophorus* is the only species that has been grown in axenic culture, on fly wings kept on nutrient agar. The nutritional requirements of these fungi have yet to be investigated. A typical fungal life cycle begins with the germination of an ascospore on the host integument. The fungus produces a *foot cell* that anchors the symbiont to the host cuticle and then produces a thallus that may consist of only a few cells. The thallus produces hairlike structures, called *appendages,* which may be unicellular or multicellular. Male and female reproductive structures and the fruiting body, or *perithecium,* occur interspersed with the appendages. In most Laboulbeniales, the symbiont is anchored to the host by a foot cell that does not appear to penetrate the cuticle. How these fungi obtain nutrients from the host, without visible signs of penetration, is not clear. The fungus may produce minute penetrating structures, or the cuticle contains sufficient nutrients for the fungus. In a few species, such as *Herpomyces sty-*

lopygae and *Trenomyces histophthorus,* penetrating structures develop from the foot cell and grow through the host integument to the hemocoel, where fungal assimilative organs are formed. In *T. histophthorus,* when the fungus reaches the hemocoel it produces spherical swellings with nodular branches that penetrate the fat bodies of the host. The fungal symbionts in all cases do not produce any visible ill effects in the host.

TRICHOMYCETES. Trichomycetes are obligate symbionts that usually live attached to the intestinal wall of marine, freshwater, and terrestrial arthropods. The fungus receives nutrients from the contents of the host's digestive tract but does not harm or benefit the host. Thus, the symbiosis is commensalistic. The host becomes infected when it swallows fungal spores. The spores germinate and the germ tubes attach to the gut lining of the host by a specialized cell, the *holdfast.* The fungal thallus consists of either branched or unbranched hyphae. *Amoebidium parasiticum* is an exceptional trichomycete in that it attaches only to the outside of aquatic crustaceans and insects (Fig. 6.7). Internal trichomycetes can be identified only by dissecting the insect host.

FUNGI AND AMBROSIA BEETLES. Associations between fungi and beetles have been known for almost 150 years. In 1836, J. Schmidberger observed an association of beetles with a glistening white material, which he could not identify, in

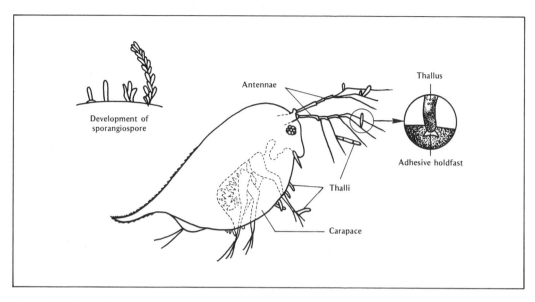

Figure 6-7. The trichomycete Amoebidium parasiticum *attached to* Daphnia *sp.* (Adapted from Moss, S. T. 1979. Commensalism of the trichomycetes. In: Insect-fungus symbiosis, 175–227, ed. L. R. Batra. Allanheld, Osmun, Montclair, N.Y.)

the galleries that beetles excavated in wood. He thought the material was sap from the wood and called it *ambrosia.* In 1844, Theodor Hartig discovered that the "ambrosia" was actually, *Monilia candida,* which was cultivated by the beetles and used as food.

Ambrosia beetles make up about thirty-six genera in two families of the insect order Coleoptera. The beetle genera *Xyleborus* and *Platypus,* each with over 200 species, are important groups of ambrosia beetles. The beetles inhabit the sapwood of weakened trees. As a rule, healthy trees and dead trees are not infected. Some beetle species attack only trees in a particular plant family, while other species infest only certain regions of a tree, such as the terminal branches. The beetles excavate galleries in the wood and feed on the fungus that grows on the gallery walls. The beetles are well adapted to live in the galleries, being small and dark with elongated bodies and short legs. Their mouth parts are delicate and covered with flexible setae. Each species of ambrosia beetle forms its own characteristic tunnel pattern. For example, in hickory trees, the *Xyleborus* beetle produces branched galleries from a single entrance into the bark (Fig. 6.8). The female beetle carries fungal spores into the tunnel and deposits them

on a substrate of chewed wood prepared by the beetle. The spores germinate to produce abundant mycelia and new spores all over the tunnel walls. The beetles graze on the chains of fungal spores. New fungal growth appears at regular intervals. The fungal garden, in general, is remarkably free from bacteria, yeast, and other contaminants. It is not known how the beetle maintains an almost pure fungus culture. The female beetles tend the fungus gardens, carefully preparing new beds into which they incorporate larval excrement. The female beetles also tend their offspring. Larvae with poor survivability are eaten or entombed in a branch tunnel. Larvae quickly die after the death of the mother beetle. The larvae are unable to maintain the fungus garden, and bacteria and other contaminants soon overgrow the galleries. Male beetles, which are wingless and short-lived, have no role in the ecology of a fungus garden, their main function being to inseminate the females. Some species of ambrosia beetles raise their offspring in communal galleries; among others, each larva develops singly in a tunnel, excavating as it grows. The larvae have strong mandibles and can excavate wood. Female beetles carry the fungus inoculum in an externally attached saclike structure, the *mycetangium* (mycangium), which is

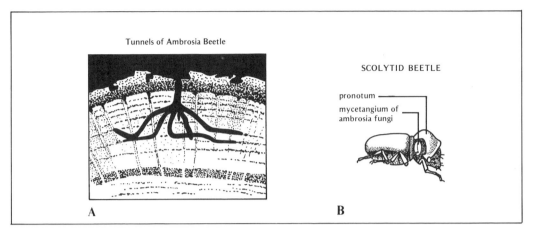

Figure 6-8. Ambrosia symbiosis. (A) Tunnels produced by the ambrosia beetle, Xyleborus, *in hickory tree. (B) Mycetangium of scolytid beetle.* (Adapted from Norris, D. M. 1979. The mutualistic fungi of Xyleborini beetles. In: Insect-fungus symbiosis, 53–63, ed. L. R. Batra. Allanheld, Osmun, Montclair, N.Y.)

located on certain parts of the beetle, such as the mandible, elytra, or thorax (Fig. 6.8). The fungus grows within the mycetangium and obtains nutrients from the host's secretions and excretions. The fungus is protected from drying out throughout the beetle's life and is thus kept available for seeding new sites during tunnel excavation.

Germinating spores of an ambrosia fungus produce germ tubes that penetrate the wood parenchyma cells adjoining the tunnel wall. The cells become filled with mycelium and the tunnel walls become covered with a velvety mass of hyphae that bear spores. Most ambrosia fungi are dimorphic, producing a yeastlike *ambrosial phase* and a *mycelial phase.* The ambrosial phase commonly occurs in the mycetangium, in the beetle galleries, and sometimes in laboratory culture. Most ambrosia fungi were originally assigned to the genus *Monilia,* but recent research has revealed the diversity of these fungi. Common genera of ambrosia fungi include *Ambrosiella, Ascoidea, Candida, Endomycopsis, Fusarium,* and *Monacrosporium.* The mycetangia of beetles often contain axenic colonies of ambrosia fungi but many beetles emerging late in the season may carry secondary, foreign fungi that develop in the tunnels. Oak wilt fungus, *Ceratocystis fagacearum,* and the Dutch elm fungus, *Ceratocystis ulmi,* are two important contaminants carried by ambrosia beetles.

The symbioses between ambrosia fungi and beetles are mutualistic. The fungal symbiont obtains several advantages from the association.

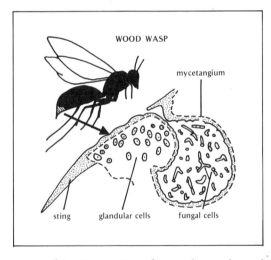

Figure 6-9. Mycetangium of a wood wasp, located next to ovipositor. (Adapted from Cooke, R. 1977. The biology of symbiotic fungi. John Wiley, London.)

Its range is increased by the beetle.

Its penetration of woody tissue is facilitated because of the mechanical injury caused by the beetles.

It uses insect remains and excretory products as nitrogen sources.

Within the mycetangium it multiplies and obtains nutrients from the host.

The beetle host also benefits from the symbiosis.

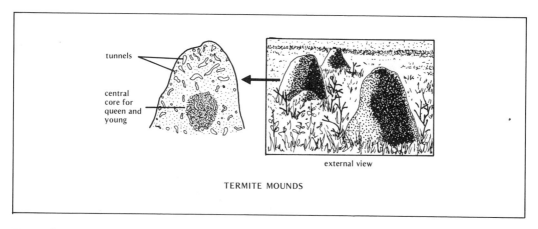

tunnels

central
core for
queen and
young

external view

TERMITE MOUNDS

Figure 6-10. Termite nests with characteristic tunnels, where fungal gardens are maintained.

The fungus produces enzymes that decompose cellulose and thus break down woody tissues. This makes the excavation process easier for the beetles.

The larvae feed exclusively on fungal conidia and adults of many species also feed on the hyphae.

The larvae obtain sterols from the fungus, which are used for the synthesis of *ecdysones,* hormones the beetle uses during its metamorphosis.

Both the beetle and the fungal symbiont show the consequences of coevolution. The behavior of the female beetle is the key to the success of the symbiosis. Survival of the larvae depends on her ability to maintain the fungal garden and to supress the growth of foreign contaminants in the galleries.

Wood wasps are similar to ambrosia beetles in their association with fungi. Female wood wasps lay eggs in the wood of conifer trees that have been weakened by disease or other insects. While the eggs are moving through the insect's ovipositor, they become coated with spores of a wood-rotting basidiomycete of the genus *Amylostereum,* or *Stereum.* The fungal spores are contained in mycetangia that are located near the base of the ovipositor (Fig. 6.9). The spores germinate in the wood and produce a mycelium that breaks down the wood and serves as a food source for the insect larvae. The larvae bore through the decaying wood, form tunnels and chambers, and develop into pupae and adults. The wasp is an obligate symbiont but the fungi

involved in these symbioses also occur as free-living fungi.

FUNGUS GARDENS OF TERMITES AND ANTS. Some species of termites and ants cultivate fungi for food. Termites use the fungi as a supplemental food source, for vitamins, whereas the ants rely entirely on the fungi for their nutrients. The fungi are grown in specially prepared beds that contain a mixture of plant material and insect excrement. Termites and ants are similar in many aspects of their symbioses with fungi. Both insects have a rigid caste system that includes workers and soldiers, and both build nests that may contain over a million insects. Termite nests are called *termitaria* and they are architectural wonders, with built-in temperature and humidity controls. Nests are produced in soil and may be up to 20 feet high (Fig. 6.10). Fungus-cultivating termites are common in Africa and Asia and differ from other termites in their lack of intestinal, symbiotic protozoa. The fungi cultivated in termitaria are members of the hyphomycete genera *Termitomyces* and *Xylaria.* White nodules in the fungus gardens are aggregates of conidiophores and conidia.

Leaf-cutting ant species of the genera *Acromyrmex, Atta,* and *Trachymyrmex* maintain fungus gardens in nests throughout Central and South America. The nests consist of chambers in which the fungus is cultivated. The fungal symbionts of most leaf-cutting ant species include the basidiomycetes *Lepiota, Rozites, Xylaria,* and *Tyridiomyces.* New nests are started by fertilized females (queens) that carry along some of the fungus from their old nest. Ant fungus gardens are developed by workers who cut leaf

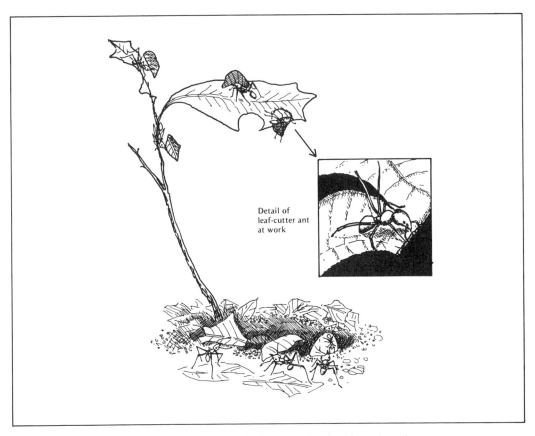

Detail of
leaf-cutter ant
at work

Figure 6-11. Worker ants of Atta *species cutting leaf sections for their fungal gardens.*

blades into large segments (10 mm in diameter) and carry them to the nest (Fig. 6.11). Young leaves are preferred over old ones and flower parts are also used. Well-worn trails are evident between ant nests and the bushes and trees that supply leaves. The leaf segments are cleaned, cut into small pieces (1 to 2 mm in diameter), and then chewed to a pulpy consistency. The ant worker mixes the spongy plant material with drops of its own exudates, which are rich in ammonia and amino acids. A small fragment of vegetation that contains a fungus from an old garden is then placed on the newly prepared bed. As the new fungus garden grows, ant workers continue to add drops of anal exudates to the bed. Fungal hyphae, which become swollen under these culture conditions, and spores are harvested periodically and used as food by the ant colony. The worker ants continuously remove secondary invaders from the nest, keeping bacteria and yeast populations at low levels. After the gardens reach an age of several months, they

are discarded and replaced with new ones. Abandoned fungus gardens quickly become overrun with foreign contaminants.

SEPTOBASIDIUM AND SCALE INSECTS. *Septobasidium* (teliomycete) is a basidiomycete that commonly occurs in warm areas of the world. Most species are obligate symbionts of scale insects. The fungus produces brown-black, flat thalli that adhere firmly to the leaves and bark of living trees. The thalli have many chambers within which the host insects live. The chambers are interconnected by tunnels and each chamber contains one insect. The fungus penetrates the insect by a few hyphae that form coiled haustoria within the hemocoel. The infected scale insect inside the fungal chamber inserts its stylet into the plant tissue and feeds on plant sap (Fig. 6.12). Nutrients that the insect obtains from the plant are also assimilated by the fungus in the hemocoel. Infected insects are small and sterile but are seldom killed. A scale insect colony also contains many uninfected individuals. In the

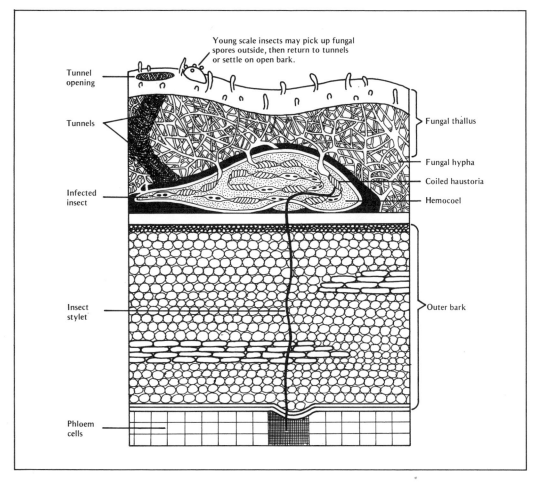

Young scale insects may pick up fungal spores outside, then return to tunnels or settle on open bark.

Tunnel opening

Tunnels

Infected insect

Insect stylet

Phloem cells

Fungal thallus

Fungal hypha

Coiled haustoria

Hemocoel

Outer bark

Figure 6-12. Mutualistic symbiosis of the fungus Septobasidium *with scale insects.* (Adapted from Cooke, R. The biology of symbiotic fungi. John Wiley, London.)

spring, as the fungus produces new mycelial growth, the uninfected female insects in the fungal chambers produce offspring called nymphs. At the same time, the fungus forms erect fruiting bodies, which produce *basidiospores* that divide by yeastlike budding. The spores stick to the bodies of the wandering nymphs. The spores germinate and the germ tubes penetrate the body hemocoel and form haustoria. Many nymphs settle down in the same fungal colony but some leave and start new colonies elsewhere. In this mutualistic association, the fungus obtains its nutrients from the scale insects and is disseminated by the nymphs while the fungus protects the insects from temperature extremes and from predation by birds and other organisms.

YEASTS AS ENDOSYMBIONTS OF INSECTS. Many insects whose diet is blood, plant sap, or wood possess yeast or yeastlike endosymbionts of the genera *Candida, Taphrina,* and *Torulopsis.* These fungi occur in the intestines, gastric cecae, fat bodies, and Malpighian tubules of insects. The symbionts are housed in specialized cells, *mycetocytes,* which are grouped together in organs called *mycetomes.* An intestinal mycetome has an epithelium that consists mainly of mycetocytes, consisting of enlarged fungus-filled cells with irregularly shaped nuclei. The cells lack microvilli on their absorptive surface, do not divide, and often are polyploid (Fig. 6.13). The mycetocytes of some insects harbor fungi only, but others contain a mixture of fungi and bacteria. Some of the symbionts can be cul-

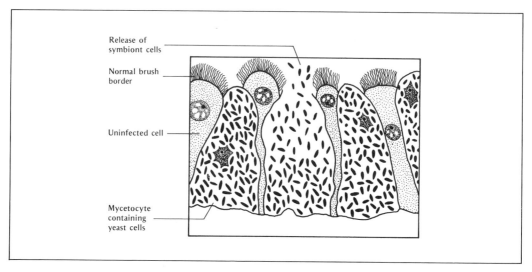

Release of
symbiont cells

Normal brush
border

Uninfected cell

Mycetocyte
containing
yeast cells

Figure 6-13. Intestinal epithelial cells of insects with mycetome-containing yeast cells. (Adapted from Cooke, R. The biology of symbiotic fungi. John Wiley, London.)

tured but most are obligate. The endosymbionts produce vitamins and essential amino acids that are used by the host insect and in return receive a stable and favorable environment in which to live. The endosymbionts are transmitted to new individuals by one of the following ways: licking and stroking of the offspring by the female parent; the outer surface of the eggs becoming coated with fungi during the egg-laying process by means of fungus-containing structures that are attached to the ovipositor; and the egg cell cytoplasm carrying the fungus.

The symbionts in a coccid insect, *Stictococcus diversisetae,* enter the ovary and infect some of the eggs produced by the insect. These eggs develop parthenogenetically and give rise to females; uninfected eggs develop into males. Loss of a chromosome is involved in the production of males. Some scientists believe that endosymbionts of infected eggs, in a manner not understood, prevent the chromosomal loss.

5. SYMBIONTS OF MAN AND OTHER VERTEBRATES. Associations of fungi and warm-blooded animals are almost universal. In nature, most animals have low-grade fungal infections. Some fungi are well known because they cause disease (*mycoses*) to domesticated animals as well as man. Specialization in this subject constitutes the field of *veterinary and medical mycology.*

Endogenous fungi make up the normal 'flora' of the skin and mucous membranes and often attack the host through injured tissue. *Exogenous fungi* live as saprophytes in the soil and sometimes produce diseases in animals. Fungal infections of vertebrates are of two main types: *systemic infection,* of deep internal tissues and vital organs such as lungs and brains; and *superficial infection,* of scalp, skin, or nails from fungi known as *dermatophytes.* Fungal parasites of vertebrates have a wide host range and are easily transmitted between humans and animals. Some fungal infections have become associated with certain occupations and life styles of humans. For example, *histoplasmosis* is prevalent among poultry farmers and cave explorers because bird droppings are a common source of fungus. Farmers, gardeners, and agricultural workers often have mild fungal infections.

Fungi that attack vertebrates have been reported from all major fungal groups, although only a few infections are caused by basidiomycetes. Following are some examples of fungal diseases of humans.

a. Candidiasis. Candidiasis is caused by a yeast-like fungus, *Candida albicans.* In humans, *Candida albicans* as well as true yeasts are residents of the mouth, alimentary canal, and vagina. The fungus normally lives as a saprophyte receiving nutrients from sloughed-off epithelial cells

of the mucous membranes. *Candida albicans* causes a disease called thrush in which the fungus invades the superficial tissue and produces soft, gray-white lesions on the gums, tonsils, and tongue. Candidiasis frequently occurs in individuals (1) undergoing antibiotic therapy, (2) on prolonged use of birth control pills, (3) during pregnancy, and (4) with diabetes mellitus. The fungus has a hyphal form during the invasive phase of the infection. The mechanisms controlling fungus morphology are not known.

b. Aspergillosis. Species of *Aspergillus* cause several types of mycoses. Aspergillosis is increasingly being recognized as an important complication in individuals suffering from malignant diseases, such as leukemia and lymphoma, and in those receiving immunosuppressive drugs. *Aspergillus fumigatus* is a virulent pathogen that causes pulmonary aspergillosis and can spread to other organs. Asthmatic symptoms and fever are common in these mycoses. *Aspergillus fumigatus* may invade the nasal sinus tissue and can cause endocarditis in patients with heart valve replacements. Recently, *fumigatotoxin,* a high molecular weight toxin, has been isolated from the fungus.

Several species of *Aspergillus* cause *avian aspergillosis.* Young birds die from appetite loss, high temperature, and diarrhea. In 1940, many herring gulls in the Boston Bay area died from infections caused by *A. fumigatus* that had developed first on rotting seaweed. Aspergilli have also been involved in *equine gutteral pouch mycosis, mycotic abortion,* and *mycotic mastitis.*

c. Histoplasmosis. Histoplasmosis is a serious and often fatal infection of the lymphatic system. The disease is widespread in warm humid regions of the world and is caused by *Histoplasma capsulatum,* a fungus that reproduces like a yeast in host tissue and produces mycelial growth in culture. The infection begins when spores of the fungus are inhaled into the lungs. After a prolonged exposure to the fungus, the individual develops chronic tuberculosislike symptoms of cough, slight fever, and pulmonary cavitation. Progressively, the disease spreads to the spleen, liver, adrenals, kidneys, nervous system, and other organs. Histoplasmosis is frequently associated with soils that become contaminated with fecal material from bats and birds, including chickens. In city parks and plazas, where birds often roost, the soil and the buildings frequently become contaminated with *H. capsulatum.* There is no satisfactory treatment for this disease.

d. Coccidiomycosis. *Coccidioides immitis* causes a mild respiratory infection that is called the San Joaquin Valley fever. The fungus causes lesions in the upper respiratory tract and progressively develops in bone, joints, and other tissues, including the central nervous system. Symptoms include fever, chills, sweating, and general weakening. *Coccidioides immitis* infections are widespread in arid parts of the United States and throughout Central America. The fungus is soil borne and produces spores that remain viable for long periods. Rodents, domesticated animals, and humans become infected from inhaling the airborne spores. The fungus is dimorphic; in the invasive phase it produces spherical cells that function as sporangia at maturity; in the saprobic phase the fungus produces arthroconidia. Most people and animals living in areas where the fungus is common have mild coccidiomycotic infections. The chance of becoming infected increases with the length of stay in the area. The disease is rarely fatal.

e. Dermatophytic Diseases: Ringworms. The fungi collectively known as *dermatophytes* are limited to the keratinized layers of skin, nails, hair, or feathers of man and animals. Diseases from dermatophytic fungi include various types of *ringworm.* Dermatophytic fungi belong to three genera: *Epidermophyton, Microsporum,* and *Trichophyton.* Species of *Epidermophyton* are confined mostly to humans and infect only the skin. These fungi are widely distributed in soil and most infections are acquired through intimate contact. Humans contract ringworm disease from infected animals and also from contaminated combs, hairbrushes, caps, and furniture. *Microsporum canis* infection is usually acquired from dogs and cats. *Trichophyton* species are transmitted to man from horses, cattle, dogs, and other animals.

Dermatophytic fungi flourish under warm and humid conditions. Most people during their lifetime become infected with at least one dermatophytic fungus. Infections are favored under conditions caused by continuous wet skin (swimmers), moist and oily skin, tight clothing and shoes that prevent evaporation of perspiration, and obesity, which results in increased folding of the skin where oil and moisture accumulate.

It is assumed that dermatophytic fungi can digest and assimilate keratinized tissue, although the mechanism of keratin digestion is not understood. Dermatophytic fungi can be grown in axenic culture, but in nature most species need a host in order to grow.

6. MYCOTOXINS. It is only since 1960 that we have learned about the significance of mycotoxins in illnesses and even death of humans and domesticated animals. In 1960, the deaths of over 100,000 turkeys within a hundred-mile radius of London led to the discovery of the mycotoxin *aflatoxin,* which was isolated from *Aspergillus flavus,* a fungus that had contaminated the peanut feed of the turkeys. Mycotoxicoses are now recognized as occurring worldwide, and a number of mycotoxin-producing fungi have been identified. Scientists in the USSR were first to appreciate the significance of mycotoxin-produced diseases. Perhaps the prolonged storage of grains and other animal feed during the long Russian winters increased the possibility of mold growth. Comsumption of moldy hay was found to be responsible for the deaths of many horses in the Ukraine in 1937. In 1954, A. K. Sarkisov published a detailed monograph in Russian on mycotoxicoses but the work remained unknown in Western countries until the mid-1960s.

All mycotoxin-produced diseases are characterized by (a) noncommunicability, (b) unresponsiveness to antibiotic drugs, and (c) geographically limited outbreaks usually linked to moldy food. The toxic response in the affected individuals occurs through organs such as the liver, kidney, lung, and nervous system. The chief mycotoxin-producing fungi belong to the genera *Aspergillus, Fusarium,* and *Penicillium.* Following are some examples of fungi and their mycotoxins.

Aspergillus flavus and *A. parasiticus* are two aflatoxin-producing fungi that are prevalent in hot humid climates. Aflatoxins have been isolated from edible nuts such as peanuts, walnuts, almonds, pecans, Brazil nuts, and pistachio nuts, and grains such as wheat, rice, corn, sorghum, cottonseed, and millets. In a 1960 British study, aflatoxin types B1, B2, G1, and G2 were identified. Since then it has been established that dairy products from cows that ingested B1 toxin contaminated hay contained M1 and M2 types of aflatoxins. The M aflatoxins are firmly bound to casein molecules. The G1 and M1 toxins are carcinogenic. Aflatoxin susceptible hosts such as rabbits, guinea pigs, rainbow trout, rats, and monkeys develop liver damage with necrotic lesions and swollen kidneys. The first signs of the disease are loss of appetite and weight followed by poor muscular coordination. A high incidence of human liver cancer in parts of Africa, Southeast Asia, and Japan is correlated with aflatoxin-contaminated food.

Penicillium vividicatum is an important mycotoxin-producing species. It usually infects stored grains and causes illness and death in humans and domesticated animals. *Penicillium citrinum* is widespread in the tropics and subtropics and its toxin produces a lethal kidney disease. Mycotoxins of *P. islandicum* have received attention from Japanese scientists because consumption of rice contaminated with this fungus produces the disease rice toxicosis. As many as four potent hepatotoxic carcinogens have been isolated from *P. islandicum.* One of these toxins, *islanditoxin,* causes severe liver damage, hemorrhage, and rapid death. *Penicillium cyclopium* is commonly associated with food grains and produces penicillic acid, cyclopiazonic acid, and two tremor-producing compounds. These fungi usually inhabit soil as saprobes and develop on the seeds only after harvest.

Fusarium graminearum infects corn and produces mycotoxins that induce vomiting in pigs. *Fusarium poa* and *F. sporotrichioides* produce the *fescue food syndrome* in cattle. The animals show symptoms of lameness, weight loss, arched back, swelling in the hind legs, separation of the hoof from the foot, abnormal horn and hoof growth, and dry gangrene in the extremities. If not checked, the disease can be fatal.

Mycotoxin research is still in a stage of infancy. There are a number of challenging questions that need additional research such as the prevalence of mycotoxins in food crops, their relationship with carcinogenicity, and ways to prevent their occurrence.

E. Summary and Perspectives

Fungi have successfully formed associations with all forms of life in most ecosystems and play a significant role in the processes of decomposition in nature. Fungal diseases in plants, animals, protoctists, and even other fungi have received considerable attention from mycologists. Unique features of fungi such as the way they obtain nutrients, their mycelial growth and spore production, and the chitinous nature of their cell wall have convinced scientists that fungi belong in a separate kingdom. In this chapter fungal associations with soil amoebae, nematodes, insects, and animals are examined in detail. Fungi have evolved morphological and physiological adaptations to trap and capture nematodes. The Laboulbeniales are microscopic ascomycetes that are obligate ectosymbionts of insects and arthropods. Some species of these fungi are so highly specialized that they grow only on certain parts of an insect host. Trichomycetes form commensalistic relationships with arthropods. Ambrosia beetles and wood wasps form mutualistic associations with wood-rotting fungi, and some species of termites and ants establish fungal gardens for food.

The basidiomycete *Septobasidium* is an obligate symbiont of scale insects, with which it forms a unique mutualistic relationship. Yeasts are endosymbionts of many insects. Fungal parasites of nematodes and insect pests have provided opportunities for scientists to develop strategies of biological control. Some fungi cause diseases of humans, such as candidiasis, histoplasmosis, coccidiomycosis, and ringworms. Mycotoxins such as aflatoxin and islanditoxin produce diseases in humans and animals. Fungal disease epidemiology is receiving increasing attention from the medical community. Cancer researchers are actively examining the role of fungal mycotoxins in certain forms of cancers.

Entomological mycology is another example of a specialty emerging from the cross-fertilization between two highly developed disciplines.

The evolutionary relationships of fungal-insect symbioses are being examined through the methods of cladistics, which is a study of genealogical relationships among species or populations.

BOX ESSAY

Roland Thaxter

PIONEER IN INSECT MYCOLOGY

In an age of commercial exploitation of science it is worth drawing attention to the career of Roland Thaxter, whose research has inspired a generation of mycologists to discover the joy of unraveling the mysteries of nature. In a large sense, scientific endeavors are like musical, artistic, and literary enterprises. Together, they enlarge and enrich our intellect and bring forth those human qualities that no monetary or material rewards can match.

Professor Roland Thaxter (1858–1932) was one of the foremost mycologists of all times. He was a native of Massachusetts, and from early childhood he loved to draw and study insects. At Harvard University, under the influence of William G. Farlow, he studied little-known fungal parasites of insects for his doctoral thesis. Farlow was so impressed with Thaxter that he invited him to join the Harvard faculty in 1891. Thaxter's years at Harvard produced several outstanding mycologists in the United States and monumental works on inconspicuous fungi associated with equally unimportant insects. His five-volume monograph on the Laboulbeniales set new standards of excellence, thoroughness, and scientific accuracy. His illustrations of these little-known fungi were both artistic and meticulous in detail. His work inspired his students, who in turn trained other scientists to investigate fungal symbionts of protoctists, nematodes, insects, and other invertebrates. Advances in insect mycology during the past 50 years have demonstrated the impact of one man's scientific talent and imagination nurtured by attention to detail, thoroughness, and scientific curiosity.

Review Questions

1. What are the characteristics of fungi that justify placing them into a separate kingdom?
2. Name several phyla of fungi.
3. Describe the morphological adaptations of predaceous fungi that enable them to trap soil amoebae.
4. What types of traps are formed by hyphomycetes to capture nematodes?
5. Name the features that endosymbiotic fungi of nematodes have in common.
6. Describe some common characteristics of insect-inhabiting fungi.
7. Name some well-known examples of entomogenous fungi.
8. Describe the features of the Laboulbeniales.
9. Describe the association between fungi and ambrosia beetles.
10. How do termites and ants grow fungi for food?
11. Describe the relationship between *Septobasidium* and scale insects.
12. Give several examples of fungal diseases of humans.
13. What are mycotoxins and what kinds of diseases do they produce? Why have mycotoxins only recently been discovered?
14. How do disease-causing fungal symbionts differ from nonsymbiotic fungi?

Further Reading

Ahrearn, D. G. 1978. Medically important yeasts. Ann. Rev. Microbiol. 32:59–68.

Barron, G. L. 1977. The nematode-destroying fungi. Canadian Biological Publications, Ontario, Canada. 140 pp. (A comprehensive, well-illustrated treatment of fungi that attack soil-inhabiting nematodes.)

Batra, L. R. (ed.). 1979. Insect-fungus symbiosis: Nutrition, mutualism, and commensalism. Second Int. Mycol. Congress. Allanheld, Osmun, Montclair, N.Y. 276 pp. (An important source of information on fungus-insect symbioses.)

Batra, S. W. T., and L. R. Batra. 1967. The fungus gardens of insects. Sci. Am. 217 (November): 112–120. (A well-illustrated introduction to fungus-gardening insects.)

Ciegler, A., and J. W. Bennett. 1980. Mycotoxins and mycotoxicosis. BioScience 30:512–515. (A good review of recent developments in the field.)

Deacon, J. W. 1980. Introduction to modern mycology. Halsted Press, New York. 197 pp. (A good introductory text on fungi.)

Madelin, M. F. 1968. Fungal parasites of invertebrates, 1. Entomogeneous fungi. In: The fungi, Vol. 3, 227–238, ed. G. C. Ainsworth and A. S. Sussman. Academic Press, New York.

Maramorosch, K. 1981. Spiroplasmas: Agents of animal and plant diseases. BioScience 31:374–379. (A comprehensive review of mycoplasmalike organisms (MLOs) that are being recognized as pathogens of plants and animals.)

Ross, I. K. 1979. Biology of the fungi: Their development, regulation, and associations. McGraw-Hill, New York. 499 pp. (Large section of the book is on fungal associations.)

Weber, N. A. 1972. The attines: The fungus-culturing ants. Amer. Scientist 60:448–456.

Wheeler, Q., and M. Blackwell (eds.). 1984. Fungus-insect relationships: Perspectives in ecology and evolution. Columbia Univ. Press, New York. 514 pp. (An important reference source for fungus-insect symbioses.)

Bibliography

Ainsworth, G. C., and P. K. C. Austwick. 1973. Fungal diseases of animals. 2d. ed. Commonwealth Agricultural Bureau, Slough, UK. 216 pp.

Ainsworth, G. C., and A. S. Sussman (eds.). 1965–1973. The Fungi: An advanced treatise. Vols. 1–4. Academic Press, New York.

Anderson, J. M., A. D. M. Rayner, and D. W. H. Walton. 1984. Invertebrate-microbial interactions. Cambridge Univ. Press, Cambridge. 349 pp. (Contains much information on insect-fungus relationships.)

Beneke, E. S., and A. L. Rogers. 1980. Human mycoses. In: Medical mycology manual. 4th ed. Burgess, Minneapolis, Minn. 237 pp.

Cooke, R. 1977. The biology of symbiotic fungi. John Wiley, London. 282 pp.

DiSalvo, A. F. 1983. Occupational Mycoses. Lea and Febiger, Philadelphia. 247 pp.

Drechsler, C. 1934. Organs of capture in some fungi preying on nematodes. Mycologia 26:135–144.

Duddington, C. L. 1957. The friendly fungi. Faber and Faber, London. 188 pp.

Duddington, C. L. 1968. Fungal parasites of invertebrates, 2. Predacious fungi. in: The fungi, Vol. 3, 239–251, ed. G. C. Ainsworth and A. S. Sussman. Academic Press, New York. (A fine treatment of fungi that attack protozoa, nematodes, and rotifers.)

Emmons, C. W., C. H. Binford, J. P. Utz, and K. W. Kwon-Chung. 1977. Medical mycology. 3d ed. Lea and Febiger, Philadelphia. 591 pp.

Kendrick, B. 1985. The fifth kingdom. Mycologue Publications, Waterloo, Ontario, Canada. 371 pp.

Sarkisov, A. M. 1954. Mycotoxicoses. State Publ. House for Agricultural Literature, Moscow. 216 pp.

Tavares, I. I. 1985. Laboulbeniales (fungi, ascomycetes). Mycologia Memoir, 9. J. Cramer, Braunschweig, West Germany. 700 pp.

Thaxter, R. 1931. Contribution towards a monograph of the Laboulbeniaceae. V. Mem. Amer. Acad. Arts Sci. 16:1–435.

CHAPTER 7

Fungal Associations
Fungi as Symbionts of Fungi, Algae, and Plants

A. Mycosymbionts of Fungi

Fungi that obtain nutrients by infecting other fungi are called mycosymbionts. Most of these symbionts are imperfect fungi and belong to the class Hyphomycetes. The terms *hyperparasitism, mycoparasitism,* and *fungicolous fungi* are often applied to interfungal associations. Mycosymbionts are generally grouped into two categories based on their mode of nutrition. *Necrotrophic mycosymbionts* obtain nutrients from hosts they kill, and *biotrophic mycosymbionts* obtain nutrients from living hosts. In the early stages of infection the host fungus seldom shows adverse effects. Following are specific examples in each of the two groups.

1. NECROTROPHIC MYCOSYMBIONTS. Common soil-inhabiting fungi such as *Ampelomyces quisqualis, Gliocladium roseum, Rhizoctonia solani,* and *Trichoderma viride* are necrotrophic mycosymbionts. They can live saprobically on dead organic matter and also attack a wide range of fungi (Fig. 7.1). Many mycosymbionts grow well in laboratory cultures. When a host fungus is introduced into a culture, the mycosymbiont grows toward it and, in some cases, its hyphae coil around the host hyphae. The mycosymbiont produces enzymelike substances that cause the host cells to disintegrate. The exact nature of these substances and their mode of action are not known. *Rhizoctonia solani* is a well-known root pathogen that kills young seedlings. In culture, *R. solani* can destroy susceptible fungal hosts such as *Rhizopus nigricans* and species of *Mucor.*

Gliocladium roseum attacks a wide range of fungi in nature as well as in laboratory cultures. The fungus curls around the host hyphae by means of short branches and kills the host cells on contact. The fungus then penetrates the dead cells and absorbs their nutrients. Cells of *Rhizopus nigricans, Rhizoctonia solani,* and *Mucor* spp. are penetrated while still alive by *G. roseum* hyphae.

Ampelomyces quisqualis attacks powdery mildew fungi. At first, the mycosymbiont grows rapidly among the host hyphae without harming them, but later it kills them. The fungus is believed to overwinter as a saprobe on mildew-infected leaves. In one study, clover powdery mildew was controlled by artificial inoculation with *A. quisqualis.*

Darluca filum attacks only spores and mycelia of rust fungi. Because of this limited host range, some scientists have speculated that the ancestor of *D. filum* was a leaf pathogen that lost its ability to infect leaf tissue and instead parasitized rust fungi.

Tuberculina maxima is often described as the purple mold of rust fungi and it is a natural enemy of the pine blister rust. This mycosymbiont infects mostly the pycniospores of the host, which in turn reduces aeciospore production. *Tuberculina maxima* also destroys rust mycelium in infected tissues of white pine. Spores of the fungus remain viable for several years. In recent years, there has been renewed interest in using *T. maxima* as a biological control agent of the pine blister rust in the western United States and Canada.

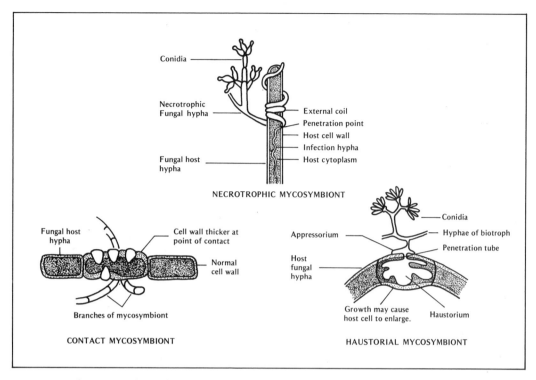

Figure 7-1. *Three types of interfungal parasitic symbiosis.* (Adapted from Cooke, R. The biology of symbiotic fungi. John Wiley, London.)

2. BIOTROPHIC MYCOSYMBIONTS

a. Contact Mycosymbionts. Contact mycosymbionts were not recognized until 1958 but are now believed to be common in nature. About six species of hyphomycetes from genera such as *Calcarisporium, Gonatobotrys,* and *Stephanoma* are involved in contact mycosymbioses (Fig. 7.1).

The mycosymbiont produces short, lateral hyphae that contact the host hyphae and cause them to release nutrients. Except for arrested growth, the infected fungus shows no adverse effects. Mycosymbionts can be grown in culture only if a growth factor, *mycotrophein,* which is present in extracts of the host fungus, is part of the culture medium. Mycotrophein is a low molecular weight substance that is active in low concentrations. Its role in the nutrition of the mycosymbiont is unknown.

b. Haustorial Mycosymbionts. Infection of species of the order Mucorales (zygomycete) by other members of the same order is common.

Spores of the mycosymbiont germinate in response to substances that diffuse from a host mycelium. The germ tube grows toward the host hyphae, and upon contact, the tip of the germ tube swells to form an *appressorium,* which is a flat, hyphal cell. Fine hyphal branches from the appressorium penetrate the host cell and then enlarge to form inflated haustoria. An internal mycelium develops from the haustoria and on maturity hyphae emerge from the host and form sporangia (Fig. 7.1).

Piptocephalis virginiana attacks a large number of *Mucor* species. Electron microscopic studies have shown that the haustorium of *P. virginiana* is similar in structure to that of fungal symbionts of plants.

B. Fungal-Algal Associations

1. MYCOPHYCOBIOSES. A *mycophycobiosis* is an obligate symbiosis between a filamentous marine fungus and a large marine alga in which the alga is the dominant partner. The fungus grows between the cells of the algal thallus but never penetrates or damages the cells. The association is different from that of a lichen because a new morphological structure is not formed and new chemical substances are not produced. Further, there is no experimental evidence of physiological interactions between the symbionts. The associations, however, appear to be mutualistic. It is thought that the fungi protect the algae from drying during low tides, when the algae are exposed. Moreover, sporelings of some marine algae such as *Ascophyllum nodosum* will not develop unless they are infected by a fungus.

Fungi that form mycophycobioses include species of *Blodgettia* and *Mycosphaerella*. *Blodgettia bornetii* is a hyphomycete that grows in the walls of tropical species of *Cladophora*. The fungus produces spores and is absent only from the growing tips of the alga. The ascomycete *Mycosphaerella ascophylli* is always associated with the brown algae *Ascophyllum nodosum* and *Pelvetia canaliculata*. Hyphae grow between cells of the cortex and medulla of the host, and the fungus undergoes sexual reproduction and produces fruiting bodies while still within the host. Jan Kohlmeyer compared these associations to ectomycorrhizas, that is, associations between fungi and plant roots in which the fungus does not penetrate the root cells.

The ascomycete *Turgidosculum complicatulum* forms associations with the green algae *Prasiola borealis* and *P. tesselata*. Hyphae of the fungus grow throughout the algal thallus and separate cells of the thallus into groups of four or into rows. These algae also occur without the fungus. *Prasiola* thalli infected with fungus are more common than uninfected forms in exposed or drier parts of the intertidal zone. The fungus protects the alga from drying. *Turgidosculum ulvae* is a common symbiont of the green alga *Ulva vexata*. The fungus grows so extensively that it separates the layers of the algal thallus.

2. LICHENS. A lichen is an association between a fungus and a photosynthetic symbiont (photobiont) that results in a stable *thallus,* or body, of specific structure. The photobiont is either an alga or a cyanobacterium. A remarkable feature of a lichen is the transformation that the symbionts, in particular the fungus, undergo during the association. A new entity, the thallus, is formed, and unique chemical compounds are

synthesized. The physiological behavior of the symbionts also changes in symbiosis. There are about 15,000 species of lichens, an indication that this type of symbiosis has been highly successful and has involved many species of fungi. Surprisingly, only about 30 different types of algae and cyanobacteria have been reported as photobionts. Taxonomically, only the fungus and alga of a lichen have Latin names, although commonly the name of the fungus is also used for that of the lichen. For example, *Cladonia cristatella* is the name of a mycobiont but, unofficially, the name of a lichen as well. The photobiont of this lichen is *Trebouxia erici*.

a. Types of Lichens. A lichen thallus usually consists of layers such as an upper and lower cortex, algal layer, and medulla. The layers differ in thickness and are better developed in some species than in others. Fungal hyphae make up most of a thallus; the photobiont cells are only a small percentage (about 7%) of the total volume.

There are three main types of thalli: crustose, foliose, and fruticose (Fig. 7.2). A crustose thallus lacks a lower cortex and is generally considered to be the most primitive type. Thalli of *Lepraria* species do not have layers but consist only of powdery granules. There are more species of crustose lichens than other types and most of them belong to the genera *Lecanora* and *Lecidea*. Many crustose lichens stick tightly to the substratum and appear to be painted on it. Some species grow inside rock crevices and bark and still manage to produce separate layers. Squamules are typical of many species of *Cladonia*. They are a specialized type of crustose thallus and are attached at only one end to the substratum. A foliose thallus has an upper and lower cortex, an algal layer, and medulla, and is usually loosely attached to the substrate by hairlike structures called *rhizines*. The thallus has many different sizes and shapes and is often divided into lobes. Common foliose genera include *Anaptychia, Cetraria, Parmelia,, Physcia,* and *Xanthoria*. Some foliose lichens, such as *Umbilicaria* (rock tripe), have thalli that are attached to the substrate by only one central point. Fruticose thalli are upright or hanging, round or flat, and often highly branched. Thalli of *Usnea* are hairlike and can reach a length of 5 meters, while those of *Evernia* are shorter and strapshaped. The layers of a fruticose thallus may surround a central, thick cord, as in *Usnea,* or a

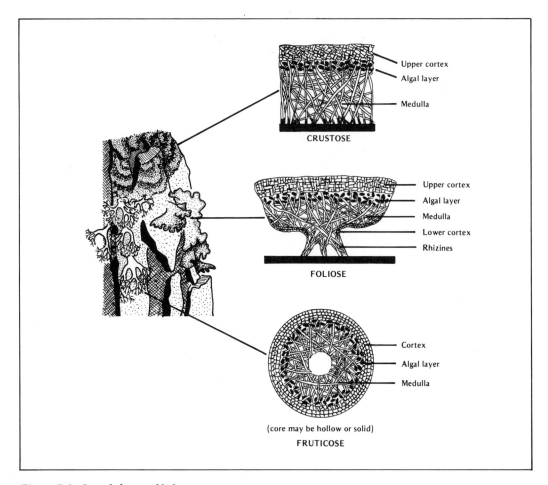

Figure 7-2. Growth forms of lichens.

hollow space as in some *Cladonia* species. Some lichens, such as *Collema,* which have *Nostoc* as a photobiont, do not form a well-organized thallus. In these cases, fungal hyphae grow inside the thick gelatinous sheaths of the photobiont, which makes up much of the thallus.

Most lichens have an ascomycete as the fungal partner, and only about a dozen are formed by basidiomycetes. The lichenized ascomycetes are one of the largest groups of the phylum Ascomycota in the kingdom Fungi. The lichenized habit obviously has had great selective benefits for fungi.

b. Distribution. Lichens grow practically everywhere, on and within rocks, on soil and tree bark, and on almost any inanimate object. They grow in deserts and in tropical rain forests, where they occur on living leaves of plants and ferns. They have been found on the shells of tortoises in the Galápagos Islands and on large weevils in New Guinea. In the dry valleys of Antarctica, endolithic lichens, such as *Buellia* and *Lecidea,* grow inside sandstone crevices. *Dermatocarpon fluviatile* and *Hydrothyria venosa* grow in freshwater streams and species of *Verrucaria* are common in the intertidal zones of rocky, ocean shores. *Verrucaria serpuloides* is a permanently submerged marine lichen that grows on stones and rocks that are 4 to 10 meters below mean low tide off the coast of the Antarctic Peninsula. Lichens abound in areas with high annual humidity, such as the fog belt zones of Chile and Baja California. Extensive lichen populations also grow in the cool, northern forests of the world where hundreds of miles of forest floor are covered with thick carpets of reindeer lichens (*Cladonia*). Trees along

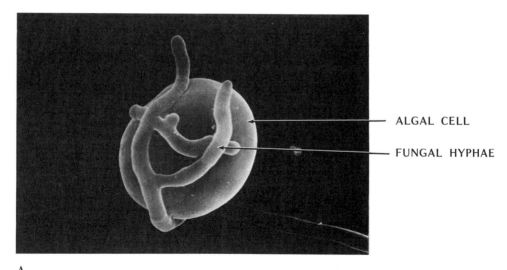

ALGAL CELL

FUNGAL HYPHAE

A

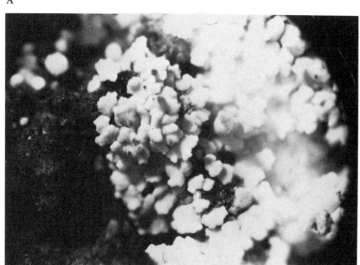

B

Figure 7-3. Synthetic lichen Cladonia cristatella. *(A) Scanning electron micrograph of algal cell enve-* *loped by fungal hyphae; early stage of lichen for-* *mation. (B) Squamules produced in soil culture.*

the coasts of northwestern United States may be blanketed with beard lichens such as *Alectoria* and *Usnea*. Lichens with organized thalli do not grow well in areas that are continuously wet, like tropical rain forests. Only poorly organized species of *Lepraria* and leaf-inhabiting lichens are found in these regions. *Lecanora conizaeoides* and *Lecanora dispersa* colonize trees and gravestones in industrial cities and towns, but most lichens cannot tolerate the polluted at-

mosphere and persistent dryness of urban areas. The sensitivity of lichens to atmospheric pollutants such as sulfur dioxide, ozone, and fluorides has made them valuable indicators of pollution in cities and industrial regions. Lichens have been used to map pollution zones around many of the world's major cities.

c. Dispersal and Reproduction. Lichens are dispersed by thallus fragments and vegetative diaspores such as *isidia* and *soredia*. Each diaspore consists of a few algal cells and fungal hyphae. Soredia are powdery granules that originate inside the thallus, as localized overgrowths of algae and hyphae, and break through the upper cortex. Insidia are cylindrical extensions of the thallus. About 39% of all foliose and fruticose species of lichens produce isidia. Diaspores are dispersed by water, wind, insects, and birds.

There is some evidence, mostly indirect, that lichen symbionts undergo sexual reproduction. Gametes produced by *Trebouxia* phycobionts in culture have fused and formed zygotes. Lichen fungi produce the same type of reproductive structures as other ascomycetes. Only some aspects of the sexual process have been seen in lichens, such as the fusion of microconidia to the tips of trichogynes.

d. Interactions between Symbionts

PHYSICAL RELATIONSHIPS. The basic unit of a lichen thallus is one algal cell with enveloping hyphae. Fungal hyphae adhere to the surface of an algal cell by means of a mucilage produced by both symbionts (Fig. 7.3). As the fungus envelopes the algal cells, it forms two types of specialized cells, appressoria and haustoria. These structures are common features of pathogenic fungi. The appressorium fastens the mycobiont tightly to the photobiont and gives rise to hyphae that grow into the algal cell and form haustoria. Hyphae penetrate the algal cell by enzymatic and physical means; that is, they partially dissolve their way through the algal wall and also push their way through. The plasma membrane of the algal cell always remains intact no matter how deeply the hyphae grow inside the cell. In some lichens, hyphae barely penetrate through the algal cell wall; in others they extend more deeply. Lichens with small haustoria are thought to be more highly evolved than those with large haustoria. Presumably, as lichens evolved, fungal penetrations into the algal cells were inhibited by stronger algal defenses. Small haustoria are difficult to see because they are obscured by the chloroplasts that fill many algal cells. An algal cell may have multiple haustoria.

The frequency of haustoria in lichens shows great variation. *Trebouxia* photobionts, which occur in about 60% of the known lichen species, commonly contain haustoria. In *Cladonia cristatella* over 50% of the algal cells have haustoria. Crustose lichens generally have the largest haustoria. Large foliose lichens such as *Sticta* and *Peltigera* possess algal cells that are closely associated with fungi but that are not penetrated by hyphae. This absence of haustoria is unusual among lichens and is thought to result from the presence of sporopollenin in walls of the *Coccomyxa* phycobionts of these lichens. Sporopollenin forms a rigid layer in the algal cell wall and appears to stop fungal penetration. Sporopollenin does not occur in the cell walls of *Trebouxia*, a fact that may explain why these algae have many haustoria. The function of haustoria is not clear but it is most likely that they absorb nutrients from the algal cells.

RECOGNITION MECHANISMS. Whether or not organisms recognize each other within a symbiotic system is a subject of much interest to scientists. In the *Rhizobium*-legume association, symbiont specificity is close and it may be regulated by a recognition mechanism that involves complementary binding of lectins and polysaccharides. The binding occurs on the external surfaces of the symbionts. In lichens, the evidence for a similar recognition mechanism is inconclusive. Although lectins have been isolated from lichens, there are conflicting reports on whether or not they function in symbiont interactions. If there is a recognition phase in lichens, it may occur later in the development of the thallus. Initial contacts between fungi and algae are nondiscriminatory and mycobiont hyphae will bind to any object, even glass beads.

SECONDARY COMPOUNDS. Lichens contain unique secondary compounds, which are commonly called *lichen acids*. Most of these compounds are products of the symbiosis, but some have been formed by the isolated fungal symbiont growing in culture. For example, one study showed that out of 50 clones of the mycobiont *Cladonia cristatella* that were grown in axenic culture, 5 clones produced the dibenzofuran didymic acid, a compound found in most natural populations of the lichen. In addition to unusual compounds such as depsides, depsidones, dibenzofurans, and usnic acids, lichens synthesize many other compounds that are the same as or similar to those formed by nonlichenized fungi. The role of the phycobiont in the synthesis of secondary compounds is not clear. The phycobiont may inhibit the decarboxylation of phenolic acid precursors produced by the mycobiont. Such precursors are used by nonlichenized fungi to synthesize a variety of compounds, such as quinones, which are toxic to algae. In a lichen, these precursors are shunted through an alternate pathway where they are esterified and coupled to form lichen secondary compounds that are not toxic to the algal symbionts.

Secondary compounds of lichens may have important ecological roles. Many have antibiotic activity and may prevent the microbial decay of lichen thalli, which may live for hundreds and even thousands of years. Lichen substances also inhibit bryophytes, other lichens, as well as seed germination and seedling development in vascular plants. The compounds protect thalli from grazing by invertebrate herbivores, such as slugs and insect larvae, and act as light screens to protect the photobionts from high light intensity. Lichen substances are also chemical weathering compounds that have a role in soil formation because of their chelating properties.

SYNTHETIC LICHENS. Some lichens have been produced in axenic laboratory cultures by recombining their separate fungal and algal symbionts. The most successful artificial syntheses have been with *Cladonia cristatella* (British soldiers) and other similar species of *Cladonia*, and with *Usnea strigosa* (beard lichen). The synthetic lichens formed in the laboratory have been identical to their natural counterparts in morphology (Fig. 7.3) and, to some extent, in chemistry.

Synthetic lichens have provided useful information concerning the nature of the lichen symbiosis and the range of specificity that exists among symbionts. Synthesis studies carried out with *Cladonia cristatella* have revealed that the mycobiont will lichenize species of *Trebouxia* other than its natural phycobiont; however it will not accept "foreign" algae, that is, those belonging to different genera. In several experiments, the mycobiont partially lichenized (formed soredia) *Friedmannia israeliensis,* a free-living alga. Other algae that were "foreign" to the *C. cristatella* mycobiont were parasitized and killed before the initial stages of lichenization took place. Compatible algae were also penetrated by the fungus but they were not killed, possibly because they possessed a defensive mechanism against the fungus. The relationship between fungus and alga in a lichen is thought to be that of controlled parasitism. The fungus parasitizes the alga but under natural conditions the parasitism is slow and infected algal cells may live for years.

e. Physiology of the Symbiosis

CARBOHYDRATE TRANSFER FROM PHOTOBIONT TO MYCOBIONT. Most of the studies that have dealt with the physiological interactions between lichen symbionts have focused on the passage of nutrients from the photobiont (autotroph) to the mycobiont (heterotroph). The reason for this is the availability of radioactive isotopes, such as $C^{14}O_2$, that can be fixed photosynthetically by the photobiont and then traced in other compounds and movement within the thallus. Studies using radioactive isotopes have revealed much information about the physiology of the lichen symbiosis.

In a lichen thallus the photobiont excretes over 90% of the carbon that it fixes photosynthetically, as a polyol or a sugar such as glucose. The polyol excreted by green symbionts is ribitol, erythritol, or sorbitol; blue-green photobionts excrete glucose. The fungus may control the rate of polyol excretion by the photobiont, according to the urease theory. This theory states that urea in a lichen thallus is hydrolyzed by the enzyme urease to produce CO_2 and NH_3. Carbon dioxide stimulates photosynthesis of the photobiont while NH_3 increases its respiration and carbohydrate release. When the lichen fungus is actively growing, it can increase the flow of nutrients from the photobiont cells by producing more urease. Lichen acids act as a feedback control because they inactivate urease. Thus, during periods of fungal growth there is a greater release of nutrients but also an increase in lichen acids, which inactivate urease and slow down photosynthesis.

What happens to the large amounts of polyols and glucose released by the photobionts? These compounds are absorbed by the mycobiont and converted to mannitol, which is a fungal storage product. Such a conversion creates a sink to which algal nutrients continue to flow. The fungus uses some of the mannitol for growth and development, but the rest is used to help it withstand the extreme conditions of its habitat. Some scientists feel that the polyol pools that accumulate in the fungus are used up during *resaturation respiration,* which occurs each time a dry lichen is rewetted. Lichens cannot control their water content and, therefore, undergo frequent cycles of wetting and drying. Each time a dry lichen is wetted, its respiration increases, remains elevated for several hours, and then returns to normal. In species of *Umbilicaria* resaturation respiration lasts for 40 minutes but in

Cetraria cucullata it may persist for 5 hours. Presumably, during resaturation respiration, the fungus respires polyols instead of proteins and other vital compounds. A lichen has to compensate for losses resulting from resaturation respiration as well as for losses that occur when carbon compounds leak out of the photobiont cells during the first few minutes of wetting.

PHOTOSYNTHESIS AND RESPIRATION. In a lichen, photosynthesis reflects the activity of the photobiont, whereas respiration is mostly that of the mycobiont, which makes up the bulk of the thallus. In *Xanthoria parietina,* the fungus constitutes about 43% of the thallus volume while the alga makes up only 6.7%; extracellular substances and air spaces make up the rest of the thallus. This wide difference in the proportions of symbionts in a thallus makes measurements of metabolic processes both difficult and complicated. Moreover, the close physical relationship between the symbionts makes it impossible to obtain pure fractions of the algal symbiont, thereby limiting the types of studies that can be performed.

The metabolic rate of a lichen is influenced by light, temperature, day length, and water content. Some lichens can adjust their metabolic responses to different environments and seasons. For example, *Caloplaca trachyphylla* and *Peltigera rufescens* can acclimate their photosynthetic rates to winter and summer temperatures, thereby achieving near maximum rates throughout the year. Similar adaptive responses may occur with light intensity, water content, day length, and season.

The net photosynthetic rate of many lichens depends on the amount of thallus water. If a thallus is saturated with water, diffusion of CO_2 to the phycobiont is much slower than when a thallus has air spaces. In some lichens, crystals of chemical substances coat the outer surface of the fungal hyphae and prevent a thallus from becoming waterlogged. Many lichens can achieve maximum rates of photosynthesis by absorbing only water vapor from the atmosphere. Lichens kept at 95% to 98% relative humidity will reach water contents of up to 70% of their dry weights and their rate of CO_2 uptake will be about 90% of the rate achieved when their thalli are wetted with liquid water. Populations of desert coastal lichens, such as those in Baja California and in Chile, are never wetted by liquid water and depend on daily fog in order to carry out photosynthesis and respiration. Photosynthesis also depends on the amount of light that reaches the algal cells. Most lichen thalli have an upper cortex, which covers the algal layer. The cortex frequently contains lichen acids and pigments that shade the photobiont.

NITROGEN FIXATION. About 8% of the known lichen species have cyanobacteria as their photobiont, usually species of *Calothrix, Fischerella (Stigonema), Nostoc,* or *Scytonema.* Cyanobacteria may be primary symbionts, as in *Collema* and *Peltigera,* or secondary symbionts as in *Lobaria, Stereocaulon,* and *Sticta.* As secondary symbionts the photobionts are housed in cephalodia, which are gall-like structures, or in separate regions of a thallus that occur on or inside a primary thallus that has a green photobiont. Most lichen-forming cyanobacteria fix atmospheric nitrogen inside specialized cells called heterocysts. The percentage of heterocysts to total vegetative cells is much greater in cephalodial cyanobacteria than in those that are primary photobionts (20% to 30% and 4%, respectively). The reason for this wide difference is not understood.

Rates of nitrogen fixation in lichens are affected by light intensity, darkness, thallus moisture, temperature, pH, desiccation of the thallus, and season. Seasonal variation in nitrogenase activity has been reported for *Peltigera canina* and *Stereocaulon paschale.*

Cyanobionts of lichens release over 95% of the nitrogen they fix as ammonia, which is converted by the mycobiont into amino acids and proteins. Hyphae near the photobiont layer of *Peltigera canina* were found to contain high concentrations of glutamic dehydrogenase, an enzyme that assimilates ammonia. The specific activity of nitrogen-assimilating enzymes of the cyanobionts, such as glutamine synthetase, is inhibited by compounds produced by the mycobiont. Thus, the cyanobionts are unable to use most of the nitrogen they fix and their growth is slowed because of nitrogen deficiency. This deficiency, in turn, causes an increase in the number of heterocysts.

Lichens with nitrogen-fixing cyanobionts are important contributors to the nitrogen economy of different ecosystems. *Lobaria oregana* is a large, foliose lichen that contains a blue-green photobiont in cephalodia. The lichen is abundant in the Douglas fir forests of the Pacific Northwest, and when its thalli fall off the trees onto the ground, they decay and release significant amounts of nitrogen (2 lbs N per acre per

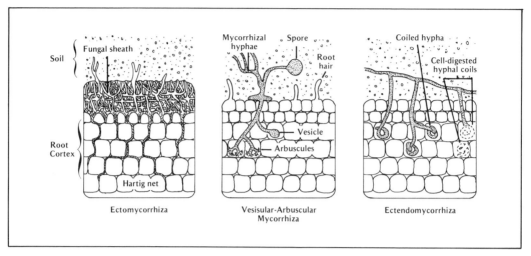

Ectomycorrhiza   Vesisular-Arbuscular   Ectendomycorrhiza
                             Mycorrhiza

Figure 7-4. *Types of mycorrhizas.* (Adapted from Deacon, J. W. 1980. Introduction to modern mycology. Halsted Press, New York.)

year). Organic nitrogen compounds may also leach out from lichen thalli during heavy rains.

C. Fungi and Plants

1. MYCORRHIZAS. Mycorrhizas are symbiotic associations between fungi and the roots of terrestrial plants. These associations were first described in 1885 by the German botanist A. B. Frank, who named the infected roots *mykorrhizen.* Frank recognized two types of mycorrhizas, which he called ectotrophic ("outside nutrition") and endotrophic ("inside nutrition"), based on whether or not a fungal sheath covered the outside of the roots. These names are no longer used because they give the false impression that mycorrhizal associations differ only in terms of how they obtain their nutrients. At present, seven types of mycorrhizas are recognized; these include *vesicular-arbuscular mycorrhiza, ectomycorrhiza, ectendomycorrhiza, arbutoid mycorrhiza, ericoid mycorrhiza, orchid mycorrhiza,* and *monotropoid mycorrhiza.*

Mycorrhizas occur in practically all terrestrial plants and are an important reason for the wellbeing of plants, especially those growing in nutrient-poor soils. Mycorrhizal plants are better able to withstand drought, high soil temperatures, toxic metals, and transplantation than plants without mycorrhiza. Fungi benefit from

the association by receiving organic compounds from the plant.

Vesicular-arbuscular mycorrhizas are the most common type of mycorrhiza and they occur in most families of angiosperms, especially in herbaceous plants, in all gymnosperms except the Pineaceae (pine family), and in ferns and liverworts. The next most common type, the ectomycorrhizas, occur only in trees and shrubs, in about 3% of the known gymnosperms and angiosperms (Fig. 7.4).

a Vesicular-arbuscular Mycorrhiza. Fungi that form vesicular-arbuscular mycorrhizas are zygomycetes that belong to the family Endogonaceae and to the genera *Acaulospora, Gigaspora, Glomus,* and *Sclerocystis.* About 80 species of these fungi are found throughout the world. The species have wide host ranges. The fungi are present in the soil or on nearby roots and they infect the developing roots. This type of mycorrhiza is difficult to recognize because there are no structural changes in the root. The fungus grows between the cortical cells of the root and also penetrates the cells. The fungus does not form an outer sheath around the root, but the hyphae do grow out from the root into the soil.

Fungal hyphae that penetrate the cells of the cortex form shrublike growths called *arbuscules* that fill most of the host cell (Fig. 7.5). Arbuscules are elaborate, branched haustoria that

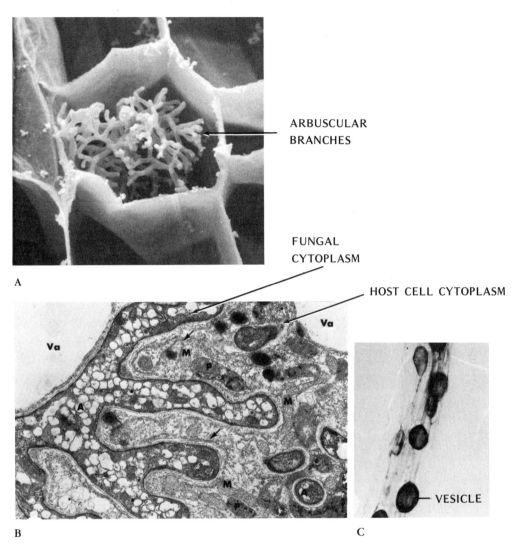

ARBUSCULAR
BRANCHES

FUNGAL
CYTOPLASM

HOST CELL CYTOPLASM

A

Va

Va

M

P

M

A

M

A

P

B

VESICLE

C

Figure 7-5. Vesicular-arbuscular mycorrhizal symbioses. (A) Arbuscules of Glomus mosseae *in root cell of* Liriodendron tulipifera. *(B) Part of a vacuolate arbuscule of* Glomus mosseae *inside root cell of yellow poplar. (C) Vesicles of* Glomus fasciculatus *in stained soybean root. (Parts (A) and (C): Merton F. Brown, University of Missouri. Part (B): Gerald Van Dyke, North Carolina State University.)*

remain surrounded by the plasma membrane of the host. These structures develop in response to fungitoxic compounds produced by the plant. An arbuscule may live for up to 15 days before it degenerates. The host cell may then become infected by new arbuscules. The fungus also produces thick-walled swellings, called *vesicles,* both within and between the plant cells (Fig. 7.5). Arbuscules and vesicles are so characteristic of these fungi that the associations they form are called *vesicular-arbuscular* mycorrhi-

zas. Vesicles store lipids; arbuscules are structures through which nutrients pass between the fungus and plant cell. An infected host cell has an enlarged nucleus and lacks starch granules. Host cells that digest the arbuscules return to their preinfected condition; that is, their nuclei return to a normal size and starch granules appear.

Vesicular-arbuscular mycorrhizas differ widely with respect to the extent of their infection in the root and the types of structures they form

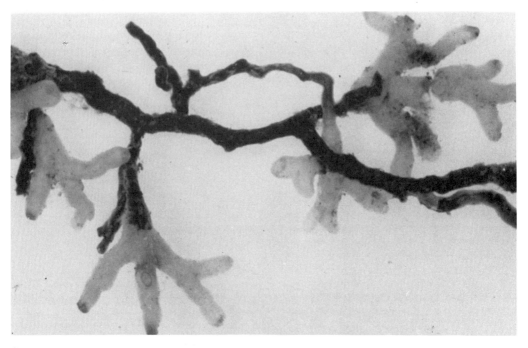

A

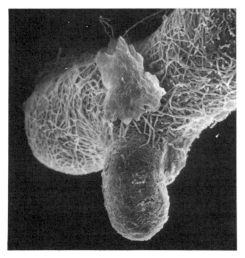

B   C

Figure 7-6. *Ectomycorrhizal symbiosis of the fungus* Pisolithus tinctorius. *(A) Ectomycorrhiza on slash pine. Note characteristic dichotomous branching. (B) and (C) Scanning electron micrographs of mantle of fungal hyphae around roots of* Eucalyptus nova-anglica. (Part (A): Donald Marx, USDA, Forest Service, Athens, Georgia. Parts (B) and (C): Gerald Van Dyke, North Carolina State University.)

within the host cells. Many of these fungi form spores and fruiting bodies of different kinds. The fungi infect only young, living root cells, generally in an area directly behind the growing root tip. As the root matures, its cells are no longer susceptible to outside fungal infection. Once infection occurs, new root tissue may be infected by hyphae that grow out from the infected plant cells or by new fungi from the soil.

Fungi that form vesicular-arbuscular mycorrhizas are obligate symbionts. They cannot live independently and receive their organic compounds from the host plant. The fungi lack enzymes, such as cellulase and pectinase, that are necessary to degrade vegetable litter and humus. The obligate nature of these fungi may explain in part why they do not grow in axenic, laboratory cultures.

Curious structures that resemble bacteria have been found in the hyphae of some vesicular-arbuscular mycorrhiza. The origin and function of these structures is not known.

The earliest known vascular plants, such as *Rhynia* and *Asteroxylon,* contained vesicular-arbuscular mycorrhizas, according to a study of the fossil remains of these plants. The presence of these associations in such early vascular plants and their widespread occurrence among present land plants indicates that mycorrhizas played an important role in helping plants colonize the terrestrial environment. The fungi may have protected the early plants from drying and helped them obtain nutrients and water from the soil.

Vesicular-arbuscular mycorrhizas are extremely common in tropical forests and in agricultural crops. These associations have immense potential as "biological fertilizers" and can be used to reclaim vast areas of barren soils in the world, particularly in the tropics, which have low levels of available phosphate. These mycorrhizas interlink different plants and are important in the transfer of nutrients between plants.

Some plant families, such as the Cyperaceae (sedges), Caryophyllaceae (carnation), Cruciferae (mustard), and Juncaceae (rush), lack mycorrhizal infections. It is not clear whether this absence of mycorrhiza results from the environmental habitats of these plants (many grow in wet habitats), which suppress fungal growth, or from a type of root epidermis and cortex that prevents fungal penetrations.

b. Ectomycorrhiza. In ectomycorrhizas the fungus forms a sheath or mantle around the root. The mantle generally is 20 to 40 μm thick, completely surrounds the root, and consists of either a tissuelike mat or a loose web of hyphae (Fig. 7.6). From the mantle, individual hyphae or strands of hyphae grow outwardly into the soil and inwardly into the root cortex. Hyphae grow through the middle lamella that separates the cortical cells and form a network between the cells called the *Hartig net,* named after its discoverer, R. E. Hartig. Hyphae do not penetrate the cortical cells or the inner core of the root where the vascular system is located. Fungal infection generally inhibits the development of root hairs and produces structural changes in the root. The changes are a result of growth hormones, auxin and cytokinins, produced by the fungus. Infected roots often assume a stumpy, corallike appearance, which is typical of many ectomycorrhizas.

In contrast to vesicular-arbuscular mycorrhizas, which involve relatively few species of zygomycetes, more than 5,000 species of basidiomycetes, mostly hymenomycetes and gasteromycetes, and even a few ascomycetes, form ectomycorrhizas. Ectomycorrhizas are common in temperate and tropical forests. A fungus may be specific to one type of tree, or, more usually, it forms associations with different trees. For example, a fungus may associate with a beech tree as well as a spruce tree. Conversely, one tree can associate with different fungi. A Douglas fir tree can form mycorrhizas with as many as 100 different fungi. Different fungi may form mycorrhizas simultaneously, on the roots of one tree, without competing with each other. Fungi present in the roots of a seedling are often replaced by other fungi as the tree matures.

Ectomycorrhizas vary considerably in shape. They may be club-shaped or nodular, simple or branched (Fig. 7.6). They start to form one to three months after a seed germinates. Differences between ectomycorrhizas include the color of the outer fungal sheath or mantle, shape of the infected roots, and the abundance and texture of the hyphae that are in the soil around the root.

Ectomycorrhizas protect the root from other parasitic organisms. The fungal sheath around the roots is a barrier through which pathogenic organisms cannot penetrate. Mycorrhizal fungi also secrete antibiotics that inhibit the growth of potential parasites. Fungitoxic compounds produced by cortical cells as a result of mycorrhizal infection also inhibit growth of pathogens.

Ectomycorrhizas develop best when plants are growing under suboptimal nutritional conditions. The fungi provide a service to plants by absorbing and making available to them the nutrients they need. Because the absorbing surface of the root is surrounded by a fungal sheath, all nutrients must pass through the fungus before they enter the root. In this way the fungus controls the amount of nutrients that pass into the root. The fungus provides phosphorus, nitrate, potassium, and other minerals to the plant. The most important mineral is phosphorus, which moves more rapidly through the fungal hyphae than soil. The sheath a fungus forms around the root stores carbon compounds and minerals that can be used by both symbionts during adverse periods. If abundant nutrients, such as nitrogen, are present in the soil, the development of ectomycorrhizas is inhibited and established associations revert to their nonsymbiotic condition. High nitrogen concentrations inhibit the fungus from producing auxin, which is necessary for mycorrhizal formation.

The plant provides sucrose to the fungus, which converts it into carbohydrates such as trehalose, mannitol, and glycogen. Since the plant cells cannot use these carbohydrates, the sheath around the roots becomes a sink to which there is a steady flow of sucrose from the plant. Most mycorrhizal fungi depend completely on the plant for their carbon supply, since they are unable to break down complex carbohydrates such as cellulose and lignin.

c. Other Mycorrhizas. There are several variant forms of ectomycorrhiza, one of which is the *ectendomycorrhiza.* In this type of association, which is common in pine seedlings, the outer fungal sheath is greatly reduced or absent. The Hartig net is present and the hyphae penetrate the host cells (Fig. 7.4). Another variant form is the *arbutoid mycorrhiza.* This type occurs in members of the heather family (Ericaceae), such as trees and shrubs of the genus *Arbutus* (Madrone), in *Pyrola* (wintergreen), and in the small, creeping plant *Arctostaphylos uva-ursi* (bearberry). This mycorrhiza has a sheath, a well-formed Hartig net, and extensive penetrations of the host cells. Ectendomycorrhizas and arbutoid mycorrhizas are produced by fungi that also form ectomycorrhizas. The kind of association produced depends on the environmental conditions and on the type of host.

The *ericoid mycorrhiza* and *monotropoid mycorrhiza* occur in the heather family and related families. Most species of these families produce *ericoid mycorrhizas,* in which neither a sheath nor a Hartig net is formed. The fungus penetrates the root cells and forms hyphal coils. When the hyphae in a host cell degenerate, the cell dies, a situation that is unlike that of other mycorrhizas in which the host cells remain alive even after their internal hyphae degenerate or are digested. The fungus that forms ericoid mycorrhizas is usually an ascomycete, such as *Pezizella,* although the basidiomycete *Clavaria* may also be involved.

Monotropoid mycorrhizas occur in the genus *Monotropa* (Ericaceae), in plants known commonly as Indian pipe and pinesap. These plants lack chlorophyll and thus cannot manufacture their own food. They live as heterotrophs and also receive nutrients from nearby trees by means of mycorrhizal connections. *Boletus,* a common forest basidiomycete, simultaneously forms ectomycorrhizas with the roots of forest trees such as beech, oak, pine, and spruce, and monotropoid mycorrhizas with species of *Monotropa.* Organic compounds pass from the tree roots through connecting fungal hyphae to the chlorophyll-less plants, which some scientists regard as *epiparasites* on the host trees. The monotropoid mycorrhiza has a thick sheath and Hartig net, and the host cells are penetrated by haustoria.

Orchid mycorrhizas are different from those of other plants. The fungi that associate with orchids are basidiomycetes and they are unique because they can break down complex carbohydrates, such as cellulose and lignin, into simpler carbon compounds, which they transport to the orchid seedlings. Other mycorrhizal fungi can use only simple carbohydrates and depend on the plant to provide them with these compounds. Orchid seeds have very little stored food and when they germinate the embryos rely on mycorrhizal fungi to supply them with carbohydrates and other nutrients during their early development. The fungi that infect the cortical cells of adult orchids are often digested by the plant cells. In this way, the plant obtains nutrients and also controls the extent of the fungal infection. Some tropical orchids that are epiphytes and some that lack chlorophyll are epiparasites on neighboring plants by means of mycorrhizal fungi, which connect the orchids to the trees.

The relationship between the fungus and the cells of a root is similar to that between a fungus and the photobiont of a lichen. From a casual inspection, mycorrhizas appear to be mutualistic associations, one in which both partners benefit. The fungus absorbs water for the plant, provides it with minerals, and also protects it from pathogenic organisms. In return, the fungus receives simple organic compounds such as sucrose and also growth factors from the plant. What appears to be an idyllic association, however, may conceal a different picture. Elias Melin, a pioneer of mycorrhizal research, has proposed that the relationship in ectomycorrhiza is one of controlled parasitism. Melin considers the mycorrhizal fungus to be a pathogen whose intrusion into the plant root is checked by fungitoxic compounds produced by the plant cells. Thus, a mutual "standoff" between fungus and plant results in a long-lasting symbiosis. If conditions change, for example, if the plant is given an optimal nutritional supply from external sources, then the plant cells kill the fungus and end the mycorrhizal association. Controlled parasitism also describes the symbiosis of lichens, where the parasitism of the photobiont cell is slow enough to allow for the development of a symbiotic relationship.

Identifying the fungi that form mycorrhizas is difficult. Ectomycorrhizal fungi that are isolated and grown in axenic cultures develop very slowly and do not have any of the characteristics of natural mycorrhizas. The cultured fungi have not revealed any unique nutritional needs or differences from other fungi to explain their symbiotic habit. The best way to identify ectomycorrhizal fungi is to see if fruiting bodies of basidiomycetes growing near the trunks of trees are connected by means of hyphal strands with the mycorrhizas of the trees.

2. PLANT DISEASES: RUSTS, SMUTS, AND POWDERY MILDEWS.

Fungi cause millions of dollars worth of damage each year to agricultural crops and to ornamental plants. Fungal diseases of plants include wilts, rots, rusts, smuts, and cankers. Since the days of Anton de Bary, plant pathologists have studied life cycles, taxonomy, biochemistry, and the genetics of fungal-host parasite relationships. In collaboration with plant geneticists, they have successfully developed disease-resistant varieties of plants. Other control measures developed include disease-free seeds and seedlings, fungicides, plant protection quarantine, and manipulation of agricultural practices. How to control fungal parasites of plants is a persistent problem for mankind, one that will never be completely solved because of the changing nature of fungi. As resistant strains of plants evolve or are developed, new strains of the parasites arise naturally to infect them. Understanding the life cycle of a fungus and the physiological and genetic basis of parasitism is the first step in any effort to control plant diseases. The late blight of potato, Dutch elm disease, and the chestnut blight are all diseases caused by fungi. In each case, a fungus practically eliminated an important plant because not enough was known about the pathogenic fungi to control them.

Rusts are basidiomycetes that cause many plant diseases. The fungi get their name from the brownish red or black spores that cover the infected areas of the plant. There are about 5,000 species of rust fungi, which infect many flowering plants, conifers, and ferns. Human concern over rusts relates to their ability to infect important crops such as wheat, corn, barley, oats, and other cereals. Up to 10% or more of a crop may be lost because of these destructive pests.

Rusts are unusual basidiomycetes because they lack fruiting bodies and have a complex life cycle during which they produce five different types of spores. Each type of spore has a particular function in the life cycle and is produced at a different time of the year. Some rusts complete their life cycle on only one plant but many need two host plants that are usually unrelated to each other. For rusts needing two hosts, eradication of the least important plant is the best way to control the spread of the disease. For example, *Puccinia graminis* (black stem rust of wheat) infects cultivated wheat and has the common barberry as its alternate host. The sexual process of the fungus takes place on the leaves of the barberry and the spores produced can infect wheat. If barberry plants are eradicated in wheat-growing areas, the life cycle of the fungus is broken and the wheat will not become infected. Unfortunately, fungal spores are very light and can travel on air currents for hundreds of miles. Thus, even a single infected barberry plant miles away from cultivated wheat crops

102

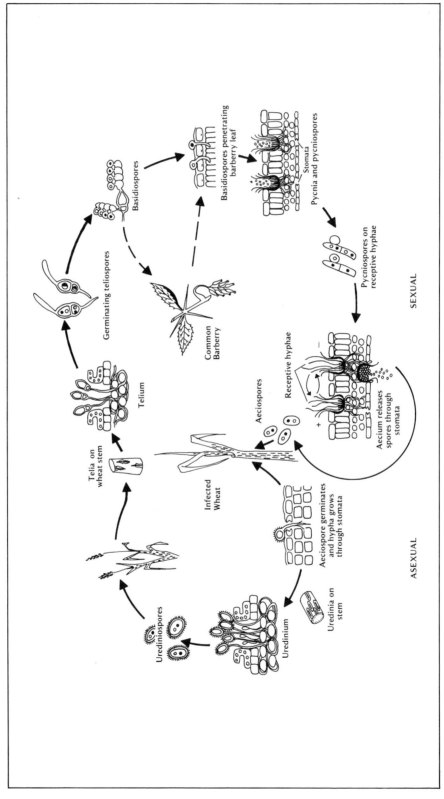

Figure 7-7. *Life cycle of wheat stem rust fungus,* Puccinia graminis tritici.

may serve as a source of infection. On barberry plants, the rust pathogen can evolve virulent races that can break down resistance in existing wheat varieties.

The life cycle of *P. graminis* begins in the spring when *basidiospores* of the fungus infect the leaves of a barberry plant (Fig. 7.7). The basidiospore produces a hyphal filament that penetrates the epidermis and forms branches in the leaf tissues. Hyphae grow inside the cells as well as between the cells. Within one to two weeks, the fungus produces *pycnia* on the upper leaf surface. Pycnia produce pycniospores that serve as gametes. The spores are contained in a thick, sugary fluid that covers the top of the pycnia. Small insects, attracted to the nectarlike fluid, pick up the sticky spores and transfer them to receptive hyphae that grow out from the pycnia. Rust fungi have plus (+) and minus (−) mating strains. Sexual compatibility occurs when *pycniospores* of one strain fuse with receptive hyphae of another strain. After fusion, the nucleus of a *pycniospore* migrates through the receptive hypha to the lower surface of the leaf where the fungus has started to form *aecia,* structures that produce another type of spore. The pycniospore nucleus pairs with a nucleus in the aecial primordium and the resulting hyphae contain both types of nuclei. These dikaryotic (two-nuclei) hyphae give rise to *aeciospores,* each of which has two nuclei. Numerous aeciospores are released from the lower surface of the barberry leaf and are carried away by the wind. If an aeciospore lands on a wheat plant, it germinates and gives rise to a dikaryotic mycelium that infects the wheat tissue. Several weeks later, the fungus produces another type of spore, the *urediniospore*. These spores are formed in such large numbers that they rupture the plant epidermis and produce rusty streaks on the stems and leaves of the infected wheat. Urediniospores infect other wheat plants and, if unchecked, can infect an entire field or crop of wheat. Continued urediniospore production in the wheat-growing regions can provide a source of inoculum for new wheat seedlings. In the autumn, the fungus produces a fifth type of spore, the *teliospore*. These black spores have thick walls and allow the fungus to survive the winter. In the spring, teliospores germinate and form basidiospores, which then infect the barberry plant and complete the life cycle of this rust.

Rust infections occur in many plants other than wheat. The infections usually result in abnormal growth of the host plants and the formation of galls. "Cedar apples" are galls on juniper plants and are caused by the rust fungus, *Gymnosporangium juniperi-virginianae.* The galls are made up of plant tissue that contains fungal hyphae and teliospores. Apple trees serve as the alternate host, where the fungus reproduces sexually.

The fungus that causes white pine blister rust, *Cronartium ribicola,* produces urediniospores and teliospores on the leaves of currants and gooseberries. The teliospores form basidiospores, which infect white pine trees. In the bark of the pine tree the fungus produces pycnia and aeciospores in such large numbers that they rupture the bark surface and form blisters. The quality of pine wood is affected adversely by the rust fungus.

Smuts are basidiomycetes that are similar to rust fungi in that they lack fruiting bodies. Unlike some rusts, they do not need an alternate host to complete their life cycle. Smuts produce large numbers of brown chlamydospores, which are often packed together into large spore balls. Smuts also produce basidiospores that behave like yeast cells; that is, they divide by budding and often fuse to form a dikaryotic stage. Each spore eventually gives rise to a dikaryotic mycelium, which infects the host plant. There are over 1,000 species of smuts. They infect many plants including agricultural crops such as corn, wheat, and oats.

Powdery mildews are ascomycetes that are obligate parasites of many plants, including lilacs, apples, grapes, and roses. These fungi form a thin mycelial layer on the surface of leaves and penetrate the epidermal cells by means of haustoria. The powdery mildews produce large numbers of asexual spores called conidia and these spores, together with the mycelium, form a powderlike coating over the infected leaves.

3. HYPOVIRULENCE. Some strains of pathogenic fungi are less virulent to their hosts than are normal strains. This reduced pathogenicity is called hypovirulence. Hypovirulent strains of *Endothia parasitica* have reduced the blight of the European chestnut (*Castanea sativa*) in many parts of Europe, and similar strains offer the first glimmer of hope of controlling the devastating fungal blight of the American chestnut (*Castanea dentata*).

Hypovirulent strains can spread, by means of hyphal fusions, among a normal pathogen pop-

BOX ESSAY

The Fungal Haustorium

Haustoria are specialized hyphal cells that are produced inside living cells of other organisms by parasitic fungi. In the case of plants, fungi penetrate the cell walls by enzymatically digesting the outer wall and then mechanically pushing through the inner wall. After the fungus has entered a cell, it forms a haustorium. Haustoria have many different shapes. Those of lichen fungi are club-shaped, and those produced by mycorrhizal fungi are branched. Haustoria of powdery mildews have fingerlike lobes and are associated with specialized structures (Fig. 7.8). The fungal haustorium is a branch of an extracellular hypha and terminates in the host cell. In some fungi the haustorium develops from an appressorium, another specialized cell of the fungus that attaches closely to the outer surface of the host cell by means of a mucilaginous substance. Haustoria of rust fungi have a narrow neck region at the site of penetration and a swollen region inside the host cell; most contain a nucleus, mitochon-

dria, and other organelles and are surrounded by a dense, granular matrix. The host cell forms a new plasma membrane, different in structure from that of the uninfected cell, around the penetrating fungus.

Although there is little direct experimental evidence to reveal the function of haustoria, it is presumed that they absorb nutrients from the penetrated cells. Whether or not haustoria kill the infected cells depends on the type of relationship the fungus has with the host. In rust and powdery mildew diseases, infected cells may be more active and live longer than uninfected cells. In lichens, haustoria kill the cells of the photobiont but only after a period of time, because the parasitism is gradual and under control.

An infected plant cell may have one or many haustoria inside its protoplast, depending on its resistance to the invading fungus. Some plant cells produce wall material around haustoria and limit their penetration into the cell.

ulation and transmit to the normal strain an agent that may be a dsRNA virus. This virus somehow brings about a reduced virulence of the pathogen. In effect, the hypovirulent strains cause a disease of the normal strains. Since the virus is present in the asexual spores (conidia) of the hypovirulent strains, the disease can spread rapidly and over great distances. The effectiveness of hypovirulent strains depends on the ability of their hyphae to fuse with those of normal strains. If there are genetic incompatibilities among the strains that prevent hyphal fusions, then the virus cannot be transmitted and the pathogen is unaffected. Thus, a disease may be controlled in some geographical areas, but not in others.

D. Summary and Perspectives

Mycosymbionts are fungi that parasitize other fungi, either killing their hosts or obtaining food from living hosts. These fungi are of interest to researchers who wish to develop biological controls for fungal pathogens. Mycophycobioses, associations between marine fungi and marine algae in which the alga is the larger partner, are being studied by scientists interested in the early stages of symbiotic evolution.

Lichens are associations of fungi mostly ascomycetes, and photobionts (photosynthetic partners), generally unicellular green algae and cyanobacteria. The lichen thallus is the result of a morphological transformation of the fungal

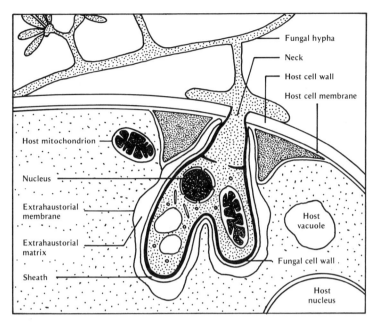

Figure 7-8. Schematic diagram of haustorium of powdery mildew fungus. (Adapted from Spencer, D. M. (ed.). 1978. The powdery mildews. Academic Press, London.)

symbiont and to a lesser extent the photobiont. The main types of thalli are crustose, foliose, and fruticose. Lichens produce secondary compounds that have a variety of roles in the symbiosis. Lichens are distributed throughout the world except in cities, where air pollution inhibits their growth. Studies on synthetic lichens have provided information about symbiont specificity, thallus development, and the physiological nature of the symbiosis. Many studies have examined carbohydrate transfer between lichen symbionts, photosynthesis and respiration of lichen thalli, and nitrogen fixation and metabolism. The photobiont excretes most of its photosynthetic products and, in the case of cyanobacteria, much of the nitrogen they fix. The fungus exerts some control over the photobiont and can regulate the amount of excreted compounds.

Mycorrhizas are mutualistic associations of fungi and the roots of terrestrial plants. There are seven different types of mycorrhizas. The most common type is the vesicular-arbuscular mycorrhiza, formed by zygomycetes, which produces arbuscules and vesicles inside root cells. Basidiomycetes form ectomycorrhizas, the sec-

ond most common type, which are characterized by a sheath or mantle of fungal tissue around the roots. Other types of mycorrhizas are ectendomycorrhiza, arbutoid mycorrhiza, ericoid mycorrhiza, monotropoid mycorrhiza, and orchid mycorrhiza. Mass-scale inoculations of pine seedlings with the ectomycorrhizal fungus *Pisolithus tinctorius* are successfully being carried out at several worldwide locations. The prospect of developing a means for the host plant to obtain minerals without having to depend on chemical fertilizers has encouraged reforestation efforts on marginally suitable lands.

The role of fungi in causing diseases of agricultural crops has been a major factor in the growth of the science of plant pathology. Rusts and smuts are obligate parasitic symbionts of plants. These fungi have complex life cycles and inflict heavy crop losses in many parts of the world. Efforts to control mildew and blight infestations of plants have played a historic role in the development of fungicides.

Hypovirulence, a reduced pathogenicity in strains of some fungi, may be caused by a virus. Hypovirulent strains of pathogenic fungi are being used to control blights of chestnut trees.

Axenic culturing of obligate parasitic fungi has had limited success. Some species of rusts and smuts have been cultured but powdery mildews have not yet been grown in axenic culture, nor have fungi that form vesicular-arbuscular mycorrhizas and many lichen fungi. Some lichen fungi and ectomycorrhizal fungi can be cultured, but their growth rates are extremely slow. The slow growth as well as the unpredictable nature of symbiotic fungi in culture has limited the types of studies that can be conducted with them. Why symbiotic fungi are difficult to grow axenically is not clear. Some species may need nutrients that are not contained in the culture medium. More likely the slow growth of these fungi is a result of their natural symbiotic adaptations. Long-lasting associations such as lichens and mycorrhizas require a balanced relationship between the symbionts, since unregulated growth of the fungus would quickly kill the host. In both lichens and ectomycorrhizas, an abundance of external nutrients inhibits the symbiotic transformation of the fungus and causes a breakdown of established associations.

Review Questions

1. Distinguish between necrotrophic and biotrophic mycosymbionts.
2. Distinguish between a mycophycobiosis and a lichen.
3. Describe the three main types of lichen thalli.
4. Describe the physical and physiological relationships between lichen symbionts.
5. How are lichens and mycorrhizas similar?
6. What role did mycorrhizas play in the early evolution of plants?
7. Distinguish between arbuscules and vesicles.
8. Why are there so many different types of mycorrhizas?
9. How do mycorrhizal partners benefit plants?
10. Describe the life cycle of *Puccinia graminis.*
11. What are smuts and powdery mildews?
12. Explain how hypovirulence can be used to control fungal blights.

Further Readings

Ahmadjian, V. 1982. The nature of lichens. Nat. Hist. 91 (March): 31–37. (Presents the view that lichens are examples of a controlled parasitism.)
Barnett, H. L., and F. L. Binder. 1973. The fungal

host-parasite relationship. Ann. Rev. Phytopathology 11:273–292.
Cooke, R. N. C., and J. M. Whipps. 1980. The evolution of modes of nutrition in fungi parasitic on terrestrial plants. Biol. Rev. 55:341–362. (A comprehensive review article.)
Hale, M. E., Jr. 1983. The biology of lichens. 3d ed. Edward Arnold, Baltimore. 190 pp. (An excellent introductory treatment of all aspects of lichens.)
Harley, J. L. 1984. The mycorrhizal associations. In: Cellular interactions. Encyclopedia of plant physiology, new series, Vol. 17, 148–186, ed. H. F. Linskens and J. Heslop-Harrison. Springer-Verlag, Berlin.
Jackson, R. M., and P. A. Mason. 1984. Mycorrhiza. Studies in Biology, no. 159. Edward Arnold, Baltimore. 59 pp.
Kohlmeyer, J., and E. Kohlmeyer. 1979. Submarine lichens and lichenlike associations. In: Marine mycology: The higher fungi, 70–78. Academic Press, New York. (Describes the known examples of mycophycobioses.)
Littlefield, L. J. 1981. Biology of the plant rusts: An introduction. Iowa State Univ. Press, Ames, Iowa. 103 pp. (A survey of the entire field of rust biology.)
Madelin, M. F. 1968. Fungi parasitic on other fungi and lichens. In: The fungi, Vol. III, 253–269, ed. G. C. Ainsworth and A. S. Sussman. Academic Press, New York.
Malloch, D. W., K. A. Pirozynski, and P. H. Raven. 1980. Ecological and evolutionary significance of mycorrhizal symbiosis in vascular plants. Proc. Natl. Acad. Sci. 77:2113–2118. (Proposes that mycorrhizal associations were instrumental in the early colonization of land by plants.)
Ruehle, J. A., and D. H. Marx. 1979. Fiber, food, fuel, and fungal symbionts. Science 206:419–422.

Bibliography

Hale, M. E. 1979. How to know the lichens. 2d ed. Wm. C. Brown, Dubuque, Iowa. 246 pp.
Harley, J. L., and S. E. Smith. 1983. Mycorrhizal symbiosis. Academic Press, London. 483 pp.
Hawksworth, D. L., and D. J. Hill. 1984. The lichen-forming fungi. Chapman and Hall, New York. 158 pp. (A capsulized treatment of lichenology.)
Horsfall, J. G., and E. B. Cowling (eds.). 1979. Plant disease: An advanced treatise. Vol. 4, How pathogens induce disease. Academic Press, New York. 466 pp.
Horsfall, J. G., and E. B. Cowling (eds.). 1980. Plant disease: An advanced treatise. Vol. 5, How plants defend themselves. Academic Press, New York. 534 pp.
Ingram, D. S., and P. H. Williams. 1982. Advances in plant pathology, Vol. 1. Academic Press, London.

220 pp. (Review articles on hypovirulence and gene-for-gene host-parasite systems.)

Lawrey, J. D. 1984. Biology of lichenized fungi. Praeger, New York. 408 pp. (An excellent account of the general field of lichenology.)

Melin, E. 1962. Physiological aspects of mycorrhizae of forest trees. In: Tree Growth, 247–263, ed. T. T. Kozlowski. Ronald Press, New York.

Misaghi, I. J. 1982. Physiology and biochemistry of plant-pathogen interactions. Plenum Press, New York. 287 pp.

Powell, C. L., and D. J. Bagyaraj. 1984. VA-Mycorrhiza. CRC Press, Boca Raton, Fla. 234 pp.

Rayner, A. D. M., and J. F. Webber. 1984. Interspecific mycelial interactions: An overview. In: The ecology and physiology of the fungal mycelium, 383–417, ed. D. H. Jennings and A. D. M. Rayner. Cambridge Univ. Press, Cambridge.

Schenck, N. C. (ed.). 1982. Methods and principles of mycorrhizal research. American Phytopathological Society, St. Paul, Minn. 244 pp.

CHAPTER 8

Protoctistan Symbiotic Associations
Parasitic and Mutualistic Protozoans

A. Introduction

Protoctists live in oceans, lakes, streams, and soils, and around the roots of plants. They also live in the alimentary canal, bloodstream, liver, lungs, brain, and other organs of animals. Spores and cysts of many protoctists float in the air. Some protoctists cause serious human illnesses, such as malaria, African sleeping sickness and Chagas' disease; others are pathogens of domesticated animals, such as horses, cattle, and poultry. The red tides that kill fish along coastal areas of the northeastern United States are caused by large populations of toxin-producing dinoflagellates.

There are an estimated 200,000 protoctist species, which is more than twice the number of chordates. The populations of some protoctists may be extremely high. For example, there may be thousands of protoctists in only 1 cc of soil and billions in the rumen of herbivores. Millions of flagellate symbionts inhabit a termite's intestine and may make up as much as one-third of the insect's weight. The White Cliffs of Dover in England consist entirely of the fossil remains of foraminiferan amoebae.

People who admire beauty, symmetry, and natural design are awed by the delicate sculpturing of the siliceous shells of radiolarians. There are approximately 45,000 species of protozoans, of which about 20,000 are fossil species of radiolaria and foraminifera. Of the 25,000 living species, about 7,000 are inhabitants of larger host animals, and the rest live in fresh water or in oceans and constitute an important source of food for large animals.

Photosynthetic protoctists are of immense im-portance in the ecology of marine organisms because they are the first link of the food chain. In this chapter, we examine symbioses that involve the nonphotosynthetic protoctists, the protozoans; in the following chapter we consider the role of photosynthetic protoctists in animal symbioses.

Recent studies in protoctistan biology have been concerned with how symbionts infect a host, overcome its defense mechanisms, and reproduce successfully. How are the host animal's antibody mechanisms and phagocytic activities of cells such as macrophages neutralized during infection? This chapter focuses on the advances in our understanding of pathogenesis of particular relationships and the conditions under which a symbiont-host relationship becomes nonpathogenic.

B. Kingdom Protoctista

1. CHARACTERISTICS. Protoctists are aerobic or anaerobic organisms that lack organized tissues and embryos. They are unicellular or multicellular. Their mode of nutrition ranges from autotrophic to absorption or ingestion and they reproduce both sexually and asexually. Protoctists may be immobile or mobile by means of pseudopodia, eukaryotic flagella, or cilia. Unicellular protoctists are known as protists.

2. CLASSIFICATION. The following is a taxonomic grouping of various organisms of the kingdom Protoctista that are discussed in chapters 8 and 9.

KINGDOM PROTOCTISTA

Phylum: Dinoflagellata (dinoflagellates); *Gymnodinium*

Phylum: Rhizopoda (amoebae); *Acanthamoeba, Entamoeba*

Phylum: Zoomastigina (animal flagellates)

   Class: Opalinida; *Opalina*

   Class: Schizopyrenida (amoeboflagellates); *Naegleria*

   Class: Kinetoplastida; *Leishmania, Trypanosoma*

   Class: Diplomonadida: *Giardia*

   Class: Trichomonadida; *Histomonas, Trichomonas*

   Class: Hypermastigida; *Trichonympha*

Phylum: Chlorophyta; *Chlorella, Cephaleuras*

Phylum: Rhodophyta; *Holmsella, Choreocolax*

Phylum: Foraminifera (forams)

Phylum: Ciliophora (ciliates)

   Class: Kinetofragminophora; *Dasytricha, Diplodinium, Entodinium, Epidinium, Isotricha*

   Class: Oligohymenophora; *Paramecium*

Phylum: Apicomplexa

   Class: Sporozoasida (gregarines and coccidians); *Monocystis, Toxoplasma, Plasmodium, Theileria*

3. A Generalized Life Cycle. The motile stage of a protozoan is called a *trophozoite.* A trophozoite feeds, grows in size, and represents the metabolically active stage in the life cycle. Food is obtained by either simple diffusion through the cell membrane or by pinocytosis and phagocytosis.

Asexual reproduction is by binary or multiple fission and is called *schizogony.* The nucleus of a trophozoite undergoes repeated mitotic divisions and each daughter nucleus becomes surrounded by a small portion of cytoplasm. The cell is now called a *schizont,* and when it ruptures it releases *merozoites,* which represent a new generation of protozoans.

In sexual reproduction, the cells that produce gametes are called *gametocytes.* The gametes may be similar or dissimilar in size and shape. In many protozoans, the zygote develops a thick wall and forms a *cyst,* which is resistant to unfavorable conditions. In a suitable environment a cyst breaks open and liberates trophozoites. In apicomplexan protozoans, the zygote is usually transformed into a *sporocyst,* the nucleus of which undergoes meiosis followed by repeated mitotic divisions. A small amount of cytoplasm surrounds each nucleus and *sporozoites* are formed, which are released when the sporocyst ruptures (Fig. 8.1).

C. Protozoans as Symbionts

1. AMOEBAE. Amoebae occur in all environments capable of supporting life. They are frequently ingested by potential hosts from drinking water or food. Many species of amoebae have become adapted to a symbiotic existence in the alimentary canal of animals. They produce no obvious pathogenic effects and are often referred to as "harmless commensals." Some amoebae, however, are human pathogens. *Entamoeba histolytica* causes dysentery and species of *Acanthamoeba* and *Naegleria* can invade the cerebrospinal fluid of mammals.

a. **Entamoeba.** *Entamoeba histolytica* infects about 400 million people worldwide, but the disease it causes, *amoebiasis,* occurs in only a relatively small number of individuals. The protozoan has been reported also from monkeys, apes, and other mammals. Infection usually begins when a host ingests fecal-contaminated food or drink, which contains cysts of the amoeba. The cysts hatch and produce amoebae, which feed on bacteria and other food particles in the large intestine. After several cycles of growth and multiplication, some amoebae encyst; that is, they become spherical, secrete a thick wall, and store glycogen. Cysts are excreted in large numbers with the feces. An infected individual may excrete an estimated 45 million cysts in one day. If the cysts are ingested by a new host, they hatch and each cyst liberates four trophozoites. Amoebic infection may occur without symptoms, or the amoebae may attack the epithelial lining of the large intestine and cause ulcers and dysentery. In more serious cases, amoebae may penetrate the intestinal wall and infect organs such as the liver and brain.

The pathogenicity of *Entamoeba histolytica* has been a subject of controversy for decades. What was previously identified as *E. histolytica* is now believed to be a composite of at least two species, *E. hartmanii* and *E. histolytica. Entamoeba hartmanii* is the smaller of the two species and is nonpathogenic. Many strains of *E. histolytica* are also thought to be nonpathogenic and can inhabit an infected individual without

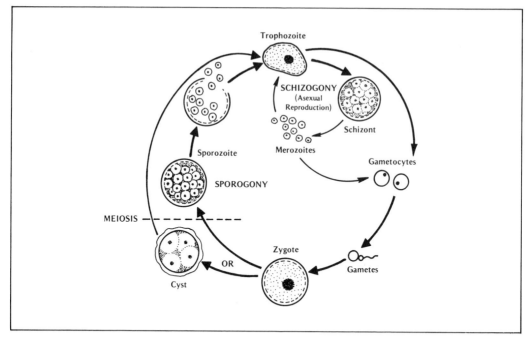

Figure 8-1. Generalized life cycle of parasitic protozoans.

BOX ESSAY

Guests and Hosts

"YOU CAN DRESS A BIOLOGIST UP,
BUT YOU CAN'T TAKE HIM OUT."

I have reluctantly come to the conclusion, after much soul-searching and not a little pain, that biologists make poor dinner guests.[1] Not that we eat with our hands, fail to bathe, or lack other social graces. The problem is that our conversations tend to encompass topics not usually included in dinner-table repartee. Just as lawyers may animatedly discuss a particular interesting case, biologists can wax poetic about cockroaches, cannibalism, and the like.

My wife has become more or less resigned

[1]Milton Love, the author of this essay, is a research associate in the Department of Biology at Occidental College, Los Angeles. The essay is reprinted from *Natural History* 88 (June 1979): 84, with permission of the author and the American Museum of Natural History.

to such behavior. She was introduced to it when, in a fit of romanticism, I named a parasite after her. She had her choice of organisms: one parasitized the gall bladders of certain fishes; the other, their urinary bladders. (She picked the urinary bladder parasite—*Davisia reginae*.)

Even she, however, eventually drew the line. During dinner one night, after discoursing on *kuru* (a viral disease found among New Guineans, caused by ritual brain cannibalism) and the pearlfish (which lives in the anus of sea cucumbers), I happened to mention *Oikopleura*, a small planktonic animal that lives in mucus "houses," which also serve as filters of food material.

"That's it!" she declared. "There will be no discussion of mucus at the dinner table." Even thoroughly inured nonbiologists have their limits.

The real problem arises when we are invited to gatherings. This was made abundantly clear during a Passover Seder my wife and I recently attended. A Seder is a ceremonial dinner celebrating the escape of Moses and his flock from Egypt. I often find the conversations at these affairs to be dull, and consequently, I spend much of these evenings in a sort of stupor.

On this occasion, the conversation swung around to the kosher dietary laws. Perking up, I pointed out that the biblical prohibition against eating pork was probably not based on the prevention of trichinosis (as is commonly believed), for trichinosis, a parasitic disease contracted by eating nematode-infected pork, is only rarely found in the Middle East.

"Trichinosis," I reflected, warming to my subject, "is an interesting disease. Infected sausage, for instance, may contain as many as one hundred thousand larval worms per ounce. A million larvae could easily be ingested during a meal."

Shamelessly, my wife (who rather enjoys these performances) egged me on: "Are the worms particularly dangerous?"

"Yes, indeed," I replied, with somewhat more enthusiasm than was perhaps warranted. "These worms burrow through the intestine and travel throughout the body, boring into muscles, brains, et cetera. There is no good cure, and in Europe of the Middle Ages, whole villages might become infected—many people dying horribly."

I vaguely noticed that the gentleman across the table had stopped eating, a gefilte-fish ball poised precariously between bowl and mouth. There was no stopping me now, however. I had their attention, and I was going to keep it.

"The whole topic of parasitism is a fascinating one. Most human parasites are very well adapted to their hosts, causing few problems. Often we never know we have them. For instance, humans can have fifty-foot-long tapeworms and have few or no symptoms. You know, Jewish women were once known as a primary host for *Diphyllobothrium latum,* the 'fish tapeworm.' The larvae are found in the muscles of fish, particularly pike, which is a major constituent of gefilte fish. Women would taste partially cooked gefilte fish before the larvae had been killed, thereby becoming infected."

The man across the table slowly pushed his untouched fish aside.

My blood surging with excitement, I continued. "However, not all tapeworms are relatively innocuous. *Echinococcus granulosus* is a species that causes very large cysts, containing ten or fifteen quarts of fluid. Humans catch the worms by swallowing their eggs, which they can get from dogs. The adult worms live in the dogs' intestines, the eggs are expelled in the feces, and the dogs often have them on their tongues after licking their rear ends. People become infected when they allow dogs to kiss them."

As one, all of the table turned toward the family schnauzer, who slunk off guiltily.

The host made a game effort to rechannel the conversation, remarking on how warm the room seemed to be. But I was unstoppable.

"Americans are just not used to thinking about parasites. We don't believe we have them. Actually, there are several types that commonly infect us. For instance, some of us here have an amoeba, *Entamoeba gingivalis,* in our mouths. It does not seem to do any harm, just sort of sits about on the gumlines, waiting for an occasional white blood cell to pop out."

At the end of the table, a woman absently poured wine on the tablecloth.

"When I taught the parasitology lab," I said, savoring the memory, "the last laboratory session was feces day, when everyone brought in their own specimens. We would find various amoebae and once in a while a worm egg. People really got into it. A woman I know had contracted amoebic dysentery in Mexico. She went off the anti-amoebic drugs she was taking, just to build up a large enough population to show the class."

Here I was shaken from my reverie by the white faces of my companions.

The conversation soon returned (on a somewhat subdued note) to the relative merits of chicken versus beef brisket as a main dish, but for the rest of the evening the group's enthusiasm for such topics flagged.

I, on the other hand, remained jovial. I had engaged in a brilliant discourse and broadened the horizons of my companions, while never once mentioning mucus at the dinner table.

producing symptoms. At one time it was believed that all cases of *E. histolytica* infection required medication, and, consequently, thousands of individuals were treated with unpleasant and even dangerous drugs. Eye disorders, allergies, migraine, rheumatoid arthritis, and liver disease were often falsely attributed to the presence of amoebae in infected persons.

Pathogenic strains of *E. histolytica* lose their virulence when they are cultured in the laboratory. Electron microscopic observations have revealed that amoebae from an infected person have a "fuzzy coat," the nature of which is not understood, that is absent in laboratory strains. Surface-active lysosomes that might be responsible for the breakdown of host cell membrane have been discovered in the amoebae. Amoebal toxins and viral symbionts may also play an important role in pathogenesis. Some scientists have observed that *E. histolytica* alone cannot produce disease and that the presence of bacterial or viral symbionts is essential for its virulence. Viruslike particles have been detected in the cytoplasm and nucleus of *E. histolytica*.

Entamoeba invadens is a common pathogen of turtles and snakes. Morphologically, the amoebae resemble *E. histolytica* and scientists have increasingly used it as a model to investigate the pathogenesis of invasive amoebiasis.

b. Acanthamoeba *and* Naegleria. Species of *Acanthamoeba* and the amoeboflagellate *Naegleria* are common in stagnant lakes and ponds and may cause a primary amoebic meningoencephalitis (PAM) in humans and other mammals. Young children swimming in such areas are most susceptible to *Naegleria* infections. Amoebae of *N. fowleri* enter the nasal passages, where they penetrate the olfactory epithelium and move along the olfactory nerve.

There are several species of *Acanthamoeba* that cause meningoencephalitis, inflammation of internal organs, corneal ulcers, gastritis, and diarrhea. In the 1920s, when acanthamoebae were isolated from diseased tissues, they were dismissed as contaminants by scientists. Culbertson and his colleagues demonstrated that intranasal inoculation of mice with *Acanthamoeba* killed the mice, and the amoebae could be observed in the brain tissues. On a worldwide basis, over 100 cases of PAM have been identified with most of them being fatal. Pathogenicity of these amoebae has been attributed

to their ability to release a phospholipolytic enzyme, which demyelinizes nerves.

2. CILIATES. Ciliates have hairlike structures called cilia and are well adapted to live in aquatic habitats. Although most ciliates are free-living, some live in the alimentary canals of vertebrates, insects, and annelids. Ciliate symbionts also inhabit organs such as the liver, as well as blood vessels and gonads of various animals.

a. Opalina ranarum: *An Amphibian Symbiont*. About 150 species of the genera *Cepedea, Opalina, Protoopalina,* and *Zeleriella* inhabit the large intestines of frogs and toads. The symbionts have large, up to 1 mm long, flat or cylindrical cells and their reproductive physiology is controlled by gonadal hormones of the host. Opalinids[2] do not produce any ill effects on the host, which may harbor thousands of these symbionts in its large intestine.

The symbionts have 2 to 200 nuclei, are covered with oblique rows of cilia,[3] and lack a mouth. Generally, adult individuals of *O. ranarum* multiply by binary fission in the rectum of frogs and toads. The symbiont life cycle is synchronized with that of the host. In the spring, at the onset of the frog's breeding period, the symbionts divide rapidly and produce small, precystic forms. These become transformed into cysts and are eliminated with feces from the host. Cysts ingested by tadpoles hatch in the duodenal region and the ciliates then migrate to the large intestines. Microgametes and macrogametes are formed from individual gametocytes. Fusion of the sex cells results in zygotes, which give rise to a new generation of ciliates (Fig. 8.2). Researchers have induced encystation of *O. ranarum* by injecting hypophysectomized and gonadectomized frogs with pregnancy urine, gonadotrophin, testosterone, and adrenalin.

b. *Rumen Ciliates: The Mutualistic Protozoans*. Mutualistic associations have evolved between ciliate populations and herbivorous mammals. The rumina of cattle and sheep con-

---

[2]Opalinids have traditionally been classified with the ciliates. Taxonomically, however, opalinids are considered as flagellates.

[3]Cilia show similarity in coordinated wave motion to spirochetes associated with termite flagellates.

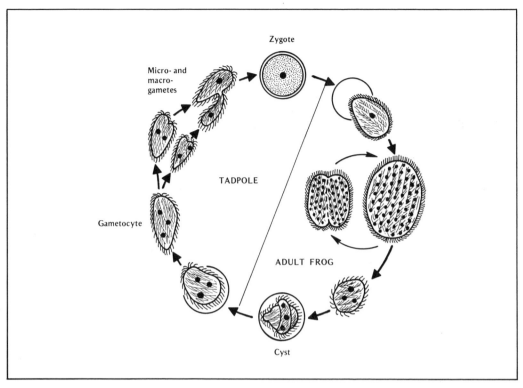

Zygote

Micro- and macro- gametes

TADPOLE

Gametocyte

ADULT FROG

Cyst

Figure 8-2. Life cycle of Opalina ranarum. (Adapted from Smyth, J. D. 1976. Introduction to animal parasitology. 2d ed. Halsted Press, New York.)

tain, in each 1 ml of gut contents, an estimated half a million ciliates consisting of 30 to 50 species.

Rumen ciliates are obligate anaerobes and new hosts acquire them through a mother's licking of her calves, or from eating grass that is dropped after being regurgitated from a rumen. The environmental conditions of the rumen are complex and have not been duplicated in the laboratory. The rumen has a temperature of 39°C, is anaerobic, contains particulate matter that is resistant to the host's digestive enzymes, and is deficient in glucose and amino acids. The alkaline or slightly acid pH of the rumen favors growth of the ciliates. The posterior portion of the stomach, the omasum, however, has an acid environment, which is lethal to the ciliate symbionts. Ciliates supply their host with about one-fifth of its protein needs, acetic, propionic, and butyric acids, and also a better balance of amino acids than do the bacterial symbionts of a rumen.

Two kinds of ciliates live in the rumen: entodiniomorphs and holotrichs (Fig. 8.3). Entodi-

niomorphs have a semirigid pellicle that covers the cell and gives each species a distinctive morphology. They also have unique internal plates, which form a type of cytoskeleton. Around the mouth of each symbiont is a band of cilia, which functions in a coordinated fashion. The number and location of ciliary bands are used to identify species of entodiniomorphs. In most species, the ciliary zone can be retracted and the pellicle drawn over it. *Diplodinium, Entodinium,* and *Epidinium* are important genera of symbiotic ciliates and they have been the subject of many physiological studies. *Entodinium bursa* is the largest of the entodiniomorphs and it consumes large quantities of plant material.

Entodiniomorphs ciliates digest cellulose and starch, as well as rumen bacteria. Many scientists have attempted, unsuccessfully, to grow rumen ciliates in the laboratory.

Rumen holotrichs resemble free-living ciliates and their number in sheep and goats has been estimated to be 160 to 200 thousand per milliliter of rumen content. Some common holotrichs are *Dasytricha ruminatium, Isotricha in-*

114

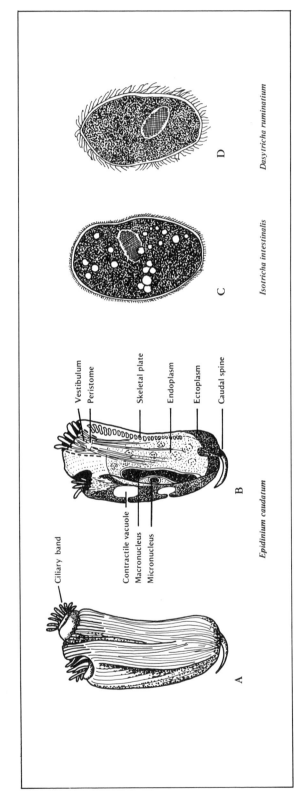

Figure 8-3. Types of mutualistic rumen ciliates. (A) and (B) External and internal morphology of a typical entodiniomorph. (C) and (D) Two examples of holotrich symbionts. (Adapted from Furness, D. N., and R. D. Butler. 1983. The cytology of sheep rumen ciliates. I. Ultrastructure of *Epidinium caudatum* Crawley. J. Protozool. 30:676–687.)

testinalis, and *I. prostoma.* Holotrich ciliates, in contrast to entodiniomorphs, can incorporate large quantities of sugar and convert it to starch for storage. Holotrich starch is indistinguishable from plant starch.

3. FLAGELLATES. Flagellated protoctists differ widely in their morphology and physiology.[4] Some have evolved a high degree of intracellular differentiation and possess unique structures whose functions have yet to be fully understood. All flagellates have flagella and can swim, which is a distinct advantage over an amoeboid life style. Some flagellates colonize the digestive, circulatory, and lymphatic systems of animals. Flagellate symbionts can be divided into two broad groups: intestinal flagellates, which inhabit the alimentary canal and genital tract of the host; and hemoflagellates, which inhabit the blood, lymph, and other tissues of vertebrate hosts and the intestines of insects.

a Intestinal Flagellates. These protoctists are obligate anaerobes and do not reproduce sexually. In some species, the trophozoite stage alternates with a resting stage, the cyst. Intestinal flagellates ingest food by phagocytosis and pinocytosis and some species have a mouthlike structure. Little is known of the nutrition and metabolism of these flagellates and only a few species have been cultured. Most species are nonpathogenic but some cause disease.

Most animals acquire flagellate symbionts from contaminated food or drink. Both the cysts and trophozoites can spread the infection. The symbionts usually occupy the lower portion of the host's alimentary canal, where digestive activities are lacking, that is, the cecum in mammals and birds and cloacal cavity in amphibians and reptiles. Below is a description of representative intestinal flagellates.

TRICHOMONADS. *Trichomonas gallinarum* causes avian trichomoniasis in pigeons, chickens, turkeys, and other domesticated birds. The symbiont inhabits the upper intestinal regions of an infected bird and may cause diarrhea, weight loss, and ruffled feathers. The trichomonad does not produce cysts. Young birds become infected by their parents. Different strains

of the flagellate have proved almost 100 percent infective in pigeons. *T. gallinarum* is used extensively to study mechanisms of pathogenicity.

Trichomonas vaginalis commonly occurs in the genital tract of humans and is sexually transmitted. Under certain conditions, still not fully understood, it becomes pathogenic on the epithelial cells of the urethra and vagina and erodes the mucosal lining. The flagellate also infects rats, mice, hamsters, and guinea pigs, and has been cultured *in vitro.*

Pathogenicity of trichomonad symbionts is correlated to their migration from the large intestine, where they are nonpathogenic, to other sites in the body. Examples of well-known pathogenic species include *T. vaginalis,* in the passages and cavities of the host's reproductive organs, and *T. gallinae,* in the upper digestive tract of birds. The pathogenicity of trichomonads is greatly reduced under axenic or monoaxenic conditions. For example, *T. gallinae,* normally a highly virulent pathogen of pigeons, loses its pathogenicity after four to five months of axenic culture. Some unknown cytoplasmic factors may become diluted when the symbiont is grown in culture. In addition, cultured flagellates have a reduced capacity to stimulate antibody production in rabbits. It is well known that pathogenic strains have either small amounts of antigens or incomplete antigens and thus trigger only a limited antibody response from the host. Although great advances have been made in understanding trichomonad pathogenesis, many intriguing problems still need attention.

GIARDIA. *Giardia lamblia* is a common inhabitant of the small intestine of man, monkeys, and pigs, especially in warmer climates. The symbiont attaches to the intestinal wall of the host by means of an adhesive disc and feeds on secretions of the intestinal mucosa. Trophozoites of this symbiont multiply by binary fission. A heavy infestation of an individual can result in the production of more than 14 billion cysts per day. Some forms of *Giardia* produce a disease called *giardiasis,* symptoms of which include diarrhea, vomiting, and poor vitamin B_{12} absorption. At least 10% of the human population in the United States carries *Giardia,* which usually is harmless and asymptomatic, although the feces from an infected person may contain large amounts of mucus and fats.

Giardia has a convex dorsal and a concave ventral side that contains an adhesive disc. The cell has two nuclei, each with a prominent nu-

---

[4]The term *flagellate* is used here in the context of eukaryotic organization where flagellum = undulipodium.

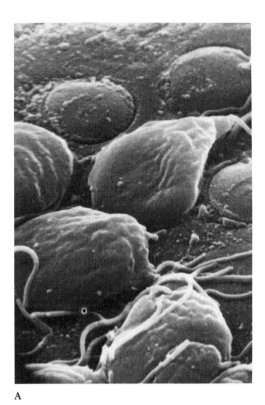

A

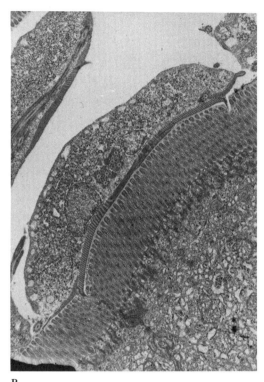

B

Figure 8-4. (A) *Trophozoites of* Giardia *attached to surface of intestinal villi of rat. Note dome-shaped lesions, which were previous sites of parasite attachment.* (B) *Electron micrograph of sectioned* Giardia *attached to microvillous border. Note the microtubular cytoskeleton of adhesive ventral disc.* (Stanley L. Erlandsen, University of Minnesota.)

cleolus. Mitochondria, smooth endoplasmic reticulum, and Golgi bodies have not been observed in the cells. The symbiont has four pairs of flagella, which are positioned to help the symbiont swim as well as to make the initial contacts with the host. The middle portion of *Giardia* contains microtubules that form a unique and distinctive cytoskeleton, which is involved in both attachment and contractibility. *Giardia* attaches to the host cell surface by means of its adhesive ventral disc. The indented margin of the disc penetrates the cell and forms an interlocking system with microvilli of the host cell (Fig. 8.4). *Giardia* microtubule cytoskeletons can be isolated by means of sonic treatments.

HISTOMONAS MELEAGRIDIS. Blackhead disease of turkey's was first identified in turkeys from Rhode Island in 1895 and by 1920 it was causing losses of up to 45% of the commercial turkey flocks in the United States. In the 1920s, Ernest Tyzzer, a Harvard University parasitologist, identified the causal agent of the disease as *Histo-monas meleagridis,* and described its biology and pathology. The pathogen first infects the intestine and then spreads to the liver and causes death of the host. Tyzzer discovered that chickens and nematodes were carriers of the flagellate. The disease declined by 1940 primarily because of improved sanitation and management practices of not putting chickens and turkeys together.

Histomonas meleagridis forms two types of cells. One type of cell is round, lacks flagella, and infects the liver and mucosal lining of the cecum. The other type of cell has one or two flagella and occurs in the intestine.

Birds that are susceptible to the cecal nematode, *Heterakis gallinarum,* have a high incidence of infection by *H. meleagridis.* Nematodes of both sexes as well as eggs of the female contain the flagellate, and when the nematodes infect poultry, the eggs hatch and produce infected larvae. Disintegration of some of the larvae by the host's digestive enzymes frees the parasite, which then invades the cecal mucosa and migrates to the liver. The symbionts cause prominent lesions on the cecal walls, a thickening of the cecal mucosa, and coagulation necrosis of liver parenchyma.

Earthworms also have a role in blackhead disease of poultry. Infected nematode eggs are often eaten by earthworms, which in turn are eaten by poultry.

Histomonas meleagridis can be cultured in the presence of bacteria from the cecum of a natural host. The cultured symbiont loses some or all of its pathogenicity but this can be restored, either fully or partially, by serial passages through poultry. Because of the unusual nature of the flagellate's life cycle and the involvement of another parasitic symbiont, a nematode, as an agent of transmission, *H. meleagridis* continues to be studied by scientists.

FLAGELLATES OF TERMITES AND ROACHES. Flagellate symbionts of the wood-eating roach *Cryptocercus punctualatus* and termites of the genera *Reticulitermes* and *Zootermopsis* have been the focus of much research. Lemuel R. Cleveland, a pioneer in this field, has devoted over four decades of his life to research on this subject. The flagellates inhabit the posterior portion of the host's alimentary canal and belong to the class Hypermastigida. The genera that have been studied extensively include *Barbulanympha, Devescovina, Joenia, Spirotrichonympha,* and *Trichonympha* (Fig. 8.5).

An estimated 30% to 50% of the total weight of a wood-eating termite consists of protoctists. Some, such as *Trichonympha,* can digest cellulose and thus allow the host insect to subsist on a diet of wood. The symbionts break down cellulose, anaerobically, to acetic acid, carbon dioxide, and hydrogen by means of the enzyme cellulase. The insect absorbs acetic acid through its hindgut wall and uses it, aerobically, as an energy source. The nature of this symbiosis is complicated because the host insects also have bacterial and fungal symbionts, which provide vitamins of the B complex to the hosts. These organisms occur in the alimentary canal, in the Malpighian tubules, or within specialized insect cells called mycetocytes. There is evidence to suggest that nitrogen-fixing bacteria also occur in the termite and provide nitrogen as well as growth factors and vitamins for the host and its flagellates. The bacteria also produce methane, the function of which is not clear.

There are over 2,000 species of termites in the world. They occur mostly in warm climates, but 41 species are known from North America. Wood-digesting, mutualistic protoctists are characteristic of the lower termite families (Mastotermitidae, Kalotermitidae, Hodotermitidae, and Rhinotermitidae). Only about 10% of termites are colonial insects, which have a high degree of social order and a distinctive "caste system" based on morphological and physiological features. Termites build complex nests, consisting of galleries and chambers, and each species produces a nest of specific shape and size that may be built underground or above ground as mounds (Fig. 6.10).

Termites undergo simple metamorphosis from egg to four nymphal instar stages and finally to the adult stage. The first nymphal instar acquires its flagellate symbionts by feeding on anal droppings of infested termites. During each molt, the entire epithelial lining of the alimentary canal is shed, resulting in the loss of intestinal flagellates. Each newly molted instar also acquires its symbionts from anal droppings. The termite digestive tract consists of three parts: (1) foregut, (2) midgut, and (3) hindgut. Most cellulose digestion takes place in the paunch region of the hindgut, which is dilated and contains the bacteria and flagellates. The flagellates digest cellulose to simpler compounds that can be used by the termites, and the host, in turn, provides an anaerobic chamber and food for the symbionts. Termite symbionts can be eliminated by starving the host or by maintaining it at high temperatures or high oxygen pressure. Termites that lose their symbionts do not survive on their normal wood diet and die within weeks. Survival of these symbiont-free termites can be prolonged by feeding them glucose or fungus-decomposed wood. Termites that are allowed to reacquire their flagellates will recover and live normally. Not all flagellates in the termite hindgut have the cellulase enzyme system. Only the cellulolytic strains are known to be mutualistic symbionts of termites, although the remaining flagellates show a great diversity of life styles.

Mixotricha paradoxa is a large wood-eating flagellate that lives in the gut of the termite *Mastotermes darwiniensis*. The symbiont earlier was thought to be a ciliate because its outer surface is covered with structures that look like cilia. Studies, however, have shown that *M. paradoxa* is covered with spirochetes and rod-shaped bacteria, and not with cilia. The spirochetes undulate in a coordinated manner and propel the

118

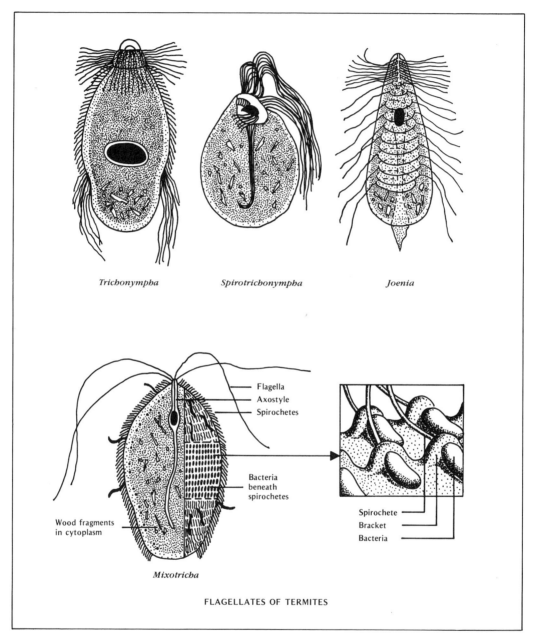

Trichonympha   *Spirotrichonympha*   *Joenia*

Flagella
Axostyle
Spirochetes

Bacteria
beneath
spirochetes

Wood fragments
in cytoplasm

Spirochete
Bracket
Bacteria

Mixotricha

FLAGELLATES OF TERMITES

Figure 8-5. *Common genera of symbiotic flagellates of termites.* (Adapted from Cleveland, L. R., and A. V. Grimstone. 1964. The fine structure of the flagellate *Mixotricha paradoxa* and its associated microorganisms. Proc. R. Soc. Lond. B. 159:668–686.)

symbiont through the intestinal fluid. The four flagella of the symbiont are believed to be used only for steering (Fig. 8.5). The association is considered to be mutualistic because the spirochetes receive nutrients from the flagellate. Lynn Margulis has suggested that the flagella of eukaryotic organisms evolved from similar spirochetelike prokaryotes that were ectosymbionts.

The flagellate symbionts of termites possess a variety of bacteria as intracellular symbionts. Some scientists believe that the bacteria are responsible for the flagellate's ability to digest cellulose.

There are many similarities between anaerobic fermentation in ruminants and cellulose degradation in termites. Both ruminants and termites ingest plant cellulose and by means of mutualistic symbionts degrade it to usable compounds.

b. Hemoflagellates. Intestinal flagellates of bloodsucking insects have successfully invaded the reticuloendothelial tissue of vertebrates. In a vertebrate host, the flagellates swim in the bloodstream or become intracellular inhabitants of macrophages, host cells that ingest and destroy microbial invaders. The symbionts have complex cells with odd-shaped mitochondria, and DNA of as yet unknown function. Hemoflagellates assume different morphological forms and in some host species the organism changes from one form to another during its life cycle. Some of the flagellates can resist the host's antibody mechanisms. Diseases produced by hemoflagellates in man and domesticated animals have cost millions of lives. African sleeping sickness, Chagas' disease in South America, and leishmaniasis in Asia and Africa are a few important human diseases. Hemoflagellates have been studied extensively by parasitologists, cell biologists, immunologists, and evolutionary biologists.

TRYPANOSOMA BRUCEI RHODESIENSIS: AFRICAN SLEEPING SICKNESS. African sleeping sickness affects thousands of people throughout equatorial Africa. About 250,000 people are infected with the disease, with about 15,000 new cases being reported each year.

Both sexes of the tsetse fly, *Glossina* sp., are vectors of the trypanosome that causes the disease. The symbiont develops into the infective form in the insect's salivary glands. When the insect feeds, it injects trypanosomes into the lymph and bloodstream of the host. The trypanosome population increases rapidly and reaches a peak in about five to seven days. The flagellates are of the *slender form* type. A week after the initial infection, two classes of host antibodies appear in the blood serum: *IgG* antibodies, which are directed against cell antigens, and *IgM* antibodies, which are active against cell surface antigens. The surface antigens of trypanosomes are believed to be glycoproteins. The antibodies cause agglutination and lysis of the trypanosomes and the disease then enters a remission phase as the number of flagellates declines sharply. During this phase, large numbers of flagellates of the *short stumpy form* appear in the blood. These forms are incapable of further development in the mammalian host. A few slender form trypanosomes develop new antigenic properties and are unaffected by the host's antibodies. These flagellates then multiply until the host's immune system manufactures new antibodies that are specific for them. The antibodies destroy most of the new trypanosomes, but again a few symbionts alter their antigens and the process is repeated. Each new antigenic variant of the symbiont is matched by the host's production of antibodies against it, but eventually the host is overwhelmed. After the insect has a blood meal from an infected individual, the stumpy form symbionts undergo several transformations and then migrate to the salivary glands, where they develop into the *infective form* (Fig. 8.6). The insect may inject as many as 40,000 trypanosomes into a host during a blood meal. The developmental stages of a symbiont last from 17 to 50 days depending on biotic and abiotic factors, and the insects remain infective for their lifetime, which is about 2 to 3 months.

Symbionts in the mammalian blood have a distinguishable coat on the outside of their cell membrane. This surface coat is present only in the infective form of the trypanosomes that emerge from the insects. Scientists have shown that antigenic variation of the trypanosome is associated with changes in its surface coat. Surface coat antigens are phenotypic expressions of trypanosome genes. Up to 10% of the trypanosome genome may consist of genes that code for antigens. A single trypanosome can produce at least 100 antigenic variants and some scientists have suggested that up to 1,000 genes may be involved. The surface coat of the trypanosome

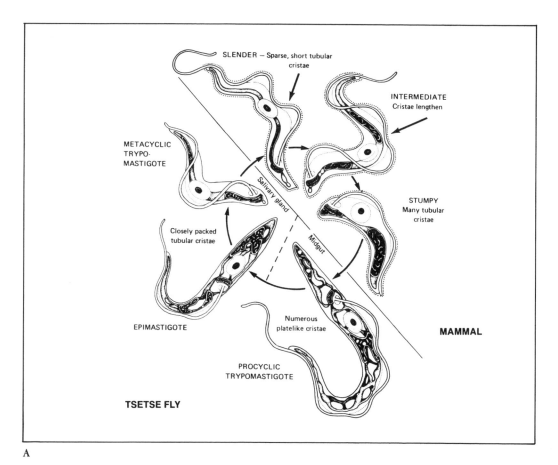

A

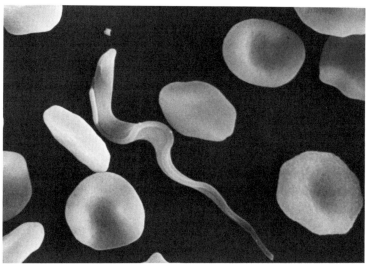

B

Figure 8-6. (A) *Schematic diagram of developmental stages of* Trypanosoma brucei *in mammals and tsetse flies, showing changes in cell surface and mitochondria. Note presence of surface coat antigens (arrows) around forms of the parasite occurring in bloodstream.* (B) *Scanning electron micrograph of trypanosome among mammalian red blood cells.* (Part (A): Adapted from Vickerman, K. 1971. Morphological and physiological considerations of extracellular blood protozoa. In: Ecology and physiology of parasites. 58–91, ed. A. M. Fallis. University of Toronto. Part (B): Steven Brentano, University of Iowa.)

is composed of about 10 million glycoprotein molecules known as *variant surface glycoproteins* (VSGs). Each VSG molecule is made up of about 500 amino acids, to which are linked two types of oligosaccharide side chains. One of the carbohydrate chains links the VSG molecule to the trypanosome cell membrane. Variations in the amino acid sequence of the VSG molecule results in new antigenic properties. Scientists are concentrating their efforts on understanding the genetic mechanisms that control the expression of VSG genes. The system of antigenic variation in trypanosomes is an innovative evolutionary adaptation that allows the symbionts to survive in the host and makes the task of developing effective vaccines against the diseases they cause a difficult one. Trypanosomes emerging from the insect host are the focus of current research in immunology.

TRYPANOSOMA CRUZI: CHAGAS' DISEASE. To live and multiply in host cells such as macrophages, whose function is to recognize and kill foreign invaders, requires a great deal of evolutionary audacity. Yet, some flagellates have successfully conquered the macrophages, one of the most hostile intracellular environments. *Trypanosoma cruzi* causes Chagas' disease in humans living in Central and South America. The flagellate enters the host through cuts in the skin or mucous membrane. The vector of *T. cruzi* is a brightly colored bug of the insect family *Reduviidae*. The insect voids the trypanosomes along with its feces during a blood meal. The flagellates appear in the blood and periodically invade the reticuloendothelial tissues, especially those of the heart. Inside the tissues, the symbiont assumes a spherical form, multiplies as an intracellular parasite for about five days, and forms a cystlike cavity (the pseudocyst). Some of the flagellates break through the pseudocyst membrane and reenter the bloodstream to infect other tissue. The symbionts that remain in the pseudocyst disintegrate and cause a lesion. Reduviid bugs ingest *T. cruzi* during the blood meal from an infected individual. The trypanosomes multiply and undergo several transformations, first in the midgut and later in the hindgut. The symbionts attach to host macrophages by means of receptor-ligand bindings and become enclosed in membrane-bound vacuoles derived from the host cell. The symbionts escape from the vacuoles into the host cytoplasm, where they multiply.

TRYPANOSOME ULTRASTRUCTURE. Trypanosomes have an elongated, flat body with a single nucleus that has a prominent, central nucleolus. The flagellum arises posteriorly, extends along the free margin of the body to form an undulating membrane, and continues anteriorly as a free flagellum. The flagellum originates from a basal body and kinetoplast, which are enclosed in a mitochondrion. The kinetoplast is disc-shaped and has parallel rows of filaments, which contain DNA. Kinetoplast DNA (K-DNA) consists of a unique network of thousands of circles, called *maxicircles* and *minicircles*. Of the total cellular DNA of *Trypanosoma*, 10% to 25% is K-DNA. Recent studies have suggested that maxicircles possess genes similar to the mitochondrial DNA of other eukaryotic cells. The minicircles make up approximately 95% of the K-DNA mass but their nature and function are unknown. Mitochondrial morphology varies a great deal in hemoflagellates. The mitochondria are usually in the form of a single or branched tube running the entire length of the cell and they possess characteristic cristae. In *Trypanosoma brucei,* the mitochondrial shape changes along with the transformation the symbiont undergoes when it passes from the vertebrate bloodstream to the insect's alimentary canal. In the bloodstream, the symbiont has a *slender form* and a long tubular mitochondrion with a double membrane but only a few cristae. In the alimentary canal of the insect, the symbiont assumes the *stumpy form* and has a well-developed spherical mitochondrion with a tubular network of prominent cristae. The mitochondrion of the slender form is inactive, that is, lacks enzymes of the Kreb's cycle and electron transport chain, and that of the stumpy form is active.

LEISHMANIA. Several species of *Leishmania* cause diseases in humans. *Leishmania tropica* is the causal agent of *oriental sore,* which is endemic to Africa, the Middle East, and India; *L. mexicana* causes *New World cutaneous leishmaniasis,* or chiclero ulcers, a disease common in Mexico, Guyana, and other countries of Central and South America; *L. braziliensis* causes an infection called *espundia,* a disease endemic to the jungles of South America, which involves mucous membranes of the mouth, nose, and pharynx; and *L. donovani* is the causal agent of *visceral leishmaniasis* in most of the tropical and subtropical countries of the world.

Leishmania has two different morphological forms depending on the type of host: (a) an oval-shaped body without an external flagellum when it occurs in a vertebrate host, and (b) an elongated body with a flagellum when it inhabits the intestines of sand flies.

In the mammalian host, *Leishmania* lives exclusively in the macrophages. The establishment of the symbiosis involves recognition, intracellular entry, surviving the host's lytic enzymes, and growth and multiplication of the symbiont.

As soon as flagellates from the insect host enter the mammalian bloodstream, a complex process of cell recognition takes place. The process is not fully understood but scientists are making progress in identifying factors that are involved. Some type of chemotaxis may take place because the flagellates are strongly attracted to macrophages and, to a lesser degree, to lymphocytes but show no affinity with red blood cells. Some scientists have suggested that the symbionts possess cell surface components that fit receptors in the macrophage cell. Macrophages permit internalization of the symbiont by phagocytosis. The flagellar end of the symbiont may play a significant role in its interaction with the host cell. Entry of the symbiont into the macrophage is accompanied by chemoluminescence and the release of hydrogen peroxide. Macrophages treated with cytochalasin B fail to phagocytize the symbionts, which suggests that the microfilament proteins actin and myosin participate in the process of symbiont entry. *Leishmania* may develop a resistance to the lysosomal enzymes of the macrophage, or it may release substances that inactivate the enzymes. Both possibilities have supportive evidence. Once inside the macrophage, the symbiont assumes its nonflagellated form, which is better suited for intracellular existence. The symbiont grows within the membrane-bound vacuole of the macrophage and multiplies slowly. Some *Leishmania* strains have developed stable associations with the host.

4. APICOMPLEXANS. Members of the phylum Apicomplexa have an apical complex that can be seen only with the electron microscope (Fig. 8.7). The complex consists of a polar ring, conoid, rhoptries, micronemes, and subpellicular microtubules. The polar ring is located at the anterior end of the symbiont, just beneath the cell membrane. The conoid is a spirally coiled structure within the polar ring. There may be

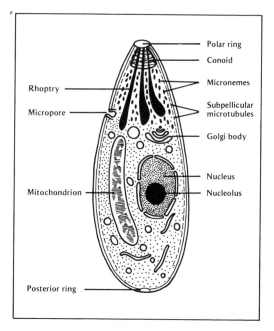

Figure 8-7. *Ultrastructural details of a typical apicomplexan trophozoite.*

two to seven rhoptries that lead to the apex of the cell. Micronemes are small convoluted structures that extend over most of the anterior end and their ducts join those of the rhoptries. Subpellicular microtubules radiate from the apex and extend most of the length of the symbiont. A mouth is located laterally, and along its edge are two concentric rings of some yet unknown material that may help the symbiont acquire nutrients. The apical complex participates in the penetration of the host cell by secreting proteolytic enzymes.

Apicomplexan life cycles require two host organisms and involve several kinds of asexual and sexual spores. The following are examples of symbionts included in the phylum.

a. Gregarines. Gregarines are extracellular symbionts that inhabit the alimentary canal and body cavity of invertebrates, especially arthropods, molluscs, and annelids. The life cycles of several *Gregarina* species are synchronized with those of their hosts. Sporocysts are produced only in the sexually mature host and are liberated along with the host sex cells, so that the next generation of larvae will become infected with the symbionts. It has been suggested that the host's hormones play a significant role in the development of the symbiont.

b. Toxoplasma gondii: *A Cosmopolitan Coccidian.* The coccidian symbiont *T. gondii* occurs in many animal species throughout the world. In man, *toxoplasmosis* is generally a mild or asymptomatic infection, but in some cases it may cause blindness and damage to the central nervous system. The symbiont invades the macrophage and circumvents its lysosomal system. Sexual stages in the life cycle begin after cats ingest cysts of the symbiont from tissues of an intermediate host such as a rodent or bird (Fig. 8-8). The cysts contain hundreds of *sporozoites,* which infect the host epithelial cells and then transform into *merozoites.* Merozoites multiply rapidly and are released into the intestinal lumen when the host cells rupture. Each merozoite can infect another epithelial cell. After several such cycles, some merozoites become *gametocytes.* Two gametocytes fuse together to form a cyst, which is then expelled with the feces.

Asexual reproduction of *T. gondii* begins when a host other than a cat ingests cysts from the meat of an infected animal. Spores liberated from the cysts penetrate the intestinal wall and enter the circulatory system and infect the macrophages. In the macrophage, spores are called *tachyzoites.* Macrophages can only phagocytize dead tachyzoites. Living tachyzoites use macrophages as an intracellular habitat and divide repeatedly by binary fission and eventually kill the host macrophage. When the host defense mechanism is mobilized, macrophages surround large numbers of symbionts and form a wall around them. This structure is called a *pseudocyst* and it contains dormant symbionts. Pseudocysts are formed in all tissues of the host but are most common in the brain, retina, and liver. Raw hamburger is believed to be an excellent source of pseudocysts. Most infections of *T. gondii* occur when undercooked or raw meat containing pseudocysts is consumed. Tachyzoites will cross placental barriers and infect the developing embryo.

TACHYZOITE INVASION OF MACROPHAGES. Early workers had reported that when tachyzoites were liberated in a host environment, they exhibited various types of movement such as gliding or rotating in a somersault fashion. Penetration of the macrophage was thought to be achieved through a collision with the host membrane. Recent evidence, however, supports the view that a symbiont's penetration is mainly the result of the host cell's phagocytic activities.

The sequence of events that results in the destruction of a pathogen in a macrophage is called the respiratory burst and involves a number of metabolic activities. When a macrophage is unable to degrade the pathogen contained in a vacuole phagosome, there is no respiratory burst. *Toxoplasma gondii* strains differ in their pathogenic abilities. A proteinaceous substance has been isolated from one strain of lysed *Toxoplasma* that enhances the virulence of another strain. *Toxoplasma* penetration of macrophages was greatly facilitated when lysozyme or hyaluronidase was added to the culture medium. The process of entry is initiated when the symbiont attaches to the macrophage by its anterior end and forms a small depression in the host cell membrane. Once inside the cell, the symbiont becomes enclosed in a *phagosome,* which is a membrane-bound vacuole. The host cell lysosomes fail to fuse with the phagosome and this suggests that *T. gondii* contains a substance that interferes with the lysosome-phagosome fusion. *Toxoplasma gondii* also changes the phagosome to allow the gathering of endoplasmic reticulum and mitochondria along the vacuole membrane. Macrophages can degrade previously killed tachyzoites, but living symbionts gain immunity to the host's destructive enzymes by altering their properties. Newly developed tachyzoites escape from the host cell by twisting through the host cell membrane.

c. Malarial Parasites. Malaria is a disease that is caused by the multiplication of the parasite *Plasmodium* in the blood and tissues of a vertebrate host. Although there are more than 100 species of *Plasmodium,* and they affect a wide range of vertebrates, each species has a narrow host range. *Plasmodium* species are obligate intracellular parasites that have several morphologically different developmental stages.

Four species of the malarial parasite infect humans: *P. falciparum, P. malariae, P. ovale,* and *P. vivax.* The malarial parasite of chimpanzees and monkeys, *P. knowlesi,* occasionally infects man and produces mild symptoms. *Plasmodium berghei* in rodents and *P. gallinaceum* and *P. lophurae* in birds have been used extensively as experimental models to explore the ultrastructural details of the symbiont-host interaction.

HISTORY OF THE DISEASE. Malaria has played a significant role in the history of man as well as in the development of modern science. More than half the world's current population lives

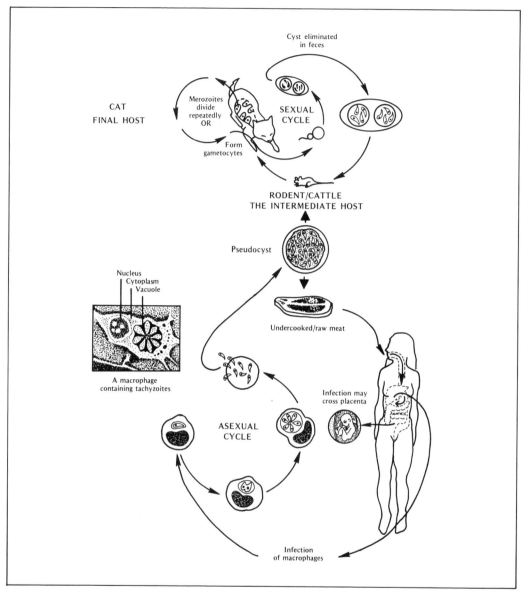

Figure 8-8. *Life cycle of* Toxoplasma gondii. (Adapted from Katz, M., D. D. Despommier, and R. W. Gwadz. 1982. Parasitic diseases. Springer-Verlag, New York.)

under the threat of malaria. Each year an estimated 150 million people contract the disease and between half a million and a million people die in Africa alone.

Discovery of the malarial parasites and understanding of the intricate details of their life cycle represent the golden age in parasitology. Ronald Ross, working with the malarial parasite of birds, and Battista Giovanni Grassi, working with human malaria, showed in 1898 that the

parasites developed in the mosquito and that humans became infected when bitten by mosquitoes. In 1880, Alphonse Laveran described the parasites in blood cells, but it wasn't until the early 1950s that the role of the liver in the disease was resolved. Both Laveran and Ross were awarded Noble Prizes for their discoveries.

The first drug to be used against malaria was quinine, which was extracted from the bark of the chinchona tree. Peruvian Indians knew of the therapeutic value of quinine, and when the European colonists learned of it the use of quinine became global.

Since the malarial symbiont is transmitted by *Anopheles* mosquitoes, the development of the pesticide DDT in the 1940s was an important breakthrough in controlling the spread of the disease. Unfortunately, as mosquito control programs spread throughout the world, DDT-resistant strains of mosquitoes began to appear.

During the Second World War, hundreds of thousands of American soldiers contracted malaria. The United States Army launched an equivalent of the Manhattan Project to combat the disease and the efforts led to the development of the antimalarial drug chloroquine. A malaria eradication program was developed, based on the use of DDT and chloroquine, and during the 1960s the incidence of malaria declined sharply in all parts of the world, with expectations that it would be totally eradicated. Then came the realization that *Plasmodium falciparum* and other species were producing strains that were resistant to chloroquine while the mosquitoes were becoming resistant to DDT. The reemergence of malaria after over two decades of relative success in controlling it has alarmed scientists, who are now trying to understand the molecular basis of the disease. Scientists are using techniques of immunology and molecular biology in attempts to develop an effective vaccine against the malarial symbiont. Much has been learned from the malarial parasite about the molecular mechanisms, genetics, immunology, and biochemistry of symbiont-host interactions. A new antimalarial drug, Qinghaosu (artemisinin), has been developed in China from the herb *Artemesia annua*. The drug has been very effective against chloroquine-resistant strains of *Plasmodium*.

LIFE CYCLE OF *PLASMODIUM*. In humans, infection begins when a female *Anopheles* mosquito penetrates the host skin and injects into the blood saliva that contains sporozoites of *Plasmodium*. The parasites migrate to the liver, where they multiply in the parenchyma cells. The sporozoites are transformed into *merozoites* and when all the liver host cells rupture, they liberate 100,000 to 300,000 merozoites. In most mammals, merozoites infect red blood cells and grow and multiply intracellularly. The mature stage of the symbiont in a red blood cell is called a *schizont*. The schizont divides to form

merozoites, which are released when the red blood cell ruptures and then infect other red blood cells. This asexual part of the life cycle is called *schizogony*. The clinical signs of malaria are associated with the cyclical blood stage of schizogony. In each cycle of schizogony, some of the invading merozoites differentiate into male (micro) and female (macro) gametocytes. The gametocytes mature in about ten days and do not develop further in the vertebrate host (Fig. 8.9).

When a female mosquito bites an infected individual, micro and macro gametocytes are ingested with the blood. In the mosquito's stomach, each microgametocyte forms eight microgametes, which fertilize the macrogametes formed by the macrogametocyte. The resulting zygotes become transformed into motile, worm-like structures called *ookinetes*, which penetrate the stomach wall and form *sporocysts*. Each sporocyst breaks open and releases *sporozoites*, which migrate to the salivary glands. The sexual part of the parasite life cycle is known as *sporogony*.

MEROZOITES: INTRACELLULAR SYMBIONTS OF ERYTHROCYTES. Erythrocytes, or red blood cells, have unique characteristics such as a complex cell membrane with numerous surface antigens and a cytoplasm that lacks internal organelles and consists mostly of one type of protein. Merozoites have many apicomplexan characteristics. The surface of a merozoite consists of a plasma membrane, glycocalyx, and a cell coat that allows the symbiont to attach to the host cell surface. The cell coat is thick and is removed during the invasion process. The coat consists of fine filaments that stand erect, like hairs, on the plasma membrane. The filament tips are either Y or T shaped and can stretch when attached to the cell. The filaments may attach to the underlying cytoskeleton of the merozoite. Two sub-plasma membranes make up a system of interconnected cisternae that lie under the apical complex. The function of the cisternae is not known. The polar ring of the apical complex is responsible for a particular merozoite shape and it forms a convergent point for rhoptries and micronemes.

It is not known how merozoites are attracted to erythrocytes, but once contact is made the parasite adheres to any part of the erythrocyte surface. The invasive process begins, however, only when the symbiont's apical complex contacts the host cell surface, which has specific re-

126

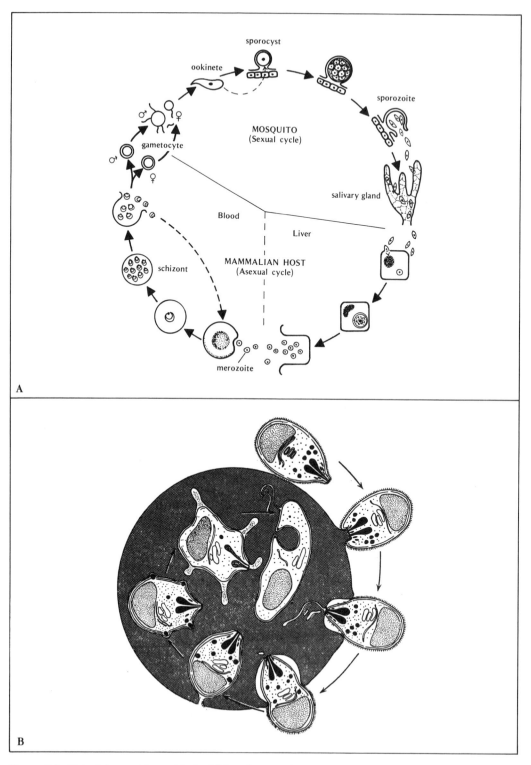

Figure 8-9. *Malarial parasitic symbiosis. (A) Devel-* *opmental stages in the life cycle of* Plasmodium. *(B)* *Stages in the invasion of an erythrocyte by a* Plas-modium *merozoite.* (Part *(B)*: Lawrence Bannister, Guy's Hospital, Medical and Dental Schools, London.)

ceptor molecules (Fig. 8.9). Merozoites can attach to a glass surface, to other merozoites, and to white blood cells. If the merozoite-erythrocyte attachment is incorrect, the merozoite detaches itself and makes another attempt. The host membrane shows a characteristic bending when the apical complex of the merozoite contacts the erythrocyte. Secretions released from rhoptries and micronemes cause the erythrocyte membrane to invaginate and form a cup into which the merozoite advances (Fig. 8.10). The surface coat of the merozoite becomes detached as the symbiont moves into the cup. Finally, the entire merozoite becomes surrounded by the host membrane and enclosed within a vacuole. It has now been firmly established that the apical complex of the parasite is responsible for the invasion of the erythrocyte. The rhoptries and micronemes are thought to be derived from the Golgi bodies and to possess lysosomic enzymes. Following the initial contact, the movement of a merozoite into the host cavity is rapid. This is interesting because merozoites lack obvious motility mechanisms. It has been suggested that suction produced by the invagination of the erythrocyte pulls the symbiont into the vacuole. In the vacuole microspheres near the cell surface of the merozoite protrude into the host cytoplasm. The contents of the microspheres are released in the vacuole and cause it to enlarge. The cisternae that lie underneath the plasma membrane of the merozoite, along with the microtubules in the cortex, disappear. The merozoite is no longer spherical and becomes transformed into a trophozoite.

d. THEILERIA PARVA. The piroplasm *Theileria parva* causes *theileriasis* in cattle and water buffalo. The symbiont is an intracellular inhabitant of erythrocytes and leukocytes. The host-symbiont relationship is unique for two reasons. First, the symbiont invades the macrophage without any vacuole membrane surrounding it. The hemoflagellate *Trypanosoma cruzi* also becomes intracytoplasmic in macrophages by escaping from the vacuole; however, *T. cruzi* eventually destroys the host cell, whereas *Theileria* becomes a permanent resident of the macrophage. Second, *T. parva* transforms the infected macrophage into a lymphoblast that is capable of carrying the symbiont indefinitely. The symbiont synchronizes its growth and duplication with that of the host cell. The schizont of *T. parva* segregates along with the host chromosomes into the two daughter cells. Recently, scientists have obtained electron microscopic evidence that suggests that the schizont of *T. parva* is closely involved with the host's mitotic apparatus. The symbiont becomes encased in microtubules during centriole replication and aster formation, and becomes compressed between the microtubules during elongation of the spindle. By the time the nuclear membrane disintegrates the schizont is fully integrated into the mitotic spindle along with the host chromosomes. During anaphase, the schizont elongates with the polar microtubules, and when the host cell constricts in the middle to produce two new cells, the schizont is also partitioned into two. Schizont-microtubule attachments have not yet been observed.

D. Summary and Perspectives

Protoctists cause diseases in millions of humans and domesticated animals. Many of these parasites have complex life cycles that involve two hosts and several types of spores.

Species of amoebae range from harmless commensals in a host's alimentary tract to virulent pathogens. *Entamoeba histolytica* causes dysentery, and species of *Acanthamoeba* cause meningoencephalitis in humans and mammals. Opalinids are commensals in the large intestine of frogs and toads.

Two kinds of ciliates, entodiniomorphs and holotrichs, live mutualistically with herbivorous mammals. The ciliates live in the rumen of the host along with bacteria.

There are two groups of flagellated protoctist symbionts: intestinal flagellates and hemoflagellates. Representative intestinal flagellates include trichomonads, *Giardia lamblia, Histomonas meleagridis,* and various genera that inhabit the alimentary canal of wood-eating termites and roaches. *Mixotricha paradoxa,* a wood-eating flagellate found in the gut of termites, is covered with spirochetes that propel the flagellate. Hemoflagellates include trypanosomes and *Leishmania,* which cause devastating diseases in man and animal such as African sleeping sickness, Chagas' disease, and leishmaniasis.

Apicomplexans include gregarines, which are extracellular symbionts of invertebrates, *Toxoplasma gondii,* a symbiont of many animal species, *Plasmodium,* a parasite that causes malaria,

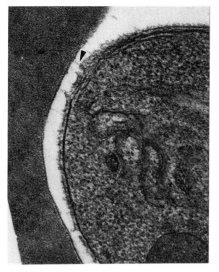

A

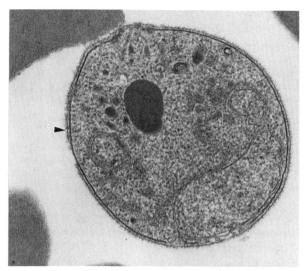

B

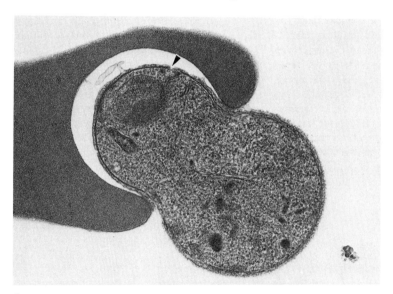

C

Figure 8-10. Electron micrographs of Plasmodium merozoites invading red blood cells. (A) A merozoite showing nonspecific type of attachment to an erythrocyte. Note the presence of surface coat (arrow). (B) Long-range attachment between parasite and host cell involving merozoite coat filaments *(arrow).* (C) *A partially depleted rhoptry is visible at the anterior end of the parasite (arrow). Note the formation of a vacuole that surrounds the symbiont.* (Lawrence Bannister, Guy's Hospital, Medical and Dental Schools, London.)

and the piroplasm *Theileria parva,* which causes theileriasis in cattle and water buffalo.

Symbiotic associations may be more complex than first imagined and may involve several layers of organization. For example, scientists now believe that *Entamoeba histolytica* cannot produce the disease amoebiasis alone; rather the presence of bacterial and viral symbionts in the cytoplasm and nucleus of the amoeba may be necessary for its virulence. Similarly, the ability of the termite symbiont *Mixotricha paradoxa* to digest cellulose may be the result of intracellular bacterial symbionts. Symbioses in the rumen of herbivorous animals and in wood-eating termites consist of complex interrelationships between bacteria, ciliates, and fungi. Similar patterns may exist in other associations.

The life cycles of the symbionts of many protoctistan associations are synchronized with each other, thereby ensuring continuity of the symbiosis from one generation to another.

Review Questions

1. Describe the life cycle of a typical protozoan.
2. List some examples of amoebic infections.
3. Explain how the life cycle of *Opalina ranarum* is synchronized with that of its host.
4. Name two types of ciliates that live in the rumen of cattle and sheep.
5. Distinguish between intestinal flagellates and hemoflagellates.
6. List some well-known examples of pathogenic flagellates.
7. Describe the nature of the symbiosis between flagellates and wood-eating roaches and termites.
8. Name some diseases of man and domesticated animals caused by hemoflagellates.
9. Explain why it is difficult to develop effective vaccines against the diseases caused by trypanosomes.
10. Describe the two different morphological forms assumed by *Leishmania* depending on the type of host.
11. Explain how a tachyzoite invades a macrophage.
12. Describe the life cycle of the malarial parasite *Plasmodium.*
13. Why is the host-symbiont relationship in the piroplasm *Theileria parva* unique?

Further Reading

Aikawa, M. 1983. Host-parasite interaction: Electron microscopic study. In: Molecular biology of parasites, 1–31, ed. J. Guardiola, L. Luzzatto, and W. Trager. Raven Press, New York.

Bannister, L. H. 1977. The invasion of red blood cells by *Plasmodium.* In: Parasite invasion, Vol. 15, 27–55, ed. A. Taylor and R. Muller. Blackwell Scientific Publications, Oxford.

Bannister, L. H. 1979. The interactions of intracellular protista and their host cells, with special reference to heterotrophic organisms. Proc. R. Soc. Lond. B. 204:141–163. (A comprehensive article dealing with some important animal parasites.)

Breznak, J. A. 1975. Symbiotic relationships between termites and their intestinal microbiota. In: Symbiosis: Symposia of the Soc. for Experimental Biology, no. 29, 559–580, ed. D. H. Jennings and D. L. Lee. Cambridge Univ. Press, Cambridge.

Breznak, J. 1982. Intestinal microbiota of termites and other xylophagous insects. Ann. Rev. Microbiol. 36:323–343.

Chang, K. P. 1983. Cellular and molecular mechanisms of intracellular symbiosis in leishmaniasis. Int. Rev. Cytology, supplement 14:267–302.

Coleman, G. S. 1978. Rumen ciliate protozoa. In: Biochemistry and physiology of protozoa, 2d ed., 381–406, ed. M. Levandowsky and S. H. Hunter. Academic Press, New York.

Davies, A. J. S., J. G. Hall, G. A. T. Targett, and M. Murray. 1980. The biological significance of the immune response with special reference to parasites and cancer. J. Parasitology 66:705–721. (A thought-provoking article considering a wide range of parasitic organisms from an evolutionary perspective.)

Donelson, J. E., and M. J. Turner. 1985. How the trypanosome changes its coat. Sci. Am. 252 (February): 44–51.

Friedman, M. J., and W. Trager. 1981. The biochemistry of resistance to malaria. Sci. Am. 244 (March): 154–164. (Considers the evolutionary significance of sickle-cell anemia and thalassemia.)

Godson, G. N. 1985. Molecular approaches to malaria vaccines. Sci. Am. 252 (May): 52–59. (*Plasmodium* defends itself against host antibodies by producing a protein with many repeated antigenic sites.)

Hungate, R. E. 1978. The rumen protozoa. In: Parasitic protozoa, Vol. 2, 655–695, ed. J. P. Kreier. Academic Press, New York.

Klayman, D. L. 1985. *Qinghaosu* (artemisinin): An antimalarial drug from China. Science 228:1049–1055.

Parkhouse, R. M. E. (ed.). 1984. Parasite evasion of the immune response. Parasitology (UK) 88:571–682. (An excellent collection of articles on the current research on the immunological basis of host-parasite symbioses.)

Vickerman, K. 1982. Parasitic protozoa: Aspects of the host-parasite interface. In: Parasites: Their world and ours, 43–52, ed. D. F. Mettrick and S. S. Desser. Elsevier Biomedical Press, Amsterdam, Netherlands.

Bibliography

Barriga, O. O. 1981. The immunology of parasitic infections: A handbook for physicians, veterinarians, and biologists. University Park Press, Baltimore. 354 pp. (Provides an excellent treatment of immunology of protoctists and helminths in addition to some basic immunology.)

Boothroyd, J. C. 1985. Antigenic variation in african trypanosomes. Ann. Rev. Microbiol. 39:475–502.

Breuer, W. V. 1985. How the malaria parasite invades its host cell, the erythrocyte. Int. Rev. Cytology 96:191–238.

Cleveland, L. R., and A. V. Grimstone. 1964. The fine structure of the flagellate *Myxotricha paradoxa* and its associated microorganisms. Proc. R. Soc. Lond. B. 159:668–686.

Cohen, S., and G. A. M. Cross (eds.). 1984. Towards the immunological control of human protozoan diseases. Phil. Trans. R. Soc. Lond. B. Biological Sciences 307:3–213. (A collection of recent review articles on the immunological research on *Trypanosoma, Entamoeba, Leishmania,* and *Plasmodium.*)

Cohen, S., and K. S. Warren (eds.). 1982. Immunology of parasitic infections. 2d ed. Blackwell Scientific Publications, Oxford. 864 pp.

Culbertson, C. G. 1971. The pathogenicity of soil amoebas. Ann. Rev. Microbiol. 24:231–254.

Erlandsen, S., and E. A. Meyer (eds.). 1984. *Giardia* and giardiasis: Biology, pathogenesis, and epidemiology. Plenum Press, New York. 407 pp.

Guardiola, J., L. Luzzatto, and W. Trager (eds.). 1983. Molecular biology of parasites. Raven Press, New York. 210 pp.

Katz, M., D. D. Despommier, and R. W. Gwadz. 1982. Parasitic diseases. Springer-Verlag, New York. 264 pp. (A brief book on medically important parasites.)

Kreier, J. P. (ed.). 1978. Parasitic protozoa. Vol. 2, Intestinal flagellates, histomonads, trichomonads, amoebae, opalinids, and ciliates. Academic Press, New York. 730 pp.

Mettrick, D. F., and S. S. Desser. 1982. Parasites: Their world and ours. Proceedings of the Fifth International Congress of Parasitology. Elsevier Biomedical Press, Amsterdam, Netherlands. 465 pp. (Many articles on the immunology, electron microscopy, and ecology of parasitic organisms.)

Nobel, E. R., and G. A. Nobel. 1982. Parasitology: The biology of animal parasites. 5th ed. Lea and Febiger, Philadelphia. 522 pp. (A standard textbook in parasitology.)

Richmond, M. H., and D. C. Smith (eds.). 1979. The cell as a habitat. Proc. R. Soc. Lond. B. 204:113–286. (An excellent collection of articles on a wide range of intracellular symbionts.)

Smyth, J. D. 1976. Introduction to animal parasitology. 2d ed. Halsted Press, New York. 466 pp. (A comprehensive, balanced treatment of animal parasitic symbioses.)

Taylor, R. A., and R. Muller (eds.). 1977. Parasite invasion. Blackwell Scientific Publications, Oxford. 155 pp.

Warren, K. S., and J. Z. Bowers (eds.). 1982. Parasitology: A global perspective. Springer-Verlag, New York. 292 pp. (Examines the role of research in parasitology and considers new approaches to old problems facing humanity.)

Warren, K. S., and E. F. Purcell. 1980. The current status and future of parasitology. Josiah Marcy, Jr., Foundation, New York. 296 pp. (Includes several articles on the future education and training of parasitologists.)

CHAPTER 9

Protoctistan Symbiotic Associations
Algae

A. Algae and Marine Invertebrates

1. INTRODUCTION. Many marine invertebrates, such as sea anemones, corals, flatworms, and protozoans, form mutualistic associations with algae. The algae, through their photosynthesis, supply nutrients to the animals, which allows them to colonize habitats they normally could not because of limited supplies of external prey. The algae appear yellow-brown and are called *zooxanthellae,* a name that does not have taxonomic status but is useful for identifying the general type of endosymbiont in a host. The most common algal symbiont of marine invertebrates is *Symbiodinium microadriaticum* (Fig. 9.1), which occurs in almost 100 species of cnidarians, protoctists, and tridacnids. Strains of this alga differ in their types of enzymes, lipids, and sterols, but only one strain occurs in each species of invertebrate. Although aposymbiotic invertebrates can be reinfected with strains of *S. microadriaticum* other than their natural ones, the foreign algae grow poorly in the unnatural host. How an animal recognizes a specific algal strain is not known, but it appears that a recognition process takes place after the algae are phagocytized. Algal strains that are compatible with the host are not digested because they either avoid or resist the host's digestive enzymes.

The number of algal cells in each animal cell varies. The hydroid *Myrionema amboinense* has one to 56 algal cells in each endodermal cell, whereas tentacle cells of the golden brown sea anemone, *Aiptasia pallida,* contain about 10 algal cells. Each algal cell is housed inside a vacuole formed by the animal cell. The algae are located in translucent parts of the animal.

Symbiodinium microadriaticum is greatly modified when it lives inside animal cells. The algal cell wall becomes thinner or is completely lost, a modification that allows for close contact between the algal cell and animal cytoplasm. The cell also loses its grooves and flagella and divides only by binary fission. The algal cells also change physiologically. In the animal, zooxanthellae excrete large amounts of glycerol, which the host uses to synthesize lipids and proteins. In some invertebrates, almost 50% of the carbon fixed by the algae is translocated, as glycerol, within 24 hours to animal tissue. Zooxanthellae also excrete glucose, alanine, and organic acids, and perhaps other compounds in amounts too small to detect with present experimental methods. When the algae are isolated from animals and grown in culture, they stop excreting substances. But when cultured algae are exposed to homogenates of their host, they once again excrete glycerol and other compounds. Although the nature of the stimulatory host-factor is not known, it seems to be a general one, since homogenates from one type of animal, such as a clam, affect zooxanthellae from another animal, such as coral, and vice versa. The homogenates are effective only if they are taken from animals that contain symbiotic algae.

In some marine animals there is a reciprocal exchange of substances between the symbionts. The animals supply the algae with acetate, which the algae use to synthesize fatty acids. The fatty acids are translocated back to the animal, which

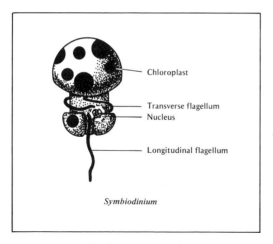

Chloroplast

Transverse flagellum
Nucleus

Longitudinal flagellum

Symbiodinium

Figure 9-1. Symbiodinium microadriaticum, *a common photosynthetic symbiont of marine invertebrates.*

uses them to synthesize waxes and other compounds.

Algal symbionts pass from one generation of invertebrates to another in different ways. Animals that reproduce asexually, by budding, transmit some of their algae directly to their offspring. In sexual reproduction, algae are contained in or on the eggs produced by the parent. In some invertebrates, offspring ingest free-living algae from their surroundings. Larvae of the jellyfish *Cassiopeia xamachana* will not develop normally unless they become infected with algae. Certain marine nudibranches obtain their algae after they eat anemones and corals, which contain symbiotic algae.

2. SEA ANEMONES AND JELLYFISH. The sea anemone *Anthopleura xanthogrammica* contains two types of symbiotic algae: zooxanthellae and zoochlorellae (green algae that are similar to the *Chlorella* symbionts of freshwater invertebrates). The relative proportion of each type of algal symbiont in the animal depends on the water temperature of the anemone's habitat. At high temperatures (26° C), zooxanthellae are more common, and at low temperatures (12° C) zoochlorellae are predominant. At intermediate temperatures the anemone has somewhat equal proportions of the symbionts. Curiously, the zoochlorellae excrete only small amounts of fixed carbon. Since in some anemones the zoochlorellae predominate, the algae must supply something other than carbon compounds,

possibly nitrogen and phosphorus, which stimulate the anemone to grow.

Invertebrates position themselves in ways to increase the exposure of their symbionts to light. For example, when tentacles of *Anthopleura pallida* are relaxed and fully stretched, the algae in their gastrodermal cells lie in a single layer and are exposed to maximum daylight. When the tentacles contract, the gastrodermal cells shrink and the algal cells lie on top of each other, the uppermost cells shading the lower ones.

Cassiopeia xamachana is a jellyfish that has been used to study how an invertebrate selects its symbiotic algae. The life cycle of *Cassiopeia* includes a sexual medusoid stage, containing algae, and an asexual polyp stage, which lacks algae. The polyps mature only after they establish an association with *S. microadriaticum.* By exposing polyps to different strains of algae, it is possible to determine which algae are able to colonize the animal. Robert K. Trench and his coworkers found that although a *Cassiopeia* polyp was able to phagocytize different strains of algae, only one strain would survive and form the symbiosis. Thus, the scientists concluded that the host recognized its symbiont only after it was phagocytized. *Cassiopeia xamachana* does not swim freely, but rather lies upside down on the sea floor, a behavioral adaptation that allows the algae in their tentacles to receive maximum daylight for photosynthesis, and gives the animal its common name, the "upside-down" jellyfish.

3. REEF-BUILDING CORALS. The symbiotic association between *S. microadriaticum* and reef-building coral animals is of great importance in tropical ecosystems and has been the subject of much study. Coral reefs support large communities of biological organisms. Coral animals are coelenterates, which are similar to sea anemones except that their polyps are small, only 10 mm in diameter, and they excrete a calcium carbonate shell around their body. As the polyps die, their shells do not decay and new polyps grow over them. After many years of this process, coral reefs are formed. Symbiotic algae live inside cells of the digestive cavity of the coral polyp. Coral animals with symbiotic algae deposit calcium carbonate around their bodies much faster than animals without algae. Pioneering studies by T. F. Goreau have revealed that the algae stimulate calcification through their photosynthetic fixation of CO_2. Removal of CO_2 by the algae increases the overall reaction rates of the calcification process. The algae supply the animal with oxygen as well as with carbon and nitrogen compounds. The animal obtains vitamins, trace elements, and other essential compounds from the digestion of captured plankton. Animal waste products, such as ammonia, are converted by the algae to amino acids, which are translocated to the animal. Such a recycling of nitrogen is an important feature in the nitrogen-poor habitats of coral.

The vast reefs that are formed in tropical waters result largely from the stimulatory effect of the zooxanthellae on the corals. The high productivity of coral reefs is noteworthy, considering that they occur in shallow, tropical waters that are low in nutrients. Pigments produced by the symbiotic corals protect both the host and the algae from harmful effects of ultraviolet radiation. How symbiotic corals evolved is not known. Although all of the reef-building corals contain symbiotic algae, many other types of corals do not form associations with algae.

4. TRIDACNID CLAMS. Among the molluscs, only tridacnid clams form symbioses with algae. These large calms grow among coral reefs in shallow, tropical waters. Like *Cassiopeia,* the calms have undergone a behavioral adaptation and they attach to the coral reef in such a way that their valves face upward and are open. This position allows their symbiotic algae to receive optimum illumination. The pigmented mantle tissue between the valve contains the algae and shades them from intense light. The algae occupy an intercellular position in the blood sinuses of the clam's mantle. Algal cells are also in other parts of the clam and commonly pass from the sinuses to the alimentary tract, where they are discharged in the clam's feces. Young clams ingest free-living algae, which move from the alimentary tract to the mantle. As with other associations, the symbiotic algae supply the clams with substantial amounts of glycerol.

5. FORAMINIFERANS AND RADIOLARIANS. Foraminiferans and radiolarians are shelled amoebae that commonly contain dinoflagellates as symbionts. The amoebae use carbon compounds excreted by the algae and also digest algal cells. Foraminiferan amoebae float or attach to objects on the ocean floor and they make thin, translucent shells that consist of calcium

carbonate. The shells have pores through which strands of cytoplasm project and inside of which the algae are contained. Algae that associate with foraminiferans also include unicellular green algae, diatoms, and several unicellular red algae. One foraminiferan may associate with different algae, sometimes simultaneously, a situation that is unlike other algal-invertebrate associations, where the symbionts are highly specific to each other.

Foraminiferan algae are enclosed in vacuoles and lack cell walls. This makes it difficult to identify the algae to species, since the cell wall is an important taxonomic trait. If grown free from the host, the algae will produce cell walls again.

The functional relationship between foraminiferans and their symbiotic algae is not fully understood. In addition to providing the host with carbon compounds, algal photosynthesis increases the foram's rate of calcification by removing excess CO_2. In return, the algae receive nitrogen, phosphorus, and vitamins from the host.

Radiolarians have silica shells or capsules that have holes through which spinelike strands of cytoplasm project. A network of the cytoplasm surrounds the outer part of the capsule and contains numerous algal cells, each within a vacuole. Algal symbionts of radiolarians are yellow-brown and they include dinoflagellates, prasinomonads, and a pyrmnesimonad. Each radiolarian has only one type of alga, but the number of algal cells in each host differs. Small radiolarians that are part of a colony may contain 30 to 50 algal cells, and large, solitary radiolarians may contain up to 5,000 algal cells. Some hosts maintain their algal population at a constant level by digesting some of their algal cells.

Colonial radiolarians are good experimental subjects for the study of algal-protoctist relationships. The colonies can be exposed to radioactive compounds and then separated into algal and host fractions. The central capsule of the radiolarian is algal-free and detaches easily from the outer network of cytoplasm, which contains the algae. The separate fractions can then be analyzed for radioactivity to determine the movement of compounds from alga to host. Studies have shown that most of the carbon compounds produced by the algae during photosynthesis pass to the host within the central capsule. The host may also obtain nutrients from prey it captures and digests. Compounds released from the digestion of prey may be used by the algae. The symbiotic algae assimilate waste products of the host, such as ammonia and carbon dioxide, thus enabling the radiolarian to conserve energy that it would have used to dispose of its wastes. Because of their symbiotic relationship with algae, radiolarians flourish in nutrient-poor waters such as those of the Sargasso Sea.

In some planktonic foraminiferans and radiolarians the dinoflagellates are moved out of their shells during the day and back into them at night. The algae are carried in and out of the shells by the cytoplasmic streaming of the host cell. This daily rhythm exposes the algae to optimum light during the day. At night the algae are brought close to the main body of the host, a position that facilitates nutrient transfer.

6. CONVOLUTA ROSCOFFENSIS.

Convoluta roscoffensis is a small, marine flatworm that lives in the intertidal zones of beaches in the Channel Islands of England and in western France. The worms are 2 mm to 4 mm long and deep green from the algae they contain; it is possible for a large worm to contain up to 40,000 algal cells. During high tide, the worms are buried in the sand but at low tide, during the daylight, they move up to the surface. During this time the algae photosynthesize until the next high tide, when the worms burrow back into the sand. When the worms are on the surface of the sand, they secrete a gelatinous substance that holds them together in colonies and protects them from drying.

The algal symbiont of *Convoluta roscoffensis* is *Platymonas convolutae,* a motile, unicellular green algae. The algae are ingested by digestive cells of the worm and become greatly modified. They lose their cell walls and flagella and develop fingerlike lobes that fit between the animal's muscle cells and greatly increase the surface area between the symbionts. The algal lobes are also near the host's cilia, which, like muscle cells, have high energy demands. Because of these close contacts, photosynthetic products of the alga pass directly to the most active regions of the animal. Despite their irregular shapes, all the algal cells face the same direction. The anterior end, which normally has four flagella, faces the inner part of the animal, while the lobed posterior end, which contains the chloroplast, faces the outside. The algal cells are separated from the host cytoplasm by membranes produced by the animal.

When the worms mature, they produce eggs that hatch into larvae, which lack algae and thus are aposymbiotic. Free-living cells of *P. convolutae* are attracted to the egg cases of the worms through an unknown chemical stimulus. The algae swim to the egg cases and attach to them. When female worms lay eggs, they also extrude algal cells and mucus that coats the eggs. Larvae that hatch from the eggs ingest the algae, which then divide until their normal population in the animal is reached.

After the worm matures and has its full complement of algae, it stops feeding and for the rest of its life depends on the algae for nutrients. Aposymbiotic larvae that do not become infected with algae will die, regardless of how much food they ingest.

The physiological relationships between *C. roscoffensis* and its symbiotic algae have been determined from various studies. The algae provide alanine, glutamate, fatty acids, and sterols to the animal. In return, the algae use uric acid, a waste product of the animal, as a source of nitrogen.

Convoluta roscoffensis can be infected with algae that are related to its normal symbiont. These foreign algae undergo the same modifications as the natural symbiont, but the animal usually does not grow as strongly with these secondary symbionts. If an animal that is infected with a secondary symbiont is exposed to cells of its natural, primary symbiont, it will discard or kill the foreign alga and replace it with its natural one.

7. SPONGES.

Marine sponges contain a variety of endosymbionts, including bacteria, dinoflagellates, diatoms, and cryptomonads. The symbioses are especially common among tropical sponges. Many sponges contain endosymbiotic cyanobacteria that are unicellular or filamentous. The cyanobacteria are either intercellular, in sponge tissue directly below the outer surface, or within vacuoles inside the host cells. The sponge obtains nutrients from the digestion of bacteria or from the excretion of compounds such as glycerol and nitrogen from the bacteria. Cyanobacterial symbionts are thought to shade sponge tissue and thus protect it from the damaging effects of intense light.

8. SEA SLUGS. An unusual type of marine relationship, which strictly speaking is not a symbiosis because it involves only part of an organism, is that between sea slugs, such as *Elysia* spp., and the chloroplasts of algae. Sea slugs ingest cytoplasm from the large cells of siphonaceous, or tubular, algae, especially *Caulerpa* and *Codium,* and the digestive cells of the animals phagocytize the chloroplasts that were in the algal cytoplasm. The chloroplasts resist digestion and continue to function for several weeks or even months within the host cells. The slugs contain so many chloroplasts that they appear green. A chloroplast may be enclosed within a vacuole or lie free in the cytoplasm of the animal. Chloroplasts photosynthesize inside the animal cell and release glucose, which the slug uses for some of its nutrient needs. Functional chloroplasts are not digested by the animal, but when a chloroplast stops photosynthesizing, it is quickly digested. How the animal cell distinguishes between active and inactive chloroplasts is not clear. Chloroplasts cannot divide inside the animal because they are unable to synthesize chlorophyll without the enzymes that are present in the algal cytoplasm. A slug constantly replenishes its chloroplast supply by feeding on new algae.

The factors that cause chloroplasts to release glucose inside the slug's digestive cells are not known. Chloroplasts that are removed from the animal stop excreting glucose or only excrete small amounts. But if homogenates from symbiotic slugs are added to the isolated chloroplasts, they will once again excrete glucose. This stimulation of excretion by unknown host factors is like that in other marine symbioses.

In a similar relationship, some planktonic foraminiferans do not readily digest the chloroplasts of diatoms and other algae on which they prey. These chloroplasts also continue to photosynthesize while they are inside the amoebae.

9. TUNICATES. Some tropical, marine tunicates contain an unusual photosynthetic symbiont, called *Prochloron,* that has characteristics of both cyanobacteria and green algae. *Prochloron* cells are green but their cell structure is prokaryotic; that is, the cell lacks a nucleus, has a wall like that of cyanobacteria, and has organelles such as polyhedral bodies. The cells, however, lack phycobiliproteins and phycobilisomes, which are universal in cyanobacteria.

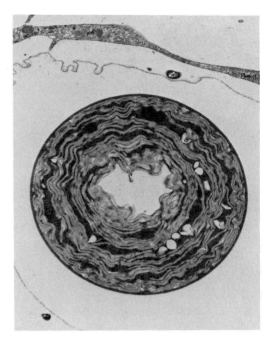

Figure 9-2. Prochloron, *a photosynthetic symbiont of tropical, marine tunicates. Note lack of nucleus and organized chloroplast.* (Rosevelt L. Pardy, University of Nebraska.)

Like green algae, *Prochloron* cells have chlorophylls a and b, whereas cyanobacteria have only chlorophyll a. *Prochloron* has such a unique mix of characteristics that it has been placed into a separate phylum, the Prochlorophyta.

All strains of *Prochloron* are unicellular and have spherical cells (Fig. 9.2). In some tunicates the symbiont cells lie within a cellulose matrix that surrounds the outer surface of the animal, while in other tunicates the symbionts are loosely attached to the cloacal wall. *Prochloron* cells always have an extracellular position in the animal.

The larvae of some tunicates have specialized pouches that carry *Prochloron* cells obtained from the parent. As the tunicate develops, it is infected at an early stage by the symbiont. Thus, the continuity between *Prochloron* and the tunicate is uninterrupted from one generation to the next.

Experimental studies on these associations are limited because *Prochloron* does not grow in laboratory culture. All studies, therefore, must be carried out on symbionts newly squeezed out of the tunicates or from dried or preserved specimens. Preliminary findings have revealed that the symbiont photosynthesizes in the tunicate and releases carbon compounds. Presumably, these compounds, as well as the oxygen produced during photosynthesis, are used by the animal.

Prochlorophytes have been found mostly in symbiosis with marine invertebrates. They may have been common in earlier periods, such as the Upper Cambrian, when the low concentrations of oxygen in the sea and atmosphere were more conducive for their free-living growth. Prochlorophytes in marine invertebrates may represent relict populations, isolated remnants of a once widespread group.

The intermediate position of *Prochloron,* between cyanobacteria and green algae, makes them interesting objects of speculation. Similarities between cells of *Prochloron* and the chloroplasts of green algae and plants suggest that prochlorophytes may have given rise to green algae through symbiotic associations.

B. Algae and Freshwater Invertebrates

1. *HYDRA. Hydra* is a common inhabitant of freshwater ponds and lakes. It remains attached to living or decaying vegetation and feeds on protists and small living animals such as shrimp. Some species of *Hydra* contain unicellular, green algae of the genus *Chlorella.* The algae divide asexually, by mitosis, and each mother cell produces four daughter cells. The animal appears green because its cells are packed with algae. The animals obtain nutrients from the algae, and can live in areas where food is scarce and survive periods when the external food supply is low. The animals ingest algal cells, as they do other particles of food. Algae that become symbiotic escape digestion and are retained inside vacuoles. When the animal divides, some of the algal cells are transmitted to the offspring, so that every individual in the animal population contains algal cells.

Green hydras live in the same habitats as brown hydras. When the natural food supply is abundant, both types of *Hydra* multiply at about the same rate. When the external food supply is scarce, however, green hydras, because of their symbiotic algae, have an advantage over brown hydras.

Hydra viridis is a common species of green

hydra (Fig. 9.3). Several strains of this species have been established in laboratories throughout the world. These strains are being studied intensively in terms of the relationships between the algae and their animal hosts. *Hydra* grows well in laboratory cultures. They are fed brine shrimp, kept in the light to allow the algae to photosynthesize, and maintained in clean, bacteria-free water. Under these conditions a hydra population doubles by asexual budding about every two days.

When *H. viridis* is exposed to high light intensity in the presence of DCMU, a chemical that inhibits photosynthesis, the animal loses its green algae and becomes aposymbiotic or bleached. Such a bleached hydra will survive and grow if it has sufficient food. *Chlorella* symbionts from several strains of *H. viridis* have been grown separately from the animal. The cultured symbionts differ from the symbiotic forms in that they do not excrete as large an amount of photosynthetic products. When the algae are placed back into the animal, they once again excrete compounds. If provided with suitable algae, bleached *Hydra* will ingest them and become green again. A foreign species of *Chlorella* was shown to infect bleached *H. viridis,* but the animal did not multiply as rapidly as normal strains.

Reinfection of aposymbiotic *H. viridis* occurs through five stages. (a) *Contact stage:* the algae attach to the membrane of a *Hydra* cell. (b) *Engulfment stage:* the animal cell ingests the algae by means of membrane projections, or microvilli. The animal cell takes in anything that contacts it, including latex spheres and foreign algae. (c) *Recognition stage:* by means of an unknown process, the digestive cells can recognize potentially symbiotic algae. Several factors may be important in the recognition process, including antigens and surface charges on the algal cells and the type of microvilli produced by the animal cell during the engulfment stage. Algae that the cell rejects are expelled from the cell; acceptable algae are retained and surrounded by vacuoles. (d) *Migration stage:* algae that are accepted by the animal are moved to the base of the digestive cell. Since the algal cells are nonmotile, their movement is controlled by the animal cell, possibly by means of microtubules. (e) *Repopulation stage:* algal cells divide asexually until the normal algal population of an animal cell is reached. Further divi-

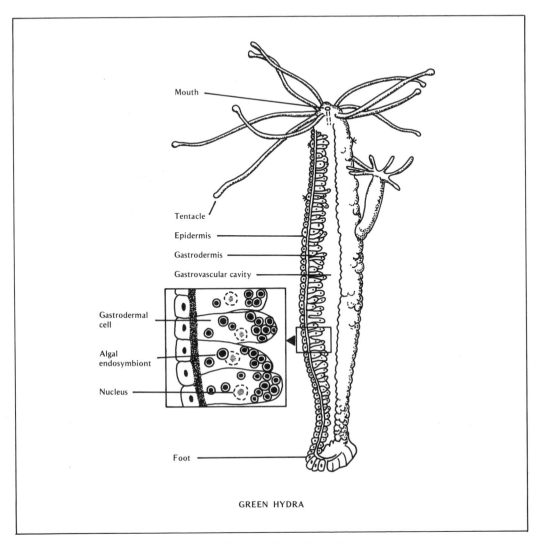

Mouth

Tentacle

Epidermis

Gastrodermis

Gastrovascular cavity

Gastrodermal cell

Algal endosymbiont

Nucleus

Foot

GREEN HYDRA

Figure 9-3. Green hydra's gastrodermal cells with endosymbiotic algae.

sion of the algal cells is closely linked to that of the host cell. When the host cell divides, so do the algal cells.

A single hydra contains about 150,000 algal cells as well as numerous bacteria that occur in vacuoles with or without algae. The bacteria may help the algae take up and store phosphate. The number of algal cells in each hydra cell varies depending on the location of the digestive cell within the animal. In the central region of the animal each digestive cell has about 19 algal cells, and in the head and tentacles each cell has about 10 algae.

Algal and bacterial symbionts are transmitted

to hydra progeny during asexual and sexual reproduction of the animal. Cells of the symbionts are attached to the outer surface of the eggs and are ingested by the young hydra when they emerge from the eggs. In some hydra strains extensions of endodermal cells carry algae into the developing eggs.

Under normal conditions symbiotic algae are not digested by hydra. There are two main reasons for this. First, cell walls of the algae contain sporopollenin, a compound that resists digestive enzymes. Second, vacuoles with algae do not fuse with lysosomes, organelles that contain digestive enzymes and normally fuse with food particles the animal ingests. But if a digestive cell takes in more algal cells than its normal algal population, the extra cells are either digested or ejected.

A bilateral movement of nutrients takes place between the symbionts of *H. viridis.* The algae supply the animal with photosynthetic products such as maltose. At pH 4.5, almost 80% of the carbon fixed by the algae is excreted as maltose, but at neutral pH's very little maltose is excreted. It is not known whether the host can change the pH of its cytoplasm in order to control maltose release by the algae. The animal rapidly hydrolyzes maltose to glucose and forms the storage compound glycogen. Algae also provide the animal with oxygen, which they produce during photosynthesis. The animal provides the algae with nutrients, including precursors of proteins and nucleic acids. These compounds, and others that are breakdown products from the hydra's food, enable the algae to survive if the animals are kept in darkness.

2. PARAMECIUM. *Paramecium bursaria* is a symbiotic protoctist with *Chlorella* as its symbiont (Fig. 9.4). Each algal cell is contained inside a vacuole that is formed by the host. The alga somehow changes the vacuole so that it does not fuse with lysosomes, and the alga, therefore, avoids being digested. When an algal cell divides, each of its four daughter cells becomes enclosed in its own vacuole. After the algal cells are ingested, they migrate from the central part of the animal cell to the periphery. Paramecia normally feed on bacteria and when the food supply is low, those animals with algae live longer than those with none. When food is plentiful, however, both types of paramecia multiply at the same rate. Algae of *P. bursaria* excrete large quantities of maltose. When fed bacteria in the dark, paramecia divide more rapidly than their symbionts and aposymbiotic paramecia develop. Aposymbiotic strains can be reinfected by different algae, including free-living strains, but only those that have been removed from a previous symbiosis grow well in the animal. During the establishment of the symbiosis, many algal cells are digested by the protoctist before enough survive to reestablish a population of about 300 algal cells in each host cell.

Many freshwater invertebrates contain symbiotic *Chlorella,* including clams of the genus *Arodanta,* species of the ciliate *Vorticella,* and a sponge, *Spongilla lacustris.* The *Chlorella* sym-

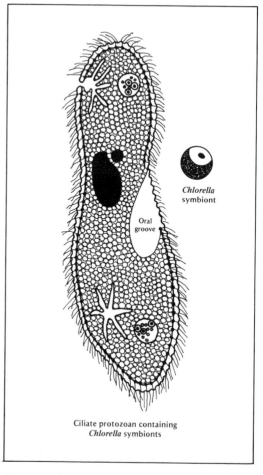

Chlorella symbiont

Oral groove

Ciliate protozoan containing *Chlorella* symbionts

Figure 9-4. Paramecium bursaria *with endosymbiotic* Chlorella.

bionts of these different invertebrates appear similar although some of the algal strains excrete glucose instead of maltose.

The presence of similar strains of *Chlorella* in unrelated invertebrates suggests that these associations developed independently, on different occasions. The associations probably began when cells of *Chlorella* were ingested by invertebrates and resisted not only the host's attempts to eject them but also the digestive enzymes of the host. Once the algae became permanent residents in the animal cell, relationships between the symbionts evolved. Why some species of invertebrates were more susceptible to algal infection than others is not known. For example, there are species of *Hydra* and *Paramecium* that resist colonization by algae.

C. Parasitic Algae

1. PARASITIC GREEN ALGAE. *Cephaleuros* is a common pathogen of vascular plants in tropical and subtropical regions. It is a member of the family Trentepohliaceae, which also includes the familiar epiphytic alga *Trentepohlia.* There are about a dozen species of *Cephaleuros,* the most common one being *C. virescens.*

Cephaleuros causes a disease called red rust on commercial plants such as citrus, coffee, and tea. The alga forms a flat, disclike, orange thallus that is usually subcuticular in leaves and has hairlike filaments that break through the cuticle; the alga also grows on fruits and stems. The damage caused by *Cephaleuros* depends on the type of host and its condition, the site of infection, season, and the number of algal thalli on the plant. Algae that penetrate leaves often kill the host cells that lie beneath the thallus, either by shading them or by toxic substances they release. The infection, however, does not extend beyond the algal thallus and the overall damage to the plant is slight. In leaves of the coffee plant, the infected host produces thick-walled cells that form a barrier between the parasite and healthy leaf tissue. Algae may girdle young stems and kill the plants. Infection of fruits like guava and mango reduces their attractiveness and marketability.

Cephaleuros does not penetrate the host cells regardless of the extent of the infection. The alga obtains water and minerals from the host and a substrate on which to grow. It also changes the carbon and nitrogen metabolism of the infected leaves, but whether it receives organic compounds from the host is not known.

Cephaleuros produces motile zoospores that spread the infection. The zoospores swim along the leaf surface or are washed by rain onto other leaves, and when they come to rest, they divide and form a thallus, which penetrates the host epidermis.

An interesting aspect of *Cephaleuros* is that it may be parasitized by a fungus to form a lichen. The fungus forms haustoria inside the living algal cells and eventually kills them. This type of parasite-parasite relationship is an example of hyperparasitism. The most common lichen formed by a fungus-*Cephaleuros* association is the genus *Strigula,* which occurs on leaf surfaces throughout the tropics. Lichenized and nonlichenized *Cephaleuros* may occupy the same habitats, but the lichenized form does not harm the leaves and grows more slowly than the unlichenized form.

Several parasitic green algae, not as common as *Cephaleuros,* cause limited damage to their hosts. *Chlorochytrium* and *Rhodochytrium* belong to the family Chlorococcaceae. The first named is an endophyte of aquatic plants, such as duckweed, and also of *Sphagnum* mosses; the second named occurs in various angiosperms, such as milkweed and ragweed. *Rhodochytrium* lacks chlorophyll and when first discovered was mistaken for a primitive fungus.

2. PARASITIC RED ALGAE. The phylum Rhodophyta consists of about 6,000 species of mostly marine organisms with complex life cycles. Over 100 species, within 50 genera, of red algae parasitize other red algae. About 80% of the parasites occur on closely related species, ones in the same family or tribe, and others parasitize unrelated hosts. Parasitic red algae occur only in the subclass Florideophyceae, especially in the orders Ceramiales and Cryptonemiales, and not in the more primitive subclass Bangiophyceae.

Parasitic red algae have the following characteristics. They are small (0.2 mm to 4 cm long), colorless or with reduced pigments, and host specific; they attack only one or two genera, to which they are closely related, and they penetrate the host tissue by means of specialized structures called rhizoids. Some parasites consist only of a rhizoidal system and do not form a vegetative thallus external to the host. Some parasites complete their life cycle 3 to 5 weeks after their spores germinate on the host; other parasites take about 18 weeks.

Red algal parasites generally do not cause disease and thus are not considered to be pathogens. Their parasitic nature, however, has been confirmed from radioisotope studies, which have demonstrated movement of organic compounds from the host to the parasite. For example, in one study, a segment of the red alga *Gracilaria,* with its attached parasite *Holmsella,* was placed in sea water that contained C^{14}-labeled bicarbonate and allowed to photosynthesize for several hours. The segment was removed from the labeled solution, washed, and either separated immediately into host and parasite plants or allowed to photosynthesize for several more hours in an unlabeled solution, before host and parasite were separated and analyzed for radioactive compounds. The results of this study clearly showed that the amount of radioactivity in the parasite increased with time, indicating a movement of compounds. The compound that moved from host to parasite was the alcohol glycoside, floridoside. When this compound reached the parasite, it was converted to mannitol and floridean starch. Similar results were obtained with the alga *Polysiphonia lanosa* and its parasite *Choreocolax polysiphoniae*. In this case, sodium mannoglycerate, another alcohol glycoside, was translocated to the parasite and converted to an unidentified sugar. Thalli of *Choreocolax* that were detached from their host fixed CO_2 more rapidly than thalli that remained on the host. This finding suggested that the host inhibited photosynthesis by the parasite.

Morphological relationships between parasites and their hosts vary considerably. Some parasites attach to the host by only one or two short rhizoids, whereas others develop extensive rhizoids in the host tissue. Parasites that penetrate deeply usually damage the host cells (for example, they become plasmolyzed), but such damage remains localized. Many parasites are linked with host cells by means of pit connections. When pit connections develop, nuclei of the parasite are transferred to the host cell. Lynda Goff and Annette Coleman discovered that the parasite *Choreocolax* inserted some of its nuclei into the cells of its host, *Polysiphonia*. The foreign nuclei caused the host cell to enlarge to twenty times its original size and to develop a thick cell wall. The number of host nuclei, mitochondria, and chloroplasts increased, as did the DNA content of the host nuclei; products of photosynthesis also accumulated in the cell.

How parasitic red algae evolved is a matter of speculation. One hypothesis is that they arose when mutated spores of a red alga produced small gametophytes that were unable to live independently but had the ability to penetrate cells of the parent plant. There are present-day red algae whose tetraspores germinate and give rise to dwarf gametophytes that attach to the parent plant. The gametophytes form rhizoids and develop pit connections with the parent cells. A second hypothesis is that the algae were originally epiphytes that became parasites. The first hypothesis seems more plausible because of the close taxonomic relationships that exist between parasites and hosts.

Some scientists believe that ascomycetes (Fungi) may have evolved from parasitic red al-

gae that lost their plastids. Ascomycetes and red algae have many common morphological and cytological features.

D. Summary and Perspectives

Symbiotic algae called zooxanthellae live in the cells of many marine invertebrates. The algae excrete glycerol and other compounds that the host uses to produce lipids and proteins; waste products of the host are used by the algae. *Anthopleura xanthogrammica*, a sea anemone, has two types of algae, zooxanthellae and zoochlorellae; the first is more prevalent when the anemone grows in warm waters and the second is more common is anemones found in cold waters. The life cycle of the upside-down jellyfish, *Cassiopeia xamachana*, cannot be completed until its asexual polyp stage becomes infected with algae. Symbiotic algae greatly enhance the activities of reef-building corals, tridacnid clams, foraminiferans, and radiolarians. A single marine flatworm, *Convoluta roscoffensis*, may contain up to 40,000 unicellular, green algal cells. The algae live inside the digestive cells of the worm and are greatly modified. A mature worm is completely dependent on the algae for food. Larvae of the worm are aposymbiotic and must be infected with algae in order to develop further. Marine sponges contain symbiotic cyanobacteria. Sea slugs obtain glucose from chloroplasts of algae they ingest. *Prochloron* occurs in tropical marine invertebrates and has characteristics of both green algae and cyanobacteria. *Hydra viridis* and *Paramecium bursaria* contain symbiotic algae that belong to the genus *Chlorella*. Infection of *H. viridis* involves different stages of contact, engulfment, recognition, migration, and repopulation. Several green algae are parasitic on vascular plants, and many small, red algae parasitize other red algae.

Symbioses between algae and invertebrates involve common features such as integration of the symbionts, regulation of the algal population, specificity, bilateral nutrient exchange, and recognition mechanisms. Integration of a symbiont into a host generally results in a reduction of parts, thin walls, loss of sexuality, and slow growth. Algal populations are regulated by the host by several mechanisms, including digestion or ejection of surplus algae and inhibition of symbiont cell division. Anemones and corals eject surplus algae, and in other symbioses algal division is restricted. Most symbioses of invertebrates with unicellular algae show a high degree of specificity; that is, a host associates with only one species of alga. There are exceptions, however, in the foraminiferans whose species associate with a wide diversity of algal symbionts. Studies of recognition systems between symbiont and host in different associations have not revealed a common pattern, a finding that is consistent with the different lines of evolutionary development among symbioses.

Review Questions

1. How is the alga *Symbiodinium microdriaticum* modified when it lives inside animal cells?
2. How do algal symbionts pass from one generation of invertebrates to another?
3. For what type of research has the jellyfish *Cassiopeia* been used?
4. Explain how symbiotic algae contribute to the formation of coral reefs.
5. Why are colonial radiolarians good subjects to study algal-protoctist relationships?
6. Describe how the algal symbiont of *Convoluta roscoffensis* becomes modified when it is ingested by the worm.
7. Why is the association between sea slugs and the chloroplasts of algae not considered to be a true symbiosis?
8. Describe the characteristics of *Prochloron*, the symbiont of tropical marine tunicates.
9. List the stages of reinfection of aposymbiotic *Hydra viridis*.
10. How does the *Chlorella* symbiont of *Paramecium bursaria* escape being digested by the host?
11. Describe several characteristics of parasitic red algae.
12. Why do so many different marine invertebrates contain the same type of symbiotic algae?

Further Reading

Chapman, R. L., and B. H. Good. 1983. Subaerial symbiotic green algae: Interactions with vascular plant hosts. In: Algal Symbiosis, 173–204, ed. L. J. Goff. Cambridge Univ. Press, Cambridge.

Cook, C. B. 1983. Metabolic interchange in algae-invertebrate symbiosis. Int. Rev. Cytology, supplement 14:177–210.

Doonan, S. A., A. E. Douglas, and G. W. Gooday. 1980. Acquisition of algae by *Convoluta roscoffensis*. In: Endocytobiology, Vol. 1, Endosymbiosis and cell biology, 293–304, ed. W. Schwemmler and H. E. A. Schenk. Walter De Gruyter, Berlin.

Douglas, A., and D. C. Smith. 1984. The green hydra symbiosis. VIII, Mechanisms in symbiont regulation. Proc. R. Soc. Lond. B. 221:291–319 (Division of the symbiotic algae is regulated by changes in pH.)

Goff, L. J. 1982. The biology of parasitic red algae. In: Progress in phycological research, Vol. 1, 289–369, ed. F. E. Round and D. J. Chapman. Elsevier Biomedical Press, Amsterdam.

Hinde, R. 1983. The retention of algal chloroplasts by molluscs. In: Algal symbiosis, 97–107, ed. L. J. Goff. Cambridge Univ. Press, Cambridge.

Holligan, P. M., and G. W. Gooday. 1975. Symbiosis in *Convoluta roscoffensis*. In: Symbiosis: Symposia of the Soc. for Experimental Biology, no. 29, 205–227, ed. D. H. Jennings and D. L. Lee. Cambridge Univ. Press, Cambridge. (Considers biochemical and physiological relationships between the symbionts.)

Karakashian, M. W. 1975. Symbiosis in *Paramecium bursaria*. In: Symbiosis: Symposia of the Soc. for Experimental Biology, no. 29, 145–173, ed. D. H. Jennings and D. L. Lee. Cambridge Univ. Press, Cambridge.

Lee, J. J. 1983. Perspectives on algal endosymbionts in larger foraminifera. Int. Rev. Cytology, supplement 14:49–77.

Lee, J. J., and M. E. McEnery. 1983. Symbiosis in foraminifera. In: Algal symbiosis, 37–68, ed. L. J. Goff. Cambridge Univ. Press, Cambridge.

Lee, J. J., A. T. Soldo, W. Reisser, M. J. Lee, K. W. Jeon, and H. D. Görtz. 1985. The extent of algal and bacterial endosymbioses in protozoa. J. Protozool. 32:391–403.

O'Brien, T. L., C. R. Wyteenbach. 1980. Some effects of temperature on the symbiotic association between zoochlorellae (Chlorophyceae) and the sea anemone *Anthopleura xanthogrammica*. Trans. Amer. Micros. Soc. 99:221–225.

Pardy, R. L. 1983. Phycozoans, phycozoology, phycozoologists? in: Algal symbiosis, 5–17, ed. L. J. Goff. Cambridge Univ. Press, Cambridge.

Pardy, R. L., R. A. Lewin, and K. Lee. 1983. The *Prochloron* symbiosis. In: Algal symbiosis, 91–96, ed. L. J. Goff. Cambridge Univ. Press, Cambridge.

Rahat, M., and V. Reich. 1984. Intracellular infection of aposymbiotic *Hydra viridis* by a foreign free-living *Chlorella* sp.: Initiation of a stable symbiosis. J. Cell. Sci. 65:265–277.

Reisser, W., and W. Wiessner. 1984. Autotrophic eukaryotic freshwater symbionts. In: Cellular interactions: Encyclopedia of plant physiology, new series, Vol. 17, 59–74, ed. H. F. Linskens and J. Heslop-Harrison. Springer-Verlag, Berlin.

Smith, D. C. 1973. Symbiosis of algae with invertebrates. Oxford Biology Readers, Oxford Univ. Press, Oxford. 16 pp. (A brief, general introduction to algal-invertebrate symbioses.)

Taylor, D. L. 1984. Autotrophic eukaryotic marine symbionts. In: Cellular interactions: Encyclopedia of plant physiology, new series, Vol. 17, 75–90, ed. H. F. Linskens and J. Heslop-Harrison. Springer-Verlag, Berlin.

Thorington, G., and L. Margulis. 1981. *Hydra viridis:* Transfer of metabolites between *Hydra* and symbiotic algae. Biol. Bull. 160:175–188. (Presents evidence of algae receiving nutrients from food eaten by the *Hydra.*)

Trench, R. K. 1979. The cell biology of plant-animal symbiosis. Ann. Rev. Physiology 30:485–531. (Considers the use of associations between unicellular algae and invertebrates as model systems with which to study basic problems of intercellular relationships.)

Wilkinson, C. R., R. Garrone, and J. Vacelet. 1984. Marine sponges discriminate between food bacteria and bacterial symbionts: Electron microscope radioautography and *in situ* evidence. Proc. R. Soc. Lond. B. 220:519–528.

Bibliography

Anderson, O. R. 1983. Radiolaria. Springer-Verlag, New York. 355 pp.

Bold, H. C., and M. J. Wynne. 1985. Introduction to the algae: Structure and reproduction. 2d ed. Prentice-Hall, Englewood Cliffs, N.J. 720 pp.

Cheng, T. C. 1967. Marine molluscs as hosts for symbioses: Advances in Marine Biology, Vol. 5, ed. F. S. Russell. Academic Press, London. 424 pp.

Goff, L. J. (ed.). 1983. Algal symbiosis. Cambridge Univ. Press, Cambridge. 216 pp.

Goreau, T. F. 1961. Problems of growth and calcium deposition in reef corals. Endeavour 20 (January):32–39.

Rosowski, J. R., and B. C. Parker. 1982. Selected papers in phycology, II. Phycological Society of America Publications, Lawrence, Kansas. 866 pp.

Vernberg, W. B. 1974. Symbiosis in the sea. Univ. of South Carolina Press, Columbia, S.C. 276 pp.

CHAPTER 10

Helminthic Symbiotic Associations
Flukes, Tapeworms, Nematodes

A. Introduction

Scientists estimate that 20% to 50% of all animal species are parasitic. Some phyla such as the Platyhelminthes, Nematoda, and Arthropoda contain large numbers of parasitic species. Hosts and parasites have evolved together, and under natural conditions many have become mutually tolerant. Host organisms can live independently but, in many cases, the parasite's association with its host is obligatory.

Animal parasites affect the health of humans and domesticated animals throughout the world. In most warm climates parasitic infections from flukes, nematodes, and arthropods greatly diminish the quality of life. Scientists estimate that three out of four people in the world are infected with parasites and many people carry multiple infections. The prevalence of parasitic diseases is extremely high among people who are poor and illiterate. Because of public health concerns, the science of *parasitology* continues to attract public support to study host-parasite relationships and to develop effective strategies to control and cure parasitic infections. Parasitology is the study of diseases caused by protozoa, helminths, and arthropods. In this chapter, we examine symbioses that involve flukes, tapeworms, and nematodes. We describe the morphological adaptations of parasites, how the host defends itself against parasitic invasion, how parasites evade a host's defenses, and coevolution of parasites and hosts.

Helminths are widely distributed parasites of plants and vertebrates and are surpassed only by insects in the variety and diversity of hosts that they parasitize. There are more than 100,000 species of helminthic parasites, but only a relatively few species are of economic and public health concern. Infections caused by helminths such as *Schistosoma,* hookworms, and filarial nematodes are the major causes of sickness in humans inhabiting tropical regions. Several species of nematodes are parasites of vegetables, fruits, and other crop plants. *Medical* and *veterinary* helminthology are specialties that deal with diseases caused by helminthic endoparasites. Recently, ecologists and parasitologists have drawn attention to the role of helminthic infections in several species of wild animals that are threatened with extinction. *Nematology,* the science of nematodes, is of interest to agriculturalists because of crop losses resulting from parasitic nematodes. Some free-living nematodes, such as *Caenorhabditis elegans,* are being used to study metazoan genetics and developmental biology because of the ease with which they can be cultured and analyzed biochemically.

Helminths are animals that have complex life cycles. They live for a long time within the host animals, and they often possess a remarkable ability to evade the host's defense mechanisms. The prevalence of helminthic infections in some areas is high, however only a few, either very young or old, hosts develop disease. Helminths in humans do not multiply and, therefore, the severity of the disease depends on the extent of the original infection. Some helminths may accumulate after repeated infections of a host.

B. Classification

In this chapter we discuss various organisms from the kingdom Animalia.

KINGDOM ANIMALIA

Phylum: Platyhelminthes (Flatworms)
  Class: Trematoda (the flukes)
        Clonorchis, Diplozoan, Fasciola,
        Paragonimus, Polystoma, and
        Schistosoma
  Class: Cestoda (the tapeworms)
        Diphyllobothrium, Echinococcus,
        Hymenolepis, Lingula, Schistocephalus,
        Spirametra, Taenia, and *Taeniarhynchus*
Phylum: Nematoda (Roundworms)
  (a) Human and Vertebrate Parasites
        Ancylostoma, Ascaris, Brugia, Dirofilaria,
        Dracunculus, Haemonchus,
        Litomosoides, Necator, Onchocerca,
        Trichinella, and *Wuchereria*
  (b) Insect Parasites
        Deladenus, Mermis, Neoaplectana,
        Romanomermis, and *Sphaerularia*
  (c) Plant Parasites
        Anguina, Bursaphalenchus, Globodera,
        Heterodera, and *Meloidogyne*

C. Trematodes: The Flukes

1. SOME FLUKE SYMBIOSES. Adult flukes are obligate endoparasites of vertebrates and include the liver fluke, the lung fluke, and the blood fluke of humans. Following sexual reproduction, the female fluke produces eggs that exit through the genital pore into the host environment and then are passed out of the host with the feces or urine. There are several larval stages, which multiply asexually in snails serving as the intermediate host. A larval stage possessing a characteristic tail emerges from a snail and either penetrates a vertebrate host immediately or encysts and attaches to grass blades, which may be ingested by a herbivorous vertebrate (Fig. 10.1).

a. **Polystoma integerrimum.** *Polystoma integerrimum* inhabits the urinary bladder of frogs throughout Europe and North America. Up to 50% of the frogs in the United States carry the fluke. The life cycle of the fluke is unusual in that its maturation is synchronized with the sex-

ual maturation of its host. When the frog spawns, the flukes become sexually mature and release their eggs. The synchronized cycles enable the parasite to produce larvae at a time when tadpoles are abundant. The fluke larvae attach to the tadpole gills and feed on mucus and other materials. The mature larvae migrate from the gills to the bladder. Experimental evidence shows that fluke maturation is regulated by hormonal activities of the host. Pituitary extracts injected into an immature frog induce sexual maturation of the parasite. The exact nature of the hormonal control of fluke development is not known. Sexual maturity of the protozoan *Opalina* is also synchronized with reproduction of its frog host.

b. **Diplozoan paradoxum.** *Diplozoan paradoxum* is a parasite of freshwater fish and has a unique life cycle that involves permanent fusions of pairs of flukes. Such fusions raise interesting questions about the biochemical, genetic, and immunological compatibility of the fused individuals. The attachments begin when the larvae pair and fuse with their posterior suckers. The reproductive openings of each worm exit on the opposite end of their partner in order to permit cross-fertilization. Larvae that do not form pairs die before becoming sexually mature.

c. **Fasciola hepatica.** *Fasciola hepatica,* the sheep liver fluke, commonly inhabits the bile duct, liver, or gall bladder of cattle, horses, pigs, and other farm animals. The parasite's life cycle, ultrastructure, nutrition, and biochemistry have been studied extensively. Upon hatching from metacercariae, young flukes penetrate the intestinal wall, move through the body cavity, and enter the liver. During their migration, the young flukes feed on muscle, intestinal, and liver parenchyma cells. When the parasites reach the bile duct, they feed on epithelial cells, blood, and liver tissue. The fluke also absorbs amino acids, glucose, and fatty acids through its tegument. Epithelial cells of the bile duct become hyperplastic during the fluke's infection. The fluke somehow stimulates production of the host tissue on which it feeds.

d. **Clonorchis sinensis.** The Chinese liver fluke, *Clonorchis sinensis,* is an important parasite of

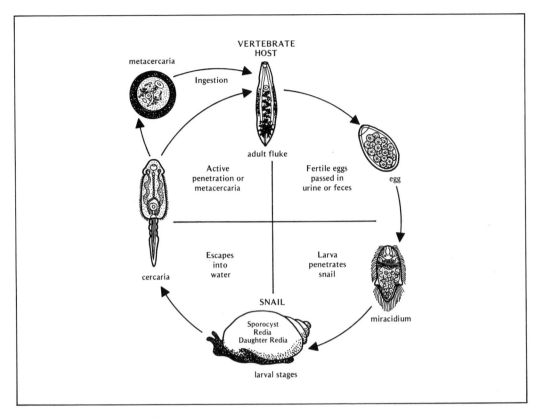

VERTEBRATE
HOST

metacercaria

Ingestion

adult fluke

Active
penetration or
metacercaria

Fertile eggs
passed in
urine or feces

egg

Escapes
into
water

Larva
penetrates
snail

cercaria

SNAIL

miracidium

Sporocyst
Redia
Daughter Redia

larval stages

Figure 10-1. A typical fluke life cycle.

man and other fish-eating mammals throughout the Far Eastern countries. The fluke makes its way to the liver and bile duct directly from the alimentary canal and feeds on epithelial and blood cells. Fish become infected with larvae, which are liberated from a snail host. People acquire these flukes by eating uncooked or smoked fish whose muscle contains metacercariae of the parasite.

e. **Paragonimus westermani.** Species of *Paragonimus* inhabit different organs of vertebrate hosts. *Paragonimus westermani,* the lung fluke, infects humans as well as cats, dogs, and rats. In humans the flukes become encapsulated in the lung tissue and produce eggs, which pass upward in the trachea to the mouth and then down the intestinal tract with the feces. This fluke is extremely prevalent in the people of China, Philippines, Thailand, and other Asian countries. Humans become infected with the fluke by eating raw crabs and crayfish. Crab juice is fre-

quently used to prepare food in Korea and the Philippines.

f. **Schistosoma.** Next to malaria, schistosomiasis is the most important parasitic disease in the world, affecting more than 200 million people. Schistosomes are blood flukes and they reside in the mesenteric blood vessels of humans. Adult schistosomes have a unique morphology and physiology. Male and female flukes are elongated and wormlike and the female fluke is held permanently in the ventral groove of the male fluke (Fig. 10.2). The presence of eggs of blood flukes in various host tissues triggers an immunological response causing the affected person to show disease symptoms, which include enlargement of the liver and spleen, bladder calcification, deformity of the ureter, and kidney disorders. Schistosomiasis is common among poor people, especially farmers who live and work under unsanitary conditions, and is increasingly being recognized as a man-made disease. The development of irrigation projects in-

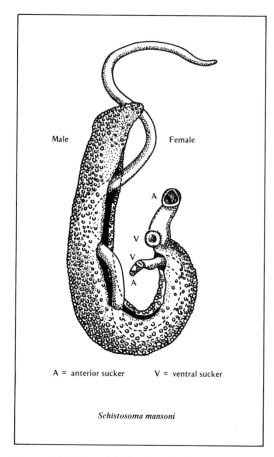

Male

Female

A

V

V

A

A = anterior sucker   V = ventral sucker

Schistosoma mansoni

Figure 10-2. An adult female of Schistosoma mansoni *in permanent copulatory union within the ventral groove of the male fluke.*

creases the range of the snail species that are the intermediate hosts of *Schistosoma.*

Three important blood flukes that infect humans are *Schistosoma mansoni, S. japonicum,* and *S. haematobium. Schistosoma mansoni* causes intestinal bilharziasis in South and Central America and the Middle East and has been studied extensively by scientists because of the ease with which it grows in laboratory mice, hamsters, and rats. *Schistosoma japonicum* affects millions of people in Asia and is also prevalent among rats, pigs, dogs, cattle, and horses. Urinary schistosomiasis is caused by *Schistosoma haematobium* and it occurs in about 40 million people of Africa and the Middle East. It is a disease of young adults between the ages of 10 and 30 years, and rarely occurs in older persons. Scientists in the Sudan have noted a cor-

relation between bladder cancer and urinary schistosomiasis.

2. FLUKE HOST-PARASITE RELATIONSHIPS. Recent advances in the biology of host-parasite relationships of flukes have focused on the structure and function of the body wall (tegument), nature of the host's immune response to the fluke, and evasion of the host's immune mechanism by the parasite.

Until the early 1960s, it was believed that all flatworms were covered with a cuticle, a tough inert structure. Ultrastructure of the fluke body covering, however, has revealed it to be a unique and biologically active structure. The term *tegument* is now applied to the outer covering of flatworms instead of cuticle, which is characteristic of roundworms.

The tegument is composed of (1) an outer region, the *distal cytoplasm,* which consists of cytoplasmic extensions of tegumental cells forming a syncytium, and (2) an inner *proximal cytoplasm,* which consists of tegumentary cells. The tegument is bound externally and internally by a plasma membrane (Fig. 10.3). The tegumentary matrix contains many mitochondria, ribosomes, vacuoles, endoplasmic reticulum, and Golgi bodies and is capable of nutrient uptake, osmoregulation, excretion, and supporting various sensory structures. The tegument contains secretory products that are believed to have an immunological role in protecting the parasite from host digestive enzymes. The tegument is also covered with a carbohydrate-protein complex called the *glycocalyx,* which is continuously produced by the tegument.

The free surface of the tegument is folded into ridges or fingerlike projections. In blood flukes the tegument has an elaborate network of channels that open to the outside. Although flukes have a well-developed digestive system, the tegumentary uptake of nutrients is significant. *In vitro* studies have shown that small molecules are absorbed through the fluke tegument and large nutrient molecules are taken in by pinocytosis.

Figure 10-3. Ultrastructure of the helminthic tegument. (A) Fluke tegument with the outer surface covered with a glycocalyx. (B) Tapeworm tegument with characteristic microtriches and glycocalyx.

147

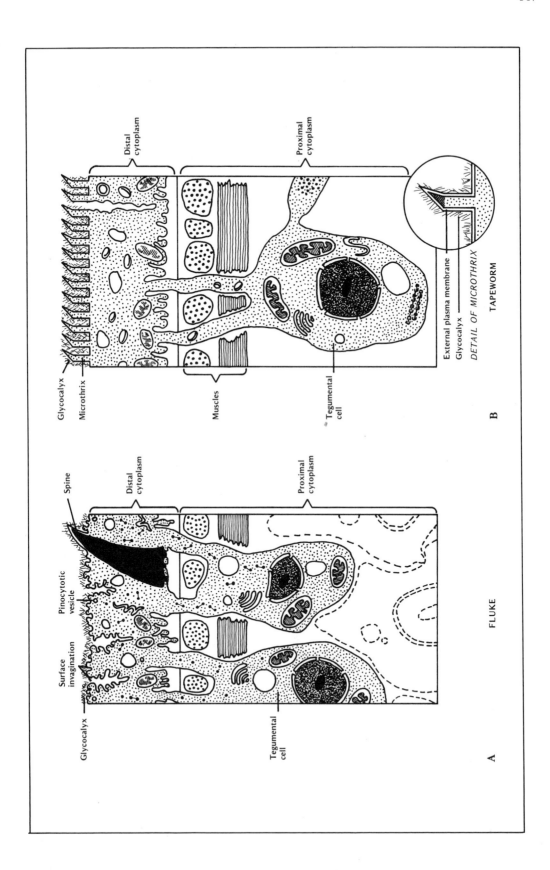

Glycocalyx

Microthrix

Distal cytoplasm

Proximal cytoplasm

Muscles

Tegumental cell

External plasma membrane
Glycocalyx
DETAIL OF MICROTHRIX

B

TAPEWORM

Spine

Distal cytoplasm

Proximal cytoplasm

Pinocytotic vesicle

Surface invagination

Glycocalyx

Tegumental cell

A

FLUKE

The tegument of the adult *Schistosoma* is unusual because it does not have a glycocalyx. Recent scanning electron microscope observations have shown that the tegument of the male schistosome has an irregular spiny surface; that of the female is mostly smooth with spines being concentrated at the tail end. When larvae of *Schistosoma mansoni* emerge from the intermediate snail host, they are surrounded by a glycocalyx. Immediately following penetration of a vertebrate host the larvae become transformed into a form called the *schistosomulum*. The larval surface membrane and associated glycocalyx are replaced by membranes derived from the host and membranous inclusions that are formed near the surface of the larval tegument within minutes of skin penetration (Fig. 10.4).

The fluke migrates from the skin to the lungs and then to the liver, and in the process loses its tegumental spines. The lung stage of the fluke is particularly resistant to the host's hormonal and cellular immune mechanisms. The schistosomulum changes or masks its original surface antigens and thus does not trigger the host's immune responses. Raymond Damian, a pioneer in parasitic immunology, proposed in 1964 the concept of *molecular mimicry* to describe how the parasitic symbiont produces surface antigens that are similar to those of the host. These antigens disguise the parasite against the host's immune response. The question of whether the hostlike antigens are synthesized by the fluke or obtained from host cells is a subject of intense research. Recent evidence confirms that schistosomula can acquire A, B, and H blood group antigens from red blood cells along with the glycoproteins of the major histocompatibility complex. These antigens and others required by the symbiont seem to play a central role in the phenomenon of molecular mimicry.

Although the first parasites successfully evade the host's immune responses, they induce the host to reject all further parasites of the same kind. This phenomenon is called *concomitant immunity*.

Adults of *Fasciola*, like *Schistosoma*, can live in the host's bile duct for long periods of time, and mechanisms of molecular mimicry fail to explain how the liver fluke evades the host's antibodies. Scientists have suggested that *Fasciola* periodically sloughs off its glycocalyx along with associated host antibodies, and replaces it with a new coat.

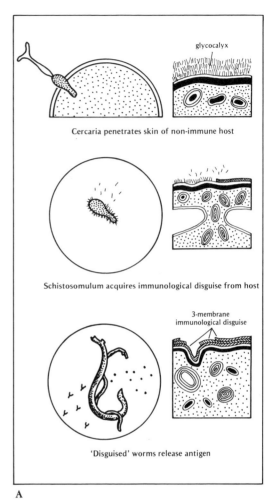

A

Figure 10-4. Tegumental changes in Schistosoma mansoni. *(A) Schematic representation of outer surface changes in cercaria following penetration of host skin. Note relative amounts of glycocalyx on cercaria, schistosomulum, and adult. Adult worms have a characteristic multilayered membrane system. (B) Electron micrograph of the fibrillar nature of the glycocalyx of a schistosome. (C) Electron micrograph of an adult schistosome and the three-membrane outer covering. (Parts (B) and (C): Dianne J. McLaren, National Institute for Medical Research, London.)*

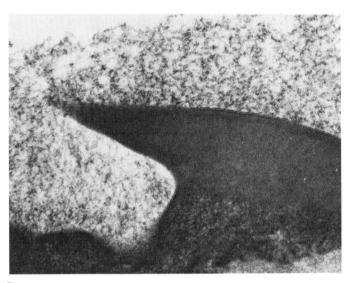

B

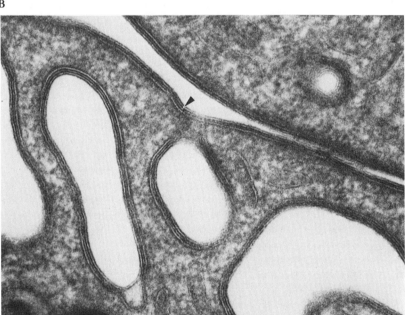

C

D. Cestodes: The Tapeworms

Tapeworms represent the 'ultimate' in biological adaptation to live in another organism. All tapeworms are obligate symbionts of vertebrates and arthropods. The adult tapeworm body consists of a head, the *scolex,* which may possess anchoring devices such as suckers and hooks, a neck or zone of proliferation, and a *body* made up of segments called *proglottids* that may number from 3 to 3,000. Sexually mature tapeworms live in the small intestines of vertebrates; their larval stages develop in the visceral organs of an alternate host, which may be a vertebrate or an arthropod. The development of larval stages in the muscles and nervous tissue of the vertebrate host produces serious disease. Some scientists view the adult tapeworms in the alimentary canal as endocommensals living in a nutrient-rich environment. Adult tapeworms are well known for their rapid growth and production of large numbers of eggs.

The extent of the disease in the host depends on the number of larvae involved, their survival, and the host immune response. Infected individuals rapidly become resistant to further infection and this immunity is passive and can be transferred to healthy hosts. Scientists have noted the similarity between the antigens of various tapeworm species. Scientists hope to collect immunogens, for developing vaccines, from a nonpathogenic tapeworm such as *Taenia hydatigena,* which is easily maintained and propagated in the laboratory.

There is increasing evidence that the establishment and survival of adult tapeworms in the host intestine is immunologically controlled. The *crowding effect,* often noted in tapeworm infections, means that the size of individual parasites is inversely related to the number of parasites in the host. Solitary tapeworms are commonly found in humans, for reasons that are not clear. It is possible that an established tapeworm secretes substances that inhibit the development of other tapeworms. Alternatively, the host may tolerate only a certain amount of parasitic load before its immunological mechanism becomes operative. Scientists believe that the competition for carbohydrates plays an important role in producing the crowding effect.

1. THE TAPEWORM TEGUMENT. Tapeworms lack a digestive system and absorb all their nutrients through their tegument, which is remarkably similar to that of the flukes. A unique feature of the tapeworm tegument are minute projections, called *microtriches,* that cover the outer surface (Fig. 10.3). Microtriches are points of attachment for the tapeworm in the host intestines, and they significantly increase the absorptive surface of the tegument. The glycocalyx of the tapeworm tegument enhances the host's amylase activities while inhibiting the host's production of trypsin, chymotrypsin, and pancreatic lipase. The glycocalyx protects the tapeworm from the host's digestive enzymes. Both the larvae and adult tapeworms live in an environment rich in amino acids, fatty acids, glycerols, acetates, and nucleotides. These molecules are absorbed by diffusion and active transport, which take place at specific sites on the tegument.

2. SOME TAPEWORM SYMBIOSES

a. **Diphyllobothrium:** *The Fish Tapeworm.* The adult form of the fish tapeworm is a common inhabitant of the alimentary canal of fish-eating mammals, birds, and fish. Fish-eating people in temperate climates often carry *D. latum* in their small intestine. The fish tapeworm is well known for its ability to absorb vitamin B_{12}, thereby causing the host to be deficient in a vitamin that is essential for the development of red blood cells. As a rule, frogs, fish, or snakes are the intermediate hosts of this tapeworm. Inhabitants of southeast Pacific Islands often use crushed frogs to prepare food and in this way acquire the tapeworm larvae. Humans infected with adult tapeworms pass eggs in the feces. Ciliated larvae emerge from the eggs and infect animals such as water fleas, where they become transformed into another larval form. Fish acquire the tapeworm by eating water fleas. The tapeworm larva penetrates the alimentary canal of the fish and develops into the final larval stage in the fish musculature.

b. **Spirometra mansonoides:** *The Beneficial Parasite.* The adult *Spirometra mansonoides* occurs in the alimentary canals of dogs and cats. From eggs passed in host feces, larvae hatch and develop in water fleas. Frogs become infected with the larval stage by eating water fleas. Mice that eat frogs become infected with the final larval stage of the tapeworm. Scientists have observed that infected mice showed increased growth and efficiency of nutrient uptake. J. F. Mueller considers *Spirometra mansonoides* to be a *beneficial parasite,* because the larger size of the infected mice makes them sluggish and therefore an easy catch for the final host, a cat.

c. **Taenia solium:** *The Pork Tapeworm.* Humans become infected with pork tapeworm when they eat poorly cooked pork. The incidence of infection is high in parts of Europe, the Middle East, East Africa, and Mexico. The parasitic larvae occur in the musculature and give pork a speckled appearance. Adult tapeworms in the human alimentary canal do not cause any noticeable symptoms.

d. **Taeniarhynchus saginatus:** *The Beef Tapeworm.* People acquire the beef tapeworm by eating raw hamburger or uncooked meat. Adult tapeworms occur in the human small intestine, where they may grow to lengths of 5 to 10 meters. Humans pass the eggs of *T. saginatus* with feces, which are ingested by cattle. The larvae hatch in the digestive tract of the cattle, penetrate the intestinal wall, enter the circulatory system, and finally become lodged in the muscles and other tissues.

e. Hymenolepis diminuta: *The Rat Tapeworm.* *Hymenolepis diminuta* has been a favorite experimental subject to investigate the nutrition, biochemistry, immunology, and developmental biology of tapeworms. The adult tapeworm commonly occurs in the alimentary canals of rats, mice, and hamsters. Insects such as beetles, fleas, cockroaches, and flies are the intermediate hosts. Rat tapeworms can live for an indefinite period and show no signs of aging. Adults of *H. diminuta* have been kept alive for more than 14 years by using serial surgical transplantations. The host shows no symptoms. The tapeworm has a diurnal migration in the intestine, moving anteriorly in the morning and returning to the posterior portion of the small intestine in the evening. The migration is correlated to the feeding behavior of the host.

f. Schistocephalus solidus: *A Case of Energy Parasitism.* *Schistocephalus solidus* inhabits the alimentary canal of ducks and has a brief life span. The birds pass tapeworm eggs, which hatch and produce swimming larvae that are ingested by water fleas. When a fish eats infected water fleas, the final larval stage of the tapeworm develops in the fish's body cavity. The larva has undifferentiated reproductive organs and significant reserves of glycogen. When a duck eats an infected fish, the tapeworm larva uses its food reserves to quickly become sexually mature and lay eggs. Scientists have shown that the larvae become sexually mature when they are incubated at 40° C in a sterile vinegar solution. It appears that the host contributes mostly body heat for the development of the tapeworms. J. Smyth has defined this type of tapeworm development as *energy parasitism.*

E. Nematodes: *The Roundworms*

Roundworms are second only to insects as the most abundant animals on earth. Most nematodes are free-living. They occur in freshwater, marine, and soil habitats and feed on microorganisms and decaying organic matter. Many nematodes are adapted for a parasitic life style in plants, fungi, and animals. Nematode parasites of plants have a syringelike mouth, which they insert into plant tissue and thereby absorb nutrients. Some nematodes are ectoparasites on plant roots; others are endoparasites and cause abnormalities of the host tissue. Scientists estimate that every kind of animal is inhabited by at least one parasitic nematode. Nematodes parasitize organs such as eyes, tongue, liver, and lungs, often causing destruction of the host tissues.

Roundworms are elongated, cylindrical helminths ranging in length from microscopic to several inches. Sexes are separate and in most species mature females lay resistant eggs. The nematode life cycle consists of four larval stages and the mature adult. Each larval stage is transformed into the next by a molting process during which the old cuticle is shed and a new one is formed. Sexually mature roundworms do not molt.

1. PARASITES OF MAN AND VERTEBRATES

a. Intestinal roundworms

ASCARIS: THE LARGE ROUNDWORMS. *Ascaris lumbricoides* is one of the largest intestinal nematodes of man and is prevalent throughout the warmer climates. An estimated 700 million people are infected with this nematode and most are asymptomatic carriers. The nematode may cause intestinal obstructions and interferes with host nutrition. Adult worms live in the small intestine and their eggs are passed with host feces to the outside. When eggs are ingested with contaminated food or drink, the larvae hatch in the small intestine, penetrate the gut lining, and are carried by the bloodstream to the liver, heart, and lungs. From the lungs the nematodes travel to the mouth and then back into the alimentary canal. It is not known why the nematode adopts such a circular path in its development in a host. During the larval migration, the parasite molts, feeds on host tissue, and grows in size. The young adult roundworms in the small intestine increase in size and become sexually mature in about two months. A mature female roundworm produces about 200,000 eggs each day. Most infected humans carry only one to a few nematodes but some may have hundreds of worms.

The adults of *A. lumbricoides* can remain in a host for up to one year. Because of its size and simple one-host life cycle, *A. lumbricoides* has been studied for its morphology, physiology, biochemistry, and host-parasite relationship.

HAEMONCHUS CONTORTUS: THE SHEEP STOMACH WORM. *Haemonchus contortus* is a bloodsucking nematode residing in the abomasum of sheep, goats, cattle, and other ruminants. Female nematodes pass their eggs with host feces onto soil where under favorable conditions the eggs hatch and the larvae undergo three developmental stages. The last larval stage does not feed and has to be ingested by a grazing host animal before it can develop further. In the host's stomach, the parasite resumes its growth, undergoes a final molt, and develops into an adult. A single sheep may carry thousands of adult worms, which attach firmly to the stomach villi and feed on blood.

A phenomenon called *self-cure* takes place in the host sheep. When an infected host receives a challenge infection, a large number of previously acquired adult worms are expelled and the host becomes resistant to reinfection. In other words, parasites acquired early in life make the host resistant to heavy infections by the same parasite. Some scientists have proposed that the host's immune response, instead of being an absolute rejection mechanism, allows the establishment of a mutualistic symbiosis. According to the *welcome mat hypothesis,* both the host and the parasite avoid mutual extinction.

Recent research on the mechanism of the self-cure phenomenon has shown that newly ingested larvae produce substances that trigger a localized allergic reaction in the host. Host animals that had expelled adults of *H. contortus* had high levels of histamines in their blood. The host animal becomes sensitized against the parasitic antigens following repeated exposure to *H. contortus.*

The self-cure phenomenon has been studied extensively in rats infected with *Nippostrongylus brasiliensis,* since both organisms can be maintained under laboratory conditions. Infected rats release eggs of the parasite in the feces by the sixth day following infection and show peak egg production on the tenth day. Within the following week egg production drops to almost zero and there is a massive expulsion of adult worms from the host intestine. A small number

of nematodes survive the expulsion phase and remain in the host intestine for a long time. The surviving worms are males and non-egg-laying females. The host animal becomes resistant to further reinfections. It is believed that the host's antibodies interfere with the acetylcholinesterase activity of the nematodes. The parasite's intestinal cells become damaged and it is unable to feed. The surviving worms are smaller and produce more and different acetylcholinesterases. They have adapted to the host's immunity and are able to survive because of their reduced antigenicity.

The phenomenon of *lactational rise* has been observed in *Haemonchus* and other gastrointestinal nematodes of herbivores. For example, the number of eggs in a lamb's feces greatly increases four to eight weeks after the young are born. Several weeks later, egg production falls to a low level. Studies have confirmed that the host's cellular immunity becomes depressed during lactation. From an evolutionary perspective, the lactational rise may be viewed as an adaptation for the survival of the nematode because it facilitates transmission of the parasite to the offspring.

ANCYLOSTOMA AND *NECTOR:* THE HOOKWORMS. An estimated one billion people who live in the warm climates of the world are infected by hookworms. The two most important hookworms are *Ancylostoma duodenale,* the Oriental hookworm of China, Japan, Asia, and North Africa, and *Necator americana,* the American hookworm prevalent in South and Central America.

The mature female hookworm resides in the small intestine of its host and each day lays about 10,000 eggs, which pass out of the host with the feces. The eggs hatch in warm, moist soil and develop into infective larvae. The larvae penetrate the human skin and enter the blood circulation, traveling to the lungs, trachea, and down the esophagus to the small intestine, where they become sexually mature. The nematodes damage the host intestine as they feed on mucus, blood, and associated tissues. Infected persons with poor nutrition often show symptoms of anemia and iron deficiency. Epidemiologists have found that humans during the first decade of their lives steadily acquire hookworms. The level of infection remains stable from the second to the fifth decade but increases again from the sixth decade. The static

BOX ESSAY

Hormones and Animal Symbioses

Hormones play important roles in some animal symbioses, regulating metabolism, reproductive physiology, life cycles, and behavior. Both symbionts may be affected by hormones but the underlying biochemical mechanisms of these hormonal interactions are not understood. In some parasitic symbioses, gigantism or weight gain by the host follows a parasitic invasion. When snails are infected by fluke larvae, they often grow faster and their reproduction is significantly reduced. Reproductive inhibition of the host is considered to be a resource management device. Experiments have shown that infected snails have higher levels of carbohydrate and protein reserves than healthy snails. Infected snails use energy resources normally intended for egg production for increased growth. Some scientists believe that the gigantism response of infected snails represents a host adaptation for prolonged life and delayed reproduction. Other examples of host weight gain caused by parasitic infection include (1) sheep infected with the liver fluke, *Fasciola hepatica,* (2) rats parasitized by the blood flagellate *Trypanosoma lewisi,* (3) obese rats infected with the tapeworm *Spirometra mansonsoides,* and (4) rats infected with the nematode *Trichinella spiralis.*

Hormones are also involved in host castration, which is an induced alteration or destruction of reproductive tissues resulting in a failure of the host to reproduce. Host castration results in changes in the host's secondary sex characteristics or in abnormal behavior. Parasitism is highly developed among barnacles and the symbionts show extreme anatomical reductions. For example, when a free-living larva of *Sacculina* attaches to a crab host, a cluster of the parasite's cells migrates to the host's thorax and grows around the visceral tissues. Eventually, a reproductive sac appears on the outside of the host. *Sacculina* changes the hormonal physiology of their host crabs. When male crabs are parasitized, their abdomen assumes the shape of female crabs and their claws become modified as egg-bearing appendages. The ovary of the parasitized female crab is completely destroyed. The parasite obtains nutrients by means of rootlike processes that radiate from the sac into the host tissue.

Similar castration effects are also known among insects. The squash bug, *Anasa,* when infected by a parasitoid fly, *Trichopida,* suffers a degeneration of its reproductive organs. Parasitoid insects undergo their larval development in the eggs, larvae, pupae, and adults of other insects and in the process of development they consume the host tissues. A single parasitoid insect infects only one host individual. It is estimated that 12% of all insect species are parasitoids of other insects.

parasitic population during much of the adult host's life is believed to result from acquired immunity.

b. Filarial Nematodes. Filarial nematodes are obligate parasites with complex life cycles involving humans and other vertebrate and arthropod vectors. The adult female worms give birth to larvae, the *microfilariae,* that invade new tissues and develop to the third larval stage. Larvae in this stage migrate to the skin or peripheral blood circulation. A bloodsucking insect becomes infected with these larvae during a meal on an infected person. *Wuchereria bancrofti* and *Brugia malayi* occur in the human lymphatic system and cause the disease *elephantiasis.* These nematodes affect about 300 million people throughout the tropics. In Africa, the same mosquito that transmits malaria is also the vector of *W. bancrofti.* The filarial worm *Onchocerca volvulus* causes skin tumors and blindness in some 20% of the people of Africa and the

Middle East. The nematode is spread to humans by bloodsucking black flies.

The so-called heartworm, *Dirofilaria immitis*, is parasitic in the heart and pulmonary artery of dogs and other mammals. The worm is transmitted by several species of mosquitoes. The disease is produced from lesions caused by dead or dying worms.

The Guinea worm, *Dracunculus medinensis*, has been known since the early recorded history of the Middle East, India, and Africa. The adult female nematode may be over 12 meters long and it emerges from a skin eruption near the ankles to lay eggs in water. Ancient physicians learned the importance of extracting the entire worm by slowly winding it on a stick. In the Bible, there is a reference to God sending "fiery serpents" among the people of Israel if they disobeyed his commandments. Many scholars believe that the "fiery serpents" were Guinea worms. The female nematode takes over one year to reach the skin of the ankles, where it causes a blister and emerges to lay eggs. The eggs are ingested by water fleas, such as copepods, where they develop into the infective larval stage. Humans become infected with Guinea worms by drinking water that contains copepods.

FILARIAL LIFE CYCLE. A typical filarial life cycle begins when humans acquire the parasite from the bite of an infected bloodsucking insect. The parasites show a high degree of host specificity as well as tissue preference. Once in the bloodstream or lymphatic vessel, the nematode larvae become sexually mature. Adult worms of *Wuchereria* and *Onchocerca* live in the lymphatic vessels for many years; the Guinea worms live for little more than a year. The mature female gives birth to microfilariae, which are the larval stages still ensheathed in the egg membrane. Microfilariae of various nematodes have a twenty-four-hour periodicity. The number of microfilariae of *Wuchereria bancrofti* reaches a peak in the peripheral blood circulation late at night. The microfilariae are acquired by night-feeding mosquitoes along with their blood meal. In the insect host, the larvae develop in the thoracic muscles, undergo two molts, and then migrate to the mouth parts. Laboratory studies on *Litomosoides carinii*, a filarial parasite of the cotton rat, have provided fundamental insights into the host-parasite relationship of filarial worms. This nematode infection of the cotton rat is prevalent throughout the open grasslands of North and South America.

FILARIAL-HOST INTERACTIONS. The development of disease from a filarial parasite varies a great deal among individuals, for reasons that are not clear. Some humans develop acute filarial disease with recurrent episodes of fever, yet the microfilariae are not always detected. Other individuals are asymptomatic but produce microfiliae, thus contributing to the spread of the infection. Some individuals are immune to filarial infections.

Scientists generally agree that symptoms of filarial disease are the results of host immune response. Obstructive lesions are produced by a delayed hypersensitive reaction to dead or dying nematodes. Elephantiasis results from the host's immune reaction to adult worms, and the skin and eye disease of onchocerciasis is a reaction to the microfilariae. Acute disease symptoms develop following treatment with antifilarial drugs, which suggests that the sudden release of dead parasites disturbs the equilibrium between the host and parasite. A harmonious relationship results when the host's immune response to the parasitic antigens is suppressed.

It is ironic that treatment to eliminate nematode parasites from a host precipitates a disease condition that does not occur in stabilized infections. The number of filarial parasites in the lymphatic system remains constant for many years, indicating that the host somehow manages to control the intensity of infection.

c. Trichinella spiralis. *Trichinella spiralis* is the largest known intracellular symbiont of vertebrates. The nematode has been studied extensively by physicians, public health officials, experimental biologists, and ecologists. The severity of the disease *trichinosis* depends on the intensity of infection in a host. Symptoms include intestinal discomfort, diarrhea and nausea from the presence of adult nematodes in the small intestine, and edema, fever, fatigue, and muscle pain caused by the migration of larvae to muscle tissue. *Trichinella spiralis* has a wide host range among carnivorous vertebrates. Human trichinosis is associated with the consumption of undercooked pork carrying parasitic larvae. The nematode has a short life cycle, does not need an intermediate host, and is easily maintained in laboratory animals.

LIFE CYCLE AND HOST-PARASITE INTERACTION. *Trichinella* infection begins with the ingestion of meat containing the *infective first-stage larvae* (Fig. 10.5). The larvae become liberated after the meat is digested and enter their intracellular habitat, a row of columnar epithelial cells in the small intestine. A larvae may occupy more than 100 epithelial cells without damaging the host cells. The columnar cell membranes fuse with each other and form a *syncytium*. It is not known how the parasite enters the columnar cells, since the larvae do not possess any special organs for cell penetration. After a few hours the larvae undergo four molts and become adult worms. The speed of the molting is unusual and might be related to the host immune response. The adult nematode increases in size and occupies an average of over 400 columnar cells. During the molting process, the symbiont undergoes extensive morphological changes in its cuticle, hypodermal gland cells, and alimentary canal. Hypodermal glands produce secretions that are believed to possess unique antigenic properties. Little information is available on how the adult worms obtain their nutrients. *Stichosomes* are unique discoid cells that occur in the anterior half of the infective larval stage. These cells produce secretory granules that are strongly antigenic. Some scientists speculate that the host immune response is directed against the secretory granules. Stichosomes of adult worms lack secretory granules.

Sexually mature worms copulate and the females give birth to larvae, the number of which varies from 200 to 1,600, depending on the host species. The newborn larvae migrate to the lymph and some move into the general circulation. Larvae collect in the lymph while a few infect organs such as the liver and lung. Recent studies have shown that striated muscle cells are the preferred intracellular habitat for newborn larvae, which are attracted to the muscles by chemical and electrical stimuli. A larva penetrates the muscle cell by a mechanical process and then doubles its volume every 24 hours until the fourth day. The larva first lives directly in the muscle cytoplasm but later becomes surrounded by three to four membranes of unknown composition. The membranes protect the parasite from the cellular and immunological response of the host. During the following three weeks, the trichinella larva develops stichosomes and a thick cuticle, and grows to 270 times its original size. The muscle cell containing the coiled larva is transformed into a *nurse cell*. Scientists have observed similarities between regenerating muscle fibers and nurse cells. The larva alters the myofibril design of the host cell, causing proliferation of the T-tubular system and increased alkaline phosphatase activities. Changes in the T-tubular system cause the muscle cell to increase its absorptive capacity. The larva-nurse cell complex extends over the life of the host animal. Muscles are specialized cells that function within a narrow range of physical and chemical conditions. Very few parasites have successfully invaded the intracellular muscle environment. *Trichinella spiralis* survives in skeletal muscle cells by transforming the structure and biochemistry of the host cell. The nurse cells lose their ability to contract like normal muscle cells. Calcification of the nurse cell is a complex host response, often developing several years after the initial infection.

TRICHINELLA: **HOST IMMUNE RESPONSE.** *Trichinella spiralis* evokes a powerful host immune response. As a new generation of larvae becomes established in the muscles, adult worms residing in the small intestines are expelled from the host. Animals previously exposed to *T. spiralis* resist further infection and the host mother transfers the immunity to her offspring. In an immune host, *T. spiralis* grows slowly and does not reproduce. Different larval and adult stages of *T. spiralis* in the same host produce different antigens. The host immune system responds by developing antibodies against the three stages of the parasite's life cycle: the adult, infective larva, and newborn larva. Multiple antigen-antibody interactions facilitate the migration of newborn larvae from the intestine to the muscle. Immunologists have shown that although the infective larvae and the young adult share common antigens, the newborn larvae do not have any antigens that are the same as those of other stages.

Trichinella was the first helminth for which an effective vaccine was developed. The vaccine was commercially used before hygienic practices to control trichinosis among pigs became prevalent.

2. NEMATODE-INSECT SYMBIOSES. Insects are the dominant form of life on earth and nematodes have successfully evolved symbioses with many of them. There are an estimated 3,000

156

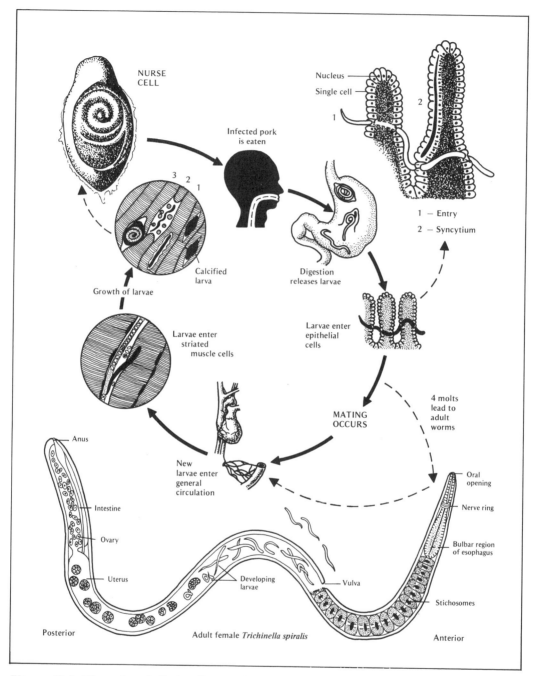

Figure 10-5. Life cycle *of* Trichinella spiralis. (Adapted from Katz, M., D. D. Despommier, and R. W. Gwadz. 1982. Parasitic diseases. Springer-Verlag, New York.)

nematode-insect associations. Nematodes of insects have intricate life cycles that are synchronized with those of their hosts. The symbionts cause changes in the insect's morphology, physiology, and behavior and reduce the reproductive potential of the host. During the past two decades there has been a renewed interest in nematode-insect symbioses because of the potential of biologically controlling insect pests by means of pathogenic symbionts.

a. **Deladenus.** The complex life cycle of *Deladenus* involves two hosts, a wood wasp and the fungus *Amylostereum.* The fungal part of the life cycle occurs in dying or dead pine trees, where the fungus grows inside galleries dug by the wasp larvae. Nematode larvae feed on the fungal hyphae and grow and molt until they become sexually mature adults. Old fungus colonies somehow trigger the parasitic behavior of the nematodes. A fertilized female nematode penetrates a wasp larva and enters its body cavity, where it increases in size and molts. The old nematode cuticle ruptures, and the hypodermal cells become transformed into absorptive cells with microvilli. When the wasp larvae pupate, the female nematode gives birth to live larvae, which migrate to the reproductive organs of the newly developed female wood wasp. The wasp disperses the nematodes by depositing them on a suitable tree along with its own eggs.

b. **Mermis nigrescens.** Mermithid nematodes are the best known obligate parasites of insects, spiders, and other invertebrates. The parasite is usually fatal to its host. The nematode larvae are found in all stages of the insect, obtaining nutrients from the host body cavity. *Mermis nigrescens* is a common parasite of grasshoppers and katydids. The female nematode is over 10 cm long and lives in the soil. After a rainfall, the nematode climbs onto vegetation and lays eggs on the foliage. The eggs remain viable for up to one year. Grasshoppers and other susceptible hosts ingest the eggs along with the foliage. The larvae hatch in the grasshopper's alimentary canal and reach the body cavity by penetrating the gut wall. The parasite increases in size from 3 mm to over 10 cm. In the last larval stage, the parasite leaves the host body and burrows in the soil. The insect host is killed during the escape process. The larvae remain dormant until the following spring, when they molt and become adults. A mature female produces over 14,000

eggs and can overwinter and resume egg laying in the following spring.

c. **Romanomermis.** The possibility of developing a biological control for mosquitoes has heightened interest in the parasitic nematode *Romanomermis,* which kills mosquito larvae (Fig. 10.6). Sexually mature adults of the nematode live for up to six months in the shallow waters of lakes, ponds, and streams. The female lays many eggs that hatch in about seven days. The larvae are strong swimmers and seek out mosquito larvae at the water surface. The nematode penetrates the cuticle of the larvae and develops in the host for five to ten days. The parasite is not carried through the pupal or adult stages of the mosquito because the infected larvae usually die or the immature nematodes leave the host. The nematode larvae return to the bottom mud, where they develop into adults and the life cycle is repeated. Once introduced into an area, the nematode becomes a permanent resident of the ecosystem. Scientists have developed methods to mass-produce the eggs of *Romanomermis* for large-scale applications.

Attempts are being made to control black flies in Africa by using parasitic nematodes that kill the insect larvae. Black flies are vectors of the filarial nematode *Onchocerca volvulus,* which causes blindness in humans.

d. **Neoaplectana.** The nematode *Neoaplectana* carries a bacterial symbiont in a specialized intestinal pouch. Nematodes are either ingested by an insect or enter the host's body through natural openings. Once inside the host alimentary canal, the nematode releases its bacterial symbionts, which move into the body cavity and multiply, causing death of the host. Nematodes develop rapidly in the dead host and become sexually mature. The females produce larvae that emerge from the host carcass carrying the bacteria and burrow into the soil, where they can survive for three years or more (Fig. 10.7). Scientists are trying to use this parasitic nematode as a way to control insect pests.

e. Fossil Symbiosis. George Poinar, an authority of nematode parasitism of insects, is a pioneer in the study of fossil symbioses. Trees such as the gum arabic tree (*Acacia arabica*) in Africa, the sandarac (*Tetraclinus articulata*) of Australia, and the algarroba (*Hymenaea courbaril*)

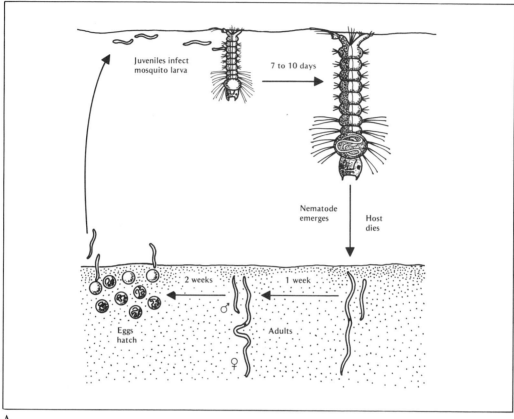

A

Figure 10-6. Romanomermis *parasitism of mosquito larvae. (A) Stages in the life cycle of* Romanomermis culicivorax. *(B) Nematodes collect at bottom of tank after emerging from mosquito larvae.*

(C) Nematodes in the thorax of mosquito larvae. (Part (A): Adapted from Nickle, W. R. (ed.). 1984. Plant and insect nematodes. Marcel Dekker, New York. Parts (B) and (C): James Peterson, USDA, Lincoln, Nebraska.)

of South America produce large volumes of resinous sap. A variety of organisms, primarily insects, becomes trapped in the resin of these trees. The organisms become entombed as the resin ages and turns amber during the course of millions of years. Electron microscopy of some amberized insects has revealed remarkable structural details of organelles such as nuclei, mitochondria, ribosomes, endoplasmic reticulum, and lipid droplets in muscle cells. Symbiotic associations have been discovered in approximately one in 100 specimens of amber fossils. Specific symbionts include entomogenous fungi, juveniles of nematodes, and parasitic mites. Amberized fossils have been found in Alaska, Canada, the Dominican Republic, Mexico, and Northern Europe and are estimated to date back 25 to 40 million years. Poinar (1982)

suggests that comparative studies of past and present symbioses should increase our understanding of the coevolutionary aspects of symbiosis. Because of the unique preservative qualities of plant resin, it may be possible to extract DNA from these fossil symbionts and conduct comparative biochemical analyses.

3. NEMATODE-PLANT SYMBIOSES. Most nematodes that attack plants are obligate parasites. The nematodes have a hollow, needlelike structure, the *stylet,* that they insert into plant tissue, like a syringe, to draw out nutrients. Most plant nematodes feed on roots but some invade parts of a plant above ground. There are over 2,000 described species of plant parasitic nematodes but only a few species are pests. Scientists are beginning to realize that serious nematode in-

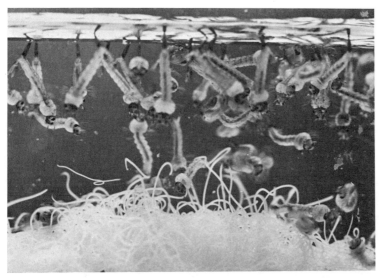

B

C

festations are the result of unsound agricultural practices such as monocropping and reliance on genetically uniform plants.

Life cycles of plant nematodes are simple, consisting of adult and juvenile stages associated with the same host. Cell proliferation, giant cell formation, suppression of cell division, and cell wall breakdown are some host responses to nematode parasitism. The infected plants are stunted and have yellow leaves, and in the case of crop plants, yields are significantly reduced. Most plant nematodes are ectoparasites, but a few species are endoparasites.

Since the end of the Second World War, soil fumigation has been the preferred method for control of nematode pests. Unfortunately, when fumigation is discontinued or suspended, the nematode populations increase to destructive proportions. Since many of the common soil fumigants are now considered to be harmful to man and other life forms, scientists are anxiously searching for alternative methods of nematode control, such as the use of biological control methods, resistant varieties of plants, crop rotation, and cultural practices that maintain the stability of the soil-microbe equilibrium.

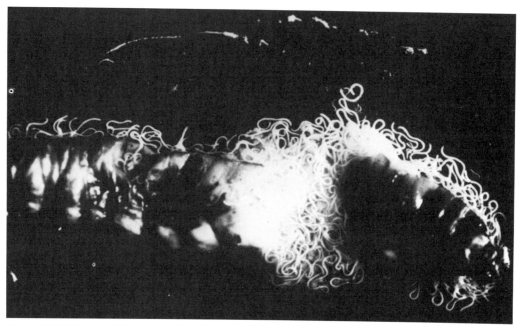

Figure 10-7. Wax moth larva parasitized by Neoaplectana carpocapsae. (W. R. Nickle, USDA, Beltsville, Maryland. Reprinted from W. R. Nickle, ed., *Plant and Insect Nematodes,* p. 643, by courtesy of Marcel Dekker, Inc.)

a. Meloidogyne: *The Root-knot Nematode.* The root-knot nematode infects a wide range of cultivated plants throughout the world. Plant nematologists have conducted extensive studies on the biology, physiology, and host-parasite interactions of the *Meloidogyne*-tomato symbiosis. Nematode juveniles penetrate the root tips of a host plant and move between the cells to the vicinity of the vascular tissues. Plant cells surrounding the nematode undergo hypertrophy and hyperplasia and form a gall. Most galls contain giant cells that are formed in response to the nematode's feeding (Fig. 10.8). The juveniles in the gall molt rapidly and become adults. The female nematode becomes enlarged and spherical and lays eggs, which are extruded in a protective gelatinous matrix. Under favorable conditions, the juveniles hatch from the eggs and infect other roots. Plants parasitized by root-knot nematodes are less resistant to infections from soil-borne fungi and bacteria and more susceptible to environmental stresses such as temperature change and drought. Yields of heavily parasitized plants are greatly reduced.

b. Heterodera *and* Globodera: *The Cyst Nematodes.* The life cycle of the cyst nematode is similar to that of the root-knot nematode. A unique feature of these parasites is that the female body containing the eggs is transformed into a resistant cyst. The infection begins when juveniles penetrate root tips of the host and migrate through the cortex. The juvenile begins the feeding process by orienting its head toward the vascular tissue. Giant cells are formed but galls do not develop. After egg production the female dies and becomes a resistant cyst, which remains in the soil for several years. The eggs in the cyst hatch when a specific stimulus is received from a host plant. Cyst nematodes cause extensive losses in crops such as sugar beet, soybean, potato, and cereals. Scientists have extensively studied the hatching factors in the hope of developing an effective cyst nematode control.

c. Anguina tritici: *The Wheat Gall Nematode.* In 1743, Turbevill Needham, an English naturalist, found support for the spontaneous generation of life by crushing wheat galls in water and observing that the contents of the gall became alive. The galls crushed by Needham contained large numbers of juveniles of *Anguina tritici* in a state of anhydrosis. Wheat galls are unknowingly sown along with wheat seed. Under warm, moist conditions, the galls break open and lib-

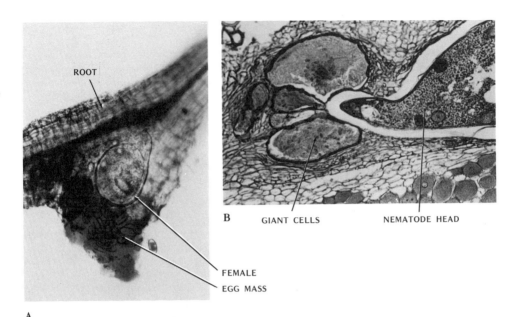

A

Figure 10-8. Plant root parasitism by the root-knot nematode Meloidogyne incognita. *(A) A root gall has formed around the female nematode. Note protruding egg mass. (B) Cross-section of an infected root, showing giant cells near the head of a feeding nematode.* (J. D. Eisenback, Virginia Polytechnic Institute and State University.)

erate juveniles, which infect nearby wheat seedlings. The juveniles first feed as ectoparasites and later penetrate the plant tissues as the wheat flowers begin to develop. After the nematodes mate, the female migrates to the ovary of the flower and lays thousands of eggs, which hatch to produce juveniles. The ovary is transformed into a resistant gall. Galls are also produced on the leaves and stems. Nematode-infected plants are usually stunted or wrinkled and have twisted leaves. *Anguina tritici* is prevalent throughout Asia, the Middle East, and Brazil. Since the seed galls float in water, they are easily separated from wheat seeds. With improved agricultural practices, the incidence of *A. tritici* is declining in many areas.

d. **Bursaphalenchus:** *Beetle-Pine Mutualistic Symbiosis. Bursaphalenchus xylophilus* is a nematode that lives in weakened or dead trees. The nematode is spread by a beetle that may carry up to 15,000 juvenile nematodes to a new location. The nematodes feed on wood tissue and are suspected of killing pine trees previously weakened by environmental stresses. Their life cycle is intertwined with that of the bark beetle. The beetle normally lays eggs under the bark, and the larvae upon hatching dig tunnels in the wood tissue, where they pupate and overwinter. As the adult beetles emerge from the pupae, the nematode juveniles develop and gather near pupal cases. The nematodes hang from the roof of the tunnel gallery by their tails. As the adult beetle emerges from the pupal case, its vigorous movements cause the nematode juveniles to attach to the beetle. The juveniles migrate swiftly to undersurfaces of the wings or to the air passages of the host. The infected beetles emerge and find other pine trees to feed on. During the feeding process, nematode juveniles leave the beetle to invade the new substrate. The beetle lays eggs in the areas softened by nematode activity. The ecological relationship between the plant-parasitic nematode and the bark beetle is thought to be mutualistic.

F. Summary and Perspectives

Helminths are metazoans with complex life cycles. Humans in many parts of the world are infected with helminthic parasites, which include flukes, tapeworms, and roundworms such as hookworms and filarial nematodes. Flukes are common in vertebrates, where they infect different organs such as the liver and urinary bladder as well as the blood. In some cases, the life cycle of a fluke is synchronized with that of its host. Schistosomes are blood flukes that cause the diseases schistosomiasis and bilharziasis. The outer body wall of a fluke, or tapeworm, is called a tegument and is a complex structure that protects the parasite from the digestive enzymes of the host and also absorbs nutrients. Some parasites disguise themselves by producing surface antigens that are similar to those of its host, a phenomenon called molecular mimicry. Tapeworms are highly specialized parasites of vertebrates and arthropods. In the tapeworm life cycle water fleas and various types of insects are common intermediate hosts. Nematodes, or roundworms, are parasites of plants, animals, and fungi. Some animals become resistant to reinfection after their initial infection by a parasite. This phenomenon is called self-cure. Hookworms infect almost a billion people in regions with warm climates. Filarial nematodes have complex life cycles that involve man and other animal vectors. Filarial diseases such as elephantiasis are the result of the host's immune responses to the parasites. The nematode *Trichinella spiralis* causes the disease trichinosis in humans and other vertebrates and results in multiple antigen-antibody interactions between host and parasite. Insects have many nematode parasites, some of which are being studied as possible biological controls of pests. Plants also are infected by different roundworms. Studies of amberized insects provide a means to compare past and present symbioses.

The use of parasites to control other parasites has had much appeal among scientists. Such biological control methods, when successful, are much better alternatives than the indiscriminate spraying of pesticides. In this respect, it is important that the life cycles of parasites be thoroughly understood in order to realize the full potential of a biological control agent.

Review Questions

1. Define *schistosomiasis*.
2. Describe the composition of the tegument of trematodes.
3. Explain the hypothesis of molecular mimicry.
4. Define *concomitant immunity*.

5. What is the function of the glycocalyx of tapeworms?
6. Name several types of tapeworms.
7. Describe the phenomenon called self-cure.
8. Describe a typical filarial life cycle.
9. How does *Trichinella spiralis* manage to survive in the skeletal muscles of its host?
10. Why has there been renewed interest in nematode-insect symbioses?
11. Name some examples of nematode-plant symbioses.

Further Reading

Damian, R. T. 1964. Molecular mimicry: Antigen sharing by parasite and host and its consequences. Am. Nat. 98:129–149.

Maggenti, A. R. 1981. Nematodes: Development as plant parasites. Ann. Rev. Microbiol. 35:135–154. (A comprehensive review on current advances in plant nematology.)

Minchella, D. J. 1985. Host life-history variation in response to parasitism. Parasitology (UK) 90:205–216.

Parkhouse, R. M. E. (ed.) 1984. Parasite evasion of the immune response. Parasitology (UK) 88:571–682. (A collection of articles on current research on the immunological basis of host-parasite symbioses.)

Poinar, G. O. 1982. Sealed in amber. Nat. Hist. 91 (June):26–32.

Taylor, A. E. R., and R. Muller (eds.). 1978. The relevance of parasitology to human welfare today. Symposia of the British Society for Parasitology, Vol. 16. Blackwell Scientific Publications, Oxford. 135 pp. (A survey of problems facing humans in medical, veterinary, and agricultural parasitology.)

Thompson, S. N. 1983. Review. Biochemical and physiological effects of metazoan endoparasites on their host species. Comp. Biochem. Physiol. 74B:183–211 (A review article on symbiotic associations of flukes, tapeworms, and nematodes.)

Wakelin, D. 1984. Immunity to parasites: How animals control parasitic infections. Edward Arnold, Baltimore. 165 pp. (An introduction to the immunological aspects of parasitism.)

Bibliography

Agabian, N., and H. Eisen (eds.) 1984. Molecular biology of host-parasite interactions. UCLA Symposia on Molecular and Cellular Biology, n.s., Vol. 13. Liss, New York.

August, J. T. (ed.). 1984. Molecular parasitology. Academic Press, New York. 293 pp.

Barriga, O. O. 1981. The immunology of parasitic infections: A handbook for physicians, veterinarians, and biologists. University Park Press, Baltimore. 354 pp.

Campbell, W. C. (ed.). 1984. *Trichinella* and trichinosis. Plenum Press, New York. 581 pp.

Crompton, D. W. T., and S. M. Joyner. 1980. Parasitic worms. Wykeham, London. 207 pp.

Katz, M., D. D. Despommier, and R. W. Gwadz. 1982. Parasitic diseases. Springer-Verlag, New York. 264 pp. (A concise treatment of parasitic diseases of humans.)

Lyon, K. M. 1979. The biology of helminth parasites. Institute of Biology, Studies in Biology, no. 102. University Park Press, Baltimore. 59 pp.

Maggenti, A. 1981. General nematology. Springer-Verlag, New York. 372 pp.

Marchalonis, J. J. (ed.). 1984. Immunobiology of parasites and parasitic infections. Contemporary topics in Immunobiology, Vol. 12. Plenum, New York. 497 pp. (A collection of articles on the use of new immunological procedures and monoclonal and hybridoma techniques to study various aspects of parasite biology including parasitic invasion, survival and escape, molecular mimicry, and host responses.)

Mettrick, D. F., and S. S. Desser. 1982. Parasites: Their world and ours. Proceedings of the Fifth International Congress of Parasitology. Elsevier Biomedical Press, Amsterdam. 465 pp.

Mueller, J. F. 1963. Parasite-induced weight gain in mice. Ann. N.Y. Acad. Sci. 113:217–233.

Needham, T. 1743. Concerning certain chalky tubulous concretions, callec malm; with some microscopical observations on the farina of red lily, and of worms discovered in smutty corn. Phil. Trans. Soc. Lond. 42:634–641.

Nickle, W. R. (ed.). 1984. Plant and insect nematodes. Marcel Dekker, New York. 925 pp.

Nobel, E. R., and G. A. Nobel. 1982. Parasitology: The biology of animal parasites. 5th ed. Lea and Febiger, Philadelphia. 522 pp.

Read, C. P. 1970. Parasitism and symbiology. Ronald Press, New York. 316 pp. (A pioneer effort to integrate parasitology with other interorganismic associations.)

Schmidt, G. D., and L. S. Roberts. 1984. Foundations of parasitology. C. V. Mosby, St. Louis. 775 pp.

Smyth, J. D. 1976. Introduction to animal parasitology. 2d ed. Halsted Press, New York. 466 pp.

Taylor, R. A., and R. Muller (eds.). 1977. Parasite invasion. Symposia of the British Society for Parasitology, Vol. 15. Blackwell Scientific Publications, Oxford. 155 pp.

Warren, K. S., and J. Z. Bowers (eds.). Parasitology: A global perspective. Springer-Verlag, New York. 292 pp.

Whitfield, P. J. 1979. The biology of parasitism: An introduction to the study of associating organisms. University Park Press, Baltimore. 277 pp. (Covers a wide range of associations; the book follows traditions established by Read's book on parasitism and symbiology.)

CHAPTER 11

Plant Symbiotic Associations

A. Parasitic Plants

Plants that parasitize other plants have been recognized for many years. The parasitic habit has developed independently in at least eleven unrelated families of dicotyledons. In some families, all the plants are parasitic; in others only one genus is parasitic. How and why this habit arose in plants, particularly in families whose other members are autotrophic, is not clear. Parasitic plants are not found in the ferns, gymnosperms, or monocotyledons. A classification of parasitic plants is given below.

There are about 3,000 species of parasitic plants, ranging from trees to small, herbaceous plants. *Holoparasites* obtain all of their nutrients from other plants, whereas *hemiparasites* still retain photosynthetic leaves and can therefore manufacture food.

1. MISTLETOES. The largest and best known group of parasitic plants are the mistletoes (Fig. 11.1). There are about 800 species and most of them occur in the tropics and subtropics. There are two families of mistletoes, the Loranthaceae and the Viscaceae. Flowers of the Loranthaceae are large and brightly colored; those of the Viscaceae are small and pale. Most mistletoes attack a wide range of host plants. For example, the tropical mistletoe *Dendrophthoe falcata* is known to attack about 350 host species that belong to unrelated families. *Viscum album,* a common mistletoe in temperate regions, also has a wide range of hosts and is an important pest in orchards and forests. A few mistletoes are specific to only one host; for example, the dwarf mistletoe *Arceuthobium douglasi* attacks only Douglas fir trees. Mistletoes can even parasitize themselves.

Most mistletoes parasitize tree branches but some, such as the giant mistletoe *Nuytsia floribunda,* have a terrestrial habit. This mistletoe forms a tree that grows as high as 10 meters and parasitizes roots of nearby grasses and other plants. It is common in West Australia and is called the Christmas tree because of the bright flowers it produces in December. The Brazilian

KINGDOM PLANTAE

Phylum:	Anthophyta (angiosperms)
Class:	Dicotyledonae (dicots)
Families:	Convolvulaceae (morning-glory family)
	Cuscuta
	Hydnoraceae (tartous family)
	Hydnora
	Loranthaceae (mistletoe family)
	Arceuthobium
	Nuytsia
	Phrygilanthus
	Orobanchaceae (broomrape family)
	Conopholis
	Epifagus
	Orobanche
	Rafflesiaceae (rafflesia family)
	Rafflesia
	Santalaceae (sandalwood family)
	Phacellaria
	Santalum
	Scrophulariaceae (figwort family)
	Alectra
	Latraea
	Striga
	Tozzia
	Viscaceae (mistletoe family)
	Dendrophthoe
	Viscum

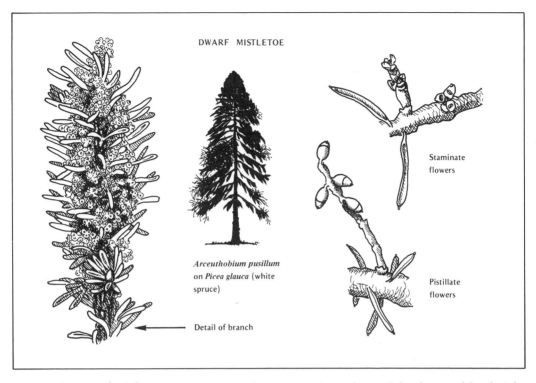

DWARF MISTLETOE

Arceuthobium pusillum
on *Picea glauca* (white
spruce)

Detail of branch

Staminate
flowers

Pistillate
flowers

Figure 11-1. Dwarf mistletoe, a common parasite of white spruce throughout the northeastern United States and Canada. Note small staminate and pistillate flowers. (Adapted from Hawksworth, F. G., and D. Wiens. 1972. Biology and classification of dwarf mistletoes (*Arceuthobium*). Agriculture Handbook, no. 401, Forest Service, US Dept. of Agriculture.)

mistletoe, *Phrygilanthus acutifolius,* is a vine and it parasitizes several different trees simultaneously by means of a large network of underground roots.

Mistletoes that grow on trees and shrubs have roots that grow along the host's branches and develop haustoria, specialized structures through which water and nutrients pass from host to parasite. The roots are not influenced by gravity and grow up and down the branches. In some mistletoes, new shoots develop from these roots and allow the parasite to compete for light with the developing canopy of the host plant.

Many mistletoes are dispersed by birds, which eat the fruits and void the seeds onto tree branches. The fruit of a mistletoe is a one-seeded berry. Some mistletoes have fruits that explode and disperse their seeds, like the dwarf mistletoe, whose seeds travel up to 15 meters and stick to any surface they contact. The dwarf mistletoe is a serious pathogen of forest trees in the western United States because of the ease with which it spreads from tree to tree.

When a mistletoe seed germinates, the emerging root forms a disclike structure that presses against the surface of the host. A column of cells arises from inside the disc, penetrates the host tissue by means of mechanical and enzymatic action, and forms a haustorium. The tissues of the haustorium fuse with the host's vascular tissue and form a passageway through which water and minerals move. A few mistletoes fuse with the host phloem tissue and in this way obtain photosynthetic products from the host. Actual penetration of a host branch by a haustorium appears to be a rapid process that follows a series of slow, preparatory stages. If a haustorium fails to penetrate the branch, the parasite forms another disclike structure from which a new haustorium develops. The parasite may make numerous attempts to penetrate the host.

In many mistletoes the haustorium stimulates

the host tissue to divide and a structure called a *placenta* forms along the haustorial surface. The shape of the placenta varies according to the species of mistletoe. Some species produce an elaborately furrowed placenta that is commonly called woodrose and is prized in Mexico for its beauty.

2. SANDALWOODS. Sandalwoods are similar to mistletoes in that they are hemiparasitic, mostly tropical plants, and may be trees, shrubs, or herbs. Most sandalwoods are root parasites but some attack tree branches. The sandalwood tree, *Santalum album,* has a fragrant wood that is used to make cabinets and carvings, and also has an oil that is used in the manufacture of perfumes and soaps. *Phacellaria* is a sandalwood that parasitizes mistletoes and other sandalwoods. The parasite grows inside the host tissues and produces small external, leafy branches that emerge from swollen nodes of the host.

3. DODDERS. Dodders are holoparasites and they consist only of twining stems and scalelike leaves. They belong to the genus *Cuscuta* and are the only parasitic members of the Convulaceae, or morning-glory family. Dodders occur throughout the world and attack a variety of hosts. When a dodder seed germinates, it gives rise to a seedling that must find a suitable host within about seven weeks or it will die. The seedling first grows upright and then in a spiral fashion until it contacts a suitable host. After contact, the dodder forms haustoria that grow toward the vascular cells of the host. When xylem cells of the host are reached, the haustorial filaments differentiate into tracheids or vessels. After the haustorial connections to the host have been established, the small roots of the dodder die and the parasite loses all connection to the earth. The tips of the dodder are phototrophic, so the parasite always grows upward along the host plant. Late in the growing season the dodder flowers and sets seeds to repeat the cycle.

4. BROOMRAPES AND FIGWORTS. Broomrapes and parasitic figworts are mostly herbaceous root parasites of temperate regions. Broomrapes are holoparasites and belong to the family Orobanchaceae; figworts are hemiparasites and are in the family Scrophulariaceae. In broomrapes and in some advanced parasitic fig-

worts, such as *Alectra, Latraea, Striga,* and *Tozzia,* the seeds germinate only in the presence of exudates from the roots of specific host plants. The exudates cause the parasite to grow toward the host and also to develop thicker roots with more root hairs. *Striga* (witchweed) species are obligate parasites of cereal crops, such as sorghum and millet, in Africa and India. Seeds of these parasites remain viable in the soil for many years and germinate only when near a suitable host. *Tozzia* lives underground much of the time, takes several years to mature, and then produces a green flowering shoot above ground. After flowering and releasing seeds, the plant dies. Common broomrapes include beechdrops (*Epifagus virginiana*), found only on roots of the American beech, squawroot (*Conopholis americana*), a parasite of the roots of red oaks, and cancer-root (*Orobanche uniflora*), a root parasite of a variety of plants.

5. *HYDNORA.* *Hydnora* is a genus of subterranean, parasitic plants and a member of a small, unique family, the Hydnoraceae, all of whose members are root parasites. The plants generally grow in remote areas of South Africa and in East African countries. *Hydnora* produces thick, coarse roots that are bright red inside and grow through the soil, and small, thin roots that parasitize the host roots. The coarse roots produce white, fleshy flowers that have a foul odor and are pollinated by beetles. *Hydnora* fruits are large and fleshy and mature underground. They have a fruity odor and are eaten by man and other animals. The dried and powdered roots of *Hydnora abyssinica,* called tartous, are used by African natives to treat diarrhea and other sicknesses. Seeds of the parasite germinate only in the presence of exudates from the host roots.

6. *RAFFLESIA.* The most exotic and spectacular parasitic plant is *Rafflesia arnoldii,* a native of the jungles of Sumatra. The plant is a holoparasite on the roots of plants and produces the largest known flower, about one meter in diameter, in the plant kingdom (Fig. 11.2). The parasite lacks stems and leaves and consists only of thin, filamentous haustoria that are so fine they have been confused with fungal hyphae. The flower of *Rafflesia* is fleshy, purplish brown with white patches, and has highly fused sexual parts. The inner parts of the petals are raised to form a circular rim or diaphragm. The flower

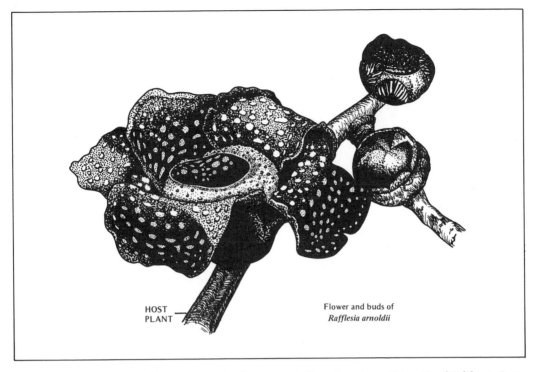

HOST
PLANT

Flower and buds of
Rafflesia arnoldii

Figure 11-2. Rafflesia arnoldii, *a parasitic plant, produces the largest known flower in the plant kingdom.* (Adapted from Kuijt, J. 1969. The biology of parasitic flowering plants. University of California Press, Berkeley.)

smells like carrion and is pollinated by flies. *Rafflesia* belongs to a small family, Rafflesiaceae, which consists of about 8 genera and 30 species of tropical plants. The fruits are large and fleshy with small seeds approximately one millimeter in diameter. How the seeds are dispersed and the conditions under which they germinate are not known. The family parasitizes only tropical members of the Vitaceae, or grape family, which form woody vines.

B. Plants and Pollinators

1. COEVOLUTION OF INSECTS AND PLANTS.

Coevolution of insects and plants over the past 200 million years has resulted in relationships that have influenced the development of both groups of organisms. An example of such a relationship is pollination in flowering plants. A primary reason why flowering plants, or angiosperms, are the dominant group of plants is their exploitation of insects as agents of pollination. Insects transport pollen from one plant to another while plants provide insects with a food source such as nectar or pollen (Fig. 11.3).

Pollination is the transfer of pollen from an anther to a stigma or similar receptive surface. If the transfer occurs between anther and stigma of the same flower, the process is known as *self-pollination*. If the transfer occurs between flowers of different plants of the same species, the process is called *cross-pollination*. Evolutionary adaptations of flowers promote cross-pollination, which results in greater hybridization and, subsequently, greater variation on which natural selection can operate. The mechanisms that plants have evolved to prevent self-pollination and to facilitate cross-pollination are many and varied. They include different maturation times for anther and stigma of the same flower, morphological adaptations of the flower, such as different style lengths in flowers of individual plants, and a self-incompatibility that is genetically determined.

Flowers of the orchid family are highly adapted to promote cross-pollination. Such ad-

Figure 11-3. Honey bee visiting flowers of heath (Erica mediterranea) *in search of nectar.* (Michael Proctor, University of Exeter.)

duced by plants to ward off predators have resulted in some insects being resistant to these compounds or even using them for their own defense against other animals. For example, larvae of the monarch butterfly feed on milkweed plants and ingest a latex from the plant that is poisonous to other insects but tolerated by the larvae. Adult monarch butterflies retain this poison and, as a consequence, are not eaten by birds.

2. FLORAL FEATURES THAT ATTRACT POLLINATORS. Insects visit flowers for several reasons, the primary one being that flowers provide a source of food, that is, pollen and nectar. Pollen is high in proteins, fats, and sugars, and probably was the original food source for many insects. Most flowers produce sufficient pollen to compensate for that eaten by insects.

When insects visit a flower, they generally become covered with pollen grains. Since they cannot clean all the grains from their bodies, they transport the pollen to the next flower they visit and bring about pollination. The anthers of most plants release their pollen through longitudinal slits, but in some plants pollen is released through terminal pores. In these flowers, bees grasp the stamens and by moving their body they cause the anthers to vibrate and release their pollen. This phenomenon is called *buzz-pollination.* Pollen released from anthers may be attracted to an insect because of electrostatic conditions around the insect and flower. Bees develop positive charges when flying to a flower, which has a negative charge, and when a bee lands on a flower the charges are dispelled.

Nectar is a sugar solution consisting of sucrose and its breakdown products glucose and fructose. The nectar is secreted from small strands of phloem that terminate in specialized nectar glands, called nectaries, located at the base of the pistil, or in protected areas such as tubes or spurs. The sugar concentration of the nectar of different plants as well as the amount of nectar produced in a flower varies and, in general, is related to the energetic needs of the pollinators. For example, flowers pollinated by birds produce more nectar than flowers pollinated by insects. Rare plants and those with specialized pollinators have flowers that produce nectar with a high sugar concentration. For example, some gentians have closed flowers

aptations have resulted in thousands of orchid hybrids that constitute one of the largest families of flowering plants.

Some insect-plant relationships are so highly evolved that neither symbiont can reproduce independently of the other. Examples of this type of relationship are the yucca and the pronuba moth, and the fig and its attendant wasp. Many plants have flowers whose morphological features are adaptations to specific types of animal pollinators. Animals have also adapted to specific types of flowers.

Many plants and animals have coevolved biochemically. Protective chemical compounds pro-

and are pollinated by large bumblebees. The insect forces open the petals in order to enter the flower. Because of the work involved in opening the flower and also the rarity of the plant, the number of flowers that a bee can visit is limited. Thus, in this insect-plant relationship, flowers with a concentrated nectar are favored over those with a less-concentrated nectar.

Flowers that bloom at night produce nectar only at night, and day-blooming flowers produce nectar only in the morning. In addition to nectar and pollen, flowers may have oil-secreting glands. The oil is gathered by pollinating insects and mixed with the pollen they collect.

Some plants produce nectaries on their stems and leaves and these glands support communities of ants or wasps that protect the plants from insect predators. Some tropical trees are called *ant trees* because they have hollow areas on their stems that house their ant protectors (Fig. 13.2). The ants feed on glycogen-rich nodules produced by the plant at the base of the leaf stalks and in return protect the plant from insect pests. The ants also remove any small, foreign plants that grow on the host plant.

Many insects use flowers as a place to lay eggs or hide from predators. The classic relationship between the *Yucca* plant and the yucca moth (*Tegeticula yuccasella*) and the fig plant and its wasp illustrates the intimate nature of some insect-plant symbioses. In both relationships, the symbionts have developed traits that are a result of their coevolution. The flowers are pollinated by the insects, which lay their eggs in the flowers. Larvae hatch from the eggs and use some of the developing seeds for food. Each relationship merits a full description.

The life cycles of the *Yucca* plant and moth are synchronized in several respects. For example, when *Yucca* flowers blossom the female moths are ready to lay their eggs, and when seeds of the flower mature the insect larvae hatch from the eggs. The female moth visits the white flowers of *Yucca* at night and by means of specialized mouth parts gathers the sticky pollen from the anther, rolls the pollen into a ball, and carries it to another flower. In the second flower the moth lays its eggs in the ovary, among the ovules or immature seeds. Then, in what seems to be almost purposeful behavior, the moth carries the ball of pollen it took from the first flower and deposits it on the stigma of the second flower, thus causing pollination and

ensuring that the ovules develop into seeds. The larvae eat only about a third of the seeds in the ovary. When the larvae are mature, they bore out of the ovary and drop to the ground, where they pupate until the next flowering season, when the adult moths emerge again.

The relationship between figs and fig wasps is similar to that of the *Yucca* and its moth in the coordinated life cycles of the two symbionts and the "purposeful" behavior of the female insects. Pollination of the fig plant is one of the most sophisticated examples of plant-insect symbioses known to scientists (Fig. 11.4). The symbiosis is an obligate mutualism between a gall wasp, *Ceratosolen arabicus,* and the sycamore fig, *Ficus sycamorus.* Fig trees have been propagated from stem cuttings for centuries throughout the mediterranean region, but in East Africa they grow wild, reproducing sexually and producing fruits throughout the year. The fig 'fruit' is actually an inflorescence termed a *syconium,* with its margins folded to a pear-shaped structure that has a small opening, the *ostiolum,* which is lined with hairlike scales. The inner fleshy surface of the inflorescence contains numerous small flowers; those close to the ostiolum are male flowers while deep within the interior are female flowers. When fertilized each female flower produces a hard, grainy seed. The female gall wasp with its ovipositor deposits an egg into the ovary of a female fig flower and induces the ovary to form a tumorlike gall, which encloses the insect egg. Female wasps enter the fig fruit through the ostiolum and frequently lose their wings, antennae, and other body parts because of the dense scales around the opening. Wasps that pass through the ostiolum are attracted to the female flowers, of which there are two types, those with long styles (1.5 mm) and those with short styles (0.8 mm). Long-styled flowers are sessile, and short-styled flowers are borne on stalks. The wasp ovipositor is about 0.8 mm long and can reach the ovary of a short-styled flower, into which it deposits an egg. On a long-styled flower, however, the female wasp inserts its ovipositor through the style, but because it cannot reach the ovary it is quickly withdrawn. Nonetheless, the wasp still places pollen collected from other flowers onto the stigma. The female wasps die shortly after they have deposited their eggs. Larvae emerge from the eggs a few days after they have been laid and feed on the gall tissue. Larvae then

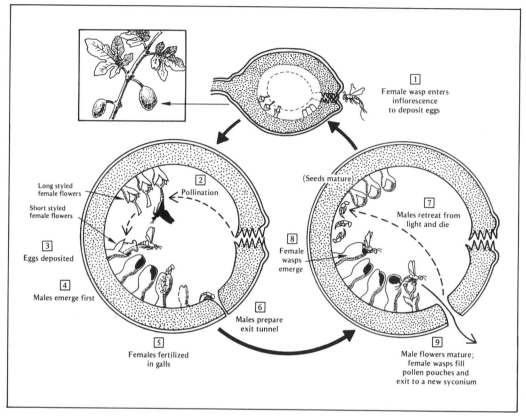

Figure 11-4. Diagrammatic representation of the mutualistic symbiosis between the sycamore fig and the gall wasp. (Adapted from Meeuse, B., and S. Morris. 1984. The sex life of flowers. Facts on File Publications, New York.)

transform into pupae from which adult wasps with strong mandibles develop and eat their way through the gall. Male wasps develop first and search for galls containing female wasps. The males puncture the galls and fertilize the females by inserting the tips of their abdomens into the gall. After they are fertilized, the female wasps emerge from the galls. The wasps then gather near the ostiolum and begin to tunnel through the syconium. Because of their sensitivity to light, the male wasps withdraw into the interior of the fig fruit and eventually die.

As the tunneling activities near completion, the male flowers mature and produce pollen. Before the female wasps leave the syconium, they fill specialized pouches on their thorax with pollen from anthers of male flowers. On emerging from a fig fruit the females search out new syconia and the entire cycle is repeated.

The symbiosis between the fig plant and wasp is obligatory and mutualistic. Neither symbiont can reproduce sexually in the absence of the other.

Insects sometimes are attracted to flowers for reasons other than food and shelter. Orchids of the genus Ophrys produce flowers that so closely resemble female bees or wasps, in morphology and fragrance, that the male insects attempt to mate with the flowers and in doing so transfer pollen from one flower to another. Similarly, flowers of the orchid Oncidium resemble enemies of some insects and when the male insects attempt to battle the flowers they bring about pollination. These forms of floral deception also occur in plants other than orchids.

Pollinators are also attracted to flowers because of their odor, color, shape, and nectar guides. Color is an important attractant over long distances, that is, greater than one meter, whereas odor is more effective at shorter distances. The various features of a flower are generally closely related to the characteristics and behavior of its pollinators. Flowers can be grouped according to the types of insects that pollinate them.

3. TYPES OF FLOWERS

a. Insect-Pollinated Flowers. Bees are the most common insect pollinators and they are also important pollinators of crop plants such as fruit trees. Bee-pollinated flowers are generally brightly colored and often have a sweet fragrance. The classic studies of the Austrian scientist Karl von Frisch, in the 1920s, showed that bees can discriminate between colors and scents. Von Frisch found that bees can distinguish the colors blue and yellow, a finding that correlates well with the fact that most bee-pollinated flowers are of those colors. Bees also detect ultraviolet light and many flowers have pigments near their nectar glands that absorb ultraviolet light and create patterns directing pollinators to the stored nectar. Bees cannot discern red flowers, except from their ultraviolet markings. The pleasant odors of bee-pollinated flowers are usually the result of essential oils, such as peppermint and lavender oils.

Nectar guides are color markings or a pattern of raised ridges on a flower that signal to an insect where the nectar is stored. In violets and orchids, the nectar guides are streaks of color on the lower petals. Bee-pollinated flowers usually have modified petals that act as platforms to allow insects to alight on the flower. Bees tend to concentrate their visits on one type of flower at a time.

Beetle-pollinated flowers are usually white or dull colored and smell like fruit or carrion. Beetles have poor vision but a keen sense of smell. They eat different parts of a flower and the flowers they normally pollinate have their ovules embedded in protective tissues away from the regions where the beetles forage.

Moth and butterfly pollinated flowers usually produce their nectar at the base of long petal tubes or spurs. The nectar is out of the reach of many insects but available to the long tongues of butterflies and moths. Some butterflies pollinate red flowers in addition to blue and yellow flowers. Since moths feed only at night, the flowers they pollinate are white or yellow and their fragrance develops strongly at night.

b. Bird and Bat Pollinated Flowers. Bird-pollinated flowers are large, usually red or yellow,

and odorless. Large flowers are necessary in order to hold the amounts of nectar birds require. Birds do not have a good sense of smell, but their vision is well developed and they can see brightly colored flowers. Hummingbirds are the most common bird pollinators in North and South America and nectar is their primary energy source. The nectar is produced in floral tubes away from the reach of smaller insects. Hummingbirds do not land on a flower but hover before it when they are drinking nectar. Many tropical plants such as hibiscus and bird of paradise have flowers that are pollinated by birds.

Bat-pollinated flowers such as those of some tropical cacti open only at night. They have dull colors and fruity odors, and are large enough to withstand the bat's foraging. As the bats drink the nectar and eat some of the floral parts, pollen collects on their fur and is thereby transferred to other flowers.

4. FLORAL CHANGES AFTER POLLINATION.

Flowers change in different ways after pollination. The changes are signals to animals that food is no longer available in these flowers and, as a consequence, the flowers are not visited by pollinators. Several hypotheses have been proposed to explain the adaptive significance of postpollination floral changes. One hypothesis is that after a certain number of insect visits to a flower, further visits may remove pollen already deposited on the stigma or damage the reproductive parts of the flower. A second hypothesis states that it is more efficient, in terms of energy expenditure of the pollinators and seed set of the flowers, for the pollinators to limit their visits to flowers that have not reached their maximum number of pollinator visits. Pollen carried by an insect to a flower that had "gone by" would be wasted and, similarly, pollen picked up from such a flower would, most likely, not be functional.

Floral changes include changes in color and orientation of the flower, the loss of odor and nectar production, the collapse of the flower, and dropping of petals. Color changes cause the flower to become less conspicuous to its pollinators. For example, a white flower that blooms at night and attracts moths becomes darker after it is pollinated and no longer contrasts well with the darkness. Similarly, the colors of nectar guides become duller and fragrant flowers lose their odor and stop producing nectar. Moth-pollinated flowers change their position after pollination and thus make it impossible for the moths to fit their proboscises into the usual channels in the flower.

C. Summary and Perspectives

Parasitic plants occur in many different families of angiosperms, especially in tropical and subtropical regions, and range in size from trees to small herbs. Mistletoes are the largest group of parasites with about 800 species. Most mistletoes have a wide host range, parasitizing tree branches by means of haustoria that fuse with the vascular tissue of the host and withdraw water and nutrients. Sandalwoods are mostly root parasites, and dodders consist mainly of thin stems that wind around the host. Broomrapes and figworts are root parasites that grow in temperate regions. Their seeds germinate only in the presence of exudates from the host plant. *Hydnora* is a root parasite that grows and flowers underground. *Rafflesia arnoldii* parasitizes the roots of other plants and produces the largest flower among the angiosperms.

Insects and flowers have coevolved for millions of years and pollination is the result of such a coevolution. Adaptations of flowers have been toward changes that promote cross-pollination, since this leads to greater hybridization and variation among the offspring. A few plant-insect relationships such as that between the *Yucca* plant and a moth and the fig plant and a wasp are so highly evolved that neither symbiont can complete its life cycle in the absence of the other one. Some plants and animals have undergone a biochemical coevolution.

Flowers attract insects with pollen and nectar as sources of food. Buzz-pollination is a means by which bees cause anthers to release their pollen. The sugar concentration of different plant nectars varies and is related to the energetic needs of the pollinators. Birds require more nectar than insects. Insects are also attracted to flowers because of their color, shape, odor, and nectar guides. Some orchids produce flowers that resemble female bees or enemies of the pollinator and this attracts male bees, which try to mate or fight with the flowers and in the process bring about pollination.

Flowers have features that relate to the behavior of their pollinators. Bee-pollinated flowers have bright colors, a sweet smell, and nectar guides, whereas flowers pollinated by beetles are white and have a fruity or carrion smell. Bird-pollinated flowers are large and odorless, those pollinated by bats open at night and have dull colors and a fruity fragrance. After pollination a flower undergoes changes that serve as signals to pollinators that food is no longer available at the flower.

The parasitic habit is found in all kingdoms, including plants. It may seem surprising that organisms that can manufacture their own food through photosynthesis should revert to parasitism and take nutrients and water from other organisms. Why this habit has evolved among separate lines of plants is unknown. In the first chapter we indicated that over a long period of time some parasitic associations have evolved into mutualistic ones. Whether this will occur among plant parasites in the future is not clear. At present, there is no evidence to suggest that plant parasites might confer a benefit on their hosts. Because the parasitism is between members of the same taxonomic group, angiosperms, the defenses of the host plant may not be sufficiently strong to slow or stop the growth of the parasite.

The relationships between flowers and their pollinators are of great interest to ecologists and evolutionary biologists, who are using pollination systems to test various hypotheses related to general ecological and evolutionary principles. The process of coevolution and the adaptive response of plants and animals to each other are seen clearly among a myriad of examples of flowering plants and their pollinators. There appears to be a continuum of relationships from plants whose flowers are accessible to many different insect pollinators to the extreme examples of the yucca and fig plants that depend on only one type of insect for pollination. Pollination biology has changed dramatically during the past fifteen years. It has gone from a purely descriptive subject to one in which computer and mathematical models are being used to simulate gametic competition in flowers and various aspects of plant-pollinator relationships.

Review Questions

1. Distinguish between holoparasites and hemiparasites.
2. Describe how a mistletoe parasitizes its host.
3. Describe how a dodder seedling finds a host.

4. Under what conditions will the seeds of broomrapes germinate?
5. Why is *Hydnora* a unique type of parasitic plant?
6. Which parasitic plant has the largest flower in the plant kingdom?
7. Define *buzz-pollination.*
8. What is the role of nectar in insect pollination?
9. Describe the relationship between the *Yucca* plant and yucca moth.
10. Describe the different types of flowers in the fig syconium and the purpose of each type.
11. Discuss how the shape, size, and color of flowers relate to their animal pollinators.
12. List some ways in which flowers change after they have been pollinated.

Further Reading

Atsatt, P., and D. O'Dowd. 1978. Mutual aid among plants. Horticulture 56 (April): 22–31. (Popular article on how plants protect themselves.)

Barth, F. G. 1985. Insects and flowers: The biology of a partnership. Princeton Univ. Press, Princeton, N.J. 297 pp. (A beautifully illustrated book about insects and pollination.)

Dafni, A. 1984. Mimicry and deception in pollination. Ann. Rev. Ecol. Syst. 15:259–278.

Faegri, K. 1971. The principles of pollination ecology. 2d ed. Pergamon Press, Oxford. 291 pp. (Considers the role of pollination in plant and animal ecology.)

Kuijt, J. 1969. The biology of parasitic flowering plants. Univ. of California Press, Berkeley. 246 pp. (The standard text for parasitic plants. Contains information on all groups of parasitic vascular plants.)

Meeuse, B., and S. Morris. 1984. The sex life of flowers. Facts on File Publications, New York. 152 pp. (Describes coadaptations between plants and their pollinators.)

Moffat, A. 1978. How plants make war. Horticulture 56 (February): 38–47.

Proctor, M., and P. Yeo. 1972. The pollination of flowers. Taplinger, New York. 418 pp. (A profusely illustrated and comprehensive treatment of pollination biology.)

Real, L. (ed.). 1983. Pollination biology. Academic Press, New York. 338 pp. (Presents modern developments in pollination biology.)

Bibliography

Calder, M., and P. Bernhardt (eds.). 1983. The biology of mistletoes. Academic Press, Sydney, New York. 348 pp.

Dawson, J. H., F. M. Ashton, W. V. Walker, J. R. Frank, and G. A. Buchanan. 1984. Dodder and its control. U.S. Department of Agriculture, Farmers Bulletin 2276.

Frisch, K. von. 1950. Bees, their vision, chemical senses and language. Cornell Univ. Press, Ithaca, New York. 119 pp.

Haustorium. Parasitic plants newsletter. 1979–1985. Nos. 1–14. Edited by J. L. Musselman. Department of Biological Sciences, Old Dominion Univ., Norfolk, VA 23508.

Hawksworth, F. G., and D. Wiens. 1972. Biology and classification of dwarf mistletoes (*Arceuthobium*). Argriculture Handbook no. 401. Forest Service, U.S. Department of Agriculture. 234 pp.

Jones, C. E., and R. J. Little. 1983. Handbook of experimental pollination biology. Scientific and Academic Editions, New York. 558 pp.

Musselman, L. J., and W. F. Mann, Jr. 1978. Root parasites of southern forests. U.S. Department of Agriculture, Forest Service, General Technical Report SO-20.

Parker, C., L. J. Musselman, R. M. Polhill, and A. K. Wilson (eds.). 1984. Proceedings of the Third International Symposium on Parasitic Weeds. International Center for Agricultural Research in the Dry Areas, Aleppo, Syria. 265 pp.

Visser, J. H. 1981. South African parasitic flowering plants. Jutta, Cape Town. 177 pp. (A well-illustrated account of a wide variety of parasitic angiosperms.)

CHAPTER 12

Behavioral and Social Symbioses

A. Introduction

Charles Darwin's book *The Expression of the Emotions in Man and Animals* (1872) was a turning point in the study of behavior. For the first time, behavior was considered from an experimental view and it was recognized that adaptive behavior of an organism fosters its survival and reproduction. Today, behavioral scientists study four aspects of behavior: (1) causation, (2) development, (3) evolutionary history, and (4) function. Observation and description are central to behavioral studies, but the problems involved in carrying out such studies are great. For example, observations cannot be made continuously during the lifetime of an organism under study. Further, behavior is complex and a scientist can study only one aspect of it at a time. Therefore, personal biases of a scientist become an important consideration in behavioral analysis. Causation is studied by examining the factors that influence behavior over relatively short periods of time in the life of an organism. These factors include environmental as well as sensory and physiological stimuli. Regulation of behavior in an interorganismic association involves internal as well as external factors. Changes in hormone levels or brain abnormalities may result in behavioral modifications of the host. Similarly, a symbiont's behavior may be influenced by developmental signals from the host.

The development of behavior results from the interplay between the genetic and environmental components of an organism's life cycle. Many behavioral patterns deteriorate or are modified because of symbiotic associations. Scientists also study the role of natural selection in the evolution of behavior within a species. Recent advances in sociobiology have brought about new insights into the evolution of social behavior. *Genetic fitness* and *natural selection* are two central tenets of sociobiology. Genetic fitness is measured as a contribution of a particular genotype to the next generation in a population. Parasitic symbioses, infectious diseases, pathogenicity, and host resistance are important factors that increase the genetic fitness of the associating species. Natural selection always maximizes fitness by increasing the chance for survival and reproduction of a species.

The functional aspect of behavior relates to an organism's adaptiveness to its environment. The processes of natural selection mold behavior to a particular environment. Examples of specific behavioral patterns include (1) migration patterns among birds, fishes, and mammals, (2) food selection among specialized feeders, and (3) symbiotic associations between different species.

Although behavioral patterns can be recognized in most symbioses, including those involving bacterial and fungal symbionts, the best-known patterns are those of animals. In this chapter, we consider cleaning symbioses of marine organisms, symbioses in which parasites modify host behavior, and symbioses of social birds and insects.

B. Behavioral Symbioses

1. CLEANING SYMBIOSES. Many marine fishes are cleaned regularly of ectoparasites and diseased or damaged tissues by specialized fish or shrimp called cleaners. The cleaners provide a valuable service by keeping fish free of parasites and disease, and, in turn, they acquire food and protection from predators. The behavioral patterns associated with these symbioses are remarkable and include the posing by host fishes

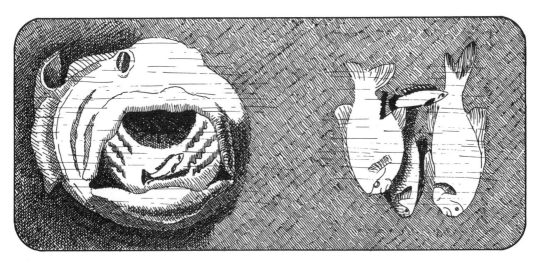

Figure 12-1. *Cleaning symbioses of marine fishes involve removal of ectoparasites by small fish called* cleaners. (Adapted from Limbaugh, C. 1961. Cleaning symbiosis. Sci. Am. 205 (August): 42–49.)

in order to expose to the cleaners that part of their body which needs attention. The posing is a means of communication between the host and the cleaner. Host poses include opening mouths and gills to allow the cleaner to enter, or assuming an unnatural vertical position (Fig. 12.1).

Cleaning symbioses have been observed in many parts of the world but are most common in tropical waters. The fish involved in this symbiosis first establish cleaning stations near prominent parts of the ocean floor, along the margins of kelp beds, or even in ship wreckage. Some stations remain constant for several years, and others are used only for a short time. Tropical cleaners have bright colors in patterns that contrast sharply with their background. The colors, along with ritualistic displays put on by the fish, attract host fish to the cleaning stations. Tropical cleaners work alone or in pairs. In contrast, cold water cleaners have dull colors and live in schools. About 50 species of fish are known to be cleaners but there are undoubtedly more. These fish generally live along the coast or in coral reefs. Deep ocean fishes, such as tuna, may also have cleaners but this has not been firmly established. Even marine turtles, iguanas, and sea urchins have a cleaning service.

Parasites that are removed from the host include small crustaceans such as copepods and isopods, as well as mats of bacteria and fungi that grow from infected host tissue. Parasitic crustaceans are usually transparent or of the same color as the host, which makes it difficult for the cleaner to remove them all. The food cleaners obtain from their hosts is only part of their diet, since they also eat small animals from the surrounding water or from the ocean floor.

Although many different types of fish have evolved the cleaning habit, some of their adaptations are similar; for example, many cleaners have a long, pointed snout that enables them to probe into small crevices of the host's skin, or gills and tweezerlike teeth that are used to remove parasites embedded in the skin.

The small wrasse or senorita, *Oxyjulis californica,* is a common cleaner off the southern California coast. It cleans a wide variety of fishes including the ocean sunfish (*Mola mola*), black sea bass (*Stereolepis gigas*), and blacksmith (*Chromis punctipinnis*). Many fish may crowd around a single cleaner and assume various poses as they wait their turn to be cleaned. Several other wrasses are also effective cleaners, such as the yellow Bahamian wrasse (*Thalassoma bifasciatum*) and the Mexican rainbow wrasse (*Thalassoma lucasanum*).

The more highly coevolved the symbiosis is between a cleaner and host, the less likely it is that the cleaner will be eaten by the host. Senoritas are rarely eaten even though they enter the mouths and gills of some large hosts. Other cleaner fish and shrimp are not as fortunate and may be eaten by their hosts. The tiny neon goby (*Elacatinus oceanops*) lives among coral reefs and cleans a variety of fishes including large groupers, parrot fishes, grunts, angels, and morays. The goby swims in and out of the mouths and gills of its hosts with impunity. Several gobies may group together to clean a large fish. The large barracuda, *Sphyraena barracuda,* grows to a length of six feet and is cleaned by a blue and yellow wrasse (*Bodianus rufus*) that is less than one inch long. Remoras are sucker fish that spend most of their lives attached to sharks and turtles and not only share in their meals but also clean them of parasites.

Several species of cleaner shrimp are common in tropical waters. The Pederson cleaning shrimp (*Periclimenes pedersoni*) is 4 cm long and has antennae that are even longer. The shrimp generally lives with an anemone, which protects the shrimp and serves as its cleaning station. When a fish approaches the anemone, the shrimp waves its antennae and body back and forth until the fish gets close enough for the shrimp to climb on it. The shrimp then crawls over the fish's body and removes parasites. In some cases, cleaner shrimps and cleaner fishes occupy the same station.

The boxer shrimp (*Stenopus hispidus*) lives in tropical waters throughout the world. It grows up to 8 cm long, has a white body with red stripes, and long, white antennae. The shrimp lives among gray corals and is easily visible to fish seeking out its services.

Experiments designed to test the importance of cleaner fish and shrimp to marine fishes have shown that cleaners control the spread of host parasites and infections. In a field experiment conducted in the Bahamas, Conrad Limbaugh removed all the cleaning organisms from two small, well-populated reefs. Limbaugh noticed that within a few weeks the number of fishes in the reefs had declined sharply and fishes that remained were infected by fungi and other parasites. A marked difference was seen between the health of fishes of reefs without cleaners and the health of those with cleaners. Another experiment was conducted with fishes in an aquarium. Fishes infected with bacteria and fungi were restored to health by the action of cleaner shrimps that were placed into the aquarium.

Some fishes change color while they are being cleaned. For example, the black surgeonfish (*Acanthurus achilles*) turns bright blue, and other fishes assume a bronze color. The adaptive value of such color changes is not clear.

Often, when fishes are being cleaned, they become alarmed and signal their alarm to their cleaners, who then quickly retreat. Cleaners working in a host's mouth during an alarm are ejected by the host or given a signal that causes them to leave.

Some fishes mimic the color patterns of cleaners. Such a disguise protects a fish from predators or allows it to approach prey without alarming them. Because the blenny (*Aspidontus taeniatus*) closely resembles the common cleaner *Labroides dimidiatus,* it can approach other fish, particularly juvenile ones, which are less wary, and bite off pieces from their fins. The blenny also removes parasites from larger fishes and some scientists feel that it may be in an intermediate stage of evolution, between a predator and a cleaner.

Much of the information on cleaning symbioses comes from observations made by divers. For this reason, studies have concentrated along coastal areas and around tropical islands and reefs. Observations of deep-sea fish populations and those of colder waters such as in the Arctic and Antarctic are limited.

Cleaning symbioses also occur with land animals. The red rock crab, *Grapsus grapsus,* is a cleaner of the marine iguana *Amblyrhynchus cristatus,* which lives on the Galápagos Islands. Various types of birds remove parasites and infected tissue from large animals such as water buffalo, crocodiles, and cattle.

2. ANEMONE-CLOWN FISH SYMBIOSIS. Fishes of the genera *Amphiprion, Dascyllus,* and *Premnas,* commonly called clown fish, form mutualistic associations with giant sea anemones that live in coral reefs throughout the Pacific Ocean. The associations are obligatory for the fish, but facultative for the anemones. The anemones eat prey they have paralyzed by means of poisonous nematocysts discharged from specialized cells in their tentacles. The clown fishes are immune to the stinging nematocysts and can nestle within the tentacles and contact them frequently without harm (Fig. 12.2). The question of how clown fish develop their immunity to the anemone has intrigued scientists.

Clown fish have to go through a period of acclimation before they become protected from the anemone. Studies designed to determine how acclimation is achieved appear to show that the mucous coating around the fish is changed by association with the anemone and the fish is no longer recognized as prey by the anemone.

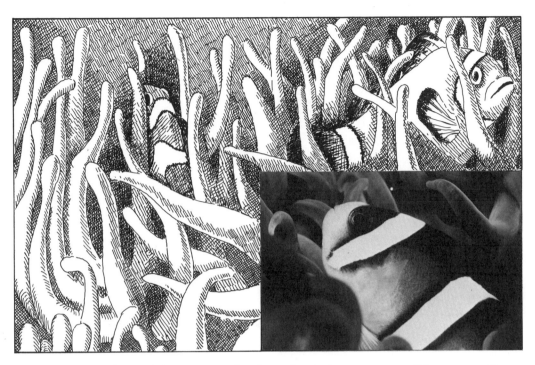

Figure 12-2. Clown fish nestled among tentacles of a giant sea anemone. The clown fish are immune to the stinging nematocysts of the anemone. (Insert: Richard M. Mariscal, Florida State University.)

The change in the mucous coat was first thought to result from internal secretions from the fish, but scientists now feel it is a result of the addition of mucus from the anemone. When the fish acquires this mucus, the anemone no longer recognizes it as foreign and thus does not discharge its nematocysts when its tentacles are contacted by the fish. In order to obtain mucus from the anemone, the fish undergoes an activity called *acclimation behavior*. The first stage of this behavior is recognition of the specific type of symbiotic anemone by the fish. The clown fish approaches the anemone cautiously, swims around its column and oral disc, and comes as close to the tentacles as possible without touching them. After frequent passes over the oral disc, the fish makes the initial contacts with the anemone by means of its tail and fins. At first, the tentacles adhere to these structures and the fish pulls away and frees itself. More extensive body contacts then follow until the fish becomes fully acclimated and can bury itself in the anemone's tentacles. The acclimation time varies but may take up to one hour. If a clown fish becomes separated from its host for longer than one hour, it must again undertake the ac-

climation behavior. The immunity a fish acquires from one anemone does not protect it against another anemone should the fish decide to move on.

Studies by Richard N. Mariscal have demonstrated the importance of the protective cover around the clown fish. When Mariscal presented pieces of a grouper and an acclimated clown fish, skin-side down, to a giant anemone the tentacles of the anemone adhered strongly to the grouper skin but not to that of the clown fish. When both pieces were presented flesh-side down, the tentacles of the anemone adhered to both pieces, which were then eaten.

Once a clown fish acclimates to its anemone, it stays with it as long as the food supply is adequate. Clown fish are brightly colored and marked, and they attract larger fish to the anemone. These fish, if they venture too close, are stung by the tentacles and eaten by the anemone. The clown fish share in the meal and also remove fragments of the prey and wastes from the anemone. The fish may also bite off and eat pieces of the anemone's tentacles that contain symbiotic algae and eat the crabs and shrimp that also live among the anemone's tentacles.

Thus, in addition to protection from predators, the clown fish obtain food.

A similar relationship exists between the Portuguese man-of-war jellyfish (*Physalia physalia*) and the horse mackerel (*Trachurus trachurus*). The fish live between the tentacles of the jellyfish and somehow avoid being stung by the nematocysts. The bright blue and silver colors of the fish, as well as its small size, attract prey for the jellyfish.

3. BEHAVIOR OF PARASITIZED HOST ANIMALS.

Many parasites, such as protozoans, helminths, and nematodes, have complex life cycles. They spend their early life in one animal species, the *intermediate host,* and reach sexual maturity in another animal species, the *definitive* or *final host.* Parasitic symbionts have a variety of physiological and ecological mechanisms to ensure their survival and reproduction. Some parasites modify the behavior of their intermediate host in such a way that the host becomes more vulnerable to predation by the definitive host species. Following are specific examples of parasites that alter host behavior.

Larvae of the nematode *Tetrameres americana* invade the muscles of grasshoppers, causing the host to become sluggish and therefore easily caught by chickens in which the nematodes live as adult worms.

Larvae of the tapeworm *Taenia multiceps* parasitize the central nervous system of animals such as sheep and cattle. The infected individual staggers in circles and becomes separated from the rest of the herd. Carnivores such as wolves and wild dogs constantly search for such weakened prey. The tapeworm becomes sexually mature in the digestive tract of the second host species.

The fluke *Leucochloridium macrostomum* commonly occurs in the rectum of European birds such as crows, sparrows, jays, and nightingales. The birds acquire the flukes in a highly unusual manner. Fluke eggs are ingested by snails, where they develop into the larval, sporocyst stage. The parasite then produces a second sporocyst generation, which migrates to the head and tentacles of the snail. These structures then enlarge and the tentacles become green, brown, or orange with prominent banding patterns. The sporocysts pulsate and this movement attracts birds, which peck on the sporocysts and ingest them. Inside the host birds, the larval cysts rupture and liberate the adult parasites.

Sporocysts acquired by a parent bird are also transferred to nestlings. In a similar manner, pulsating sporocysts of the fluke *Ptychogonimus megastoma* are ingested by shore crabs and the fluke larvae become encysted in the crab tissue. Fishes acquire the parasite by ingesting the infected crabs.

The lancelet fluke *Dicrocelium dendriticum* inhabits the gallbladder of sheep, deer, rabbits, and other grazing mammals. Eggs of the fluke are eaten by land snails, in which the larvae develop. The mature larvae exit the snail in a *slime ball,* which is eaten by ants. The fluke forms cysts in the brain of the infected ant and radically transforms its behavior. The affected ant climbs to the tip of a blade of grass and hangs on tightly. This abnormal behavior increases the chances of the ant being transferred to a grazing host.

Acanthocephalan worms are endoparasites that inhabit the alimentary tracts of vertebrates, where they attain sexual maturity. The intermediate host of these parasites is an arthropod, whose behavior is transformed after infection. The altered behavior significantly increases the chances of the parasite being acquired by the final host through preferential predation of the infected intermediate host. There are several well-known examples of this phenomenon.

Polymorphus paradoxus is an endoparasite inhabiting the alimentary canals of beavers, mallards, and muskrats. Aquatic crustaceans such as amphipods are the intermediate hosts of the parasite. Healthy amphipods normally move away from light and rarely appear at the surface of a lake or pond. When disturbed, they dive and burrow into the mud at the bottom. Parasitized amphipods, however, are attracted to light and are commonly observed clinging to vegetation on the surface of the water. This altered behavior increases the likelihood of the amphipods being eaten by a final host species. Similar anomalous behavior of amphipods infected with *Polymorphus marilis* allows ducks to catch the crustaceans.

Songbirds preferentially feed on pill bugs or sow bugs (isopods) that have become infected with the acanthocephalan *Plagiorhynchus cylindraceus.* Songbirds normally do not feed on pill bugs but the infected bugs may make up over 40% of a bird's diet. The behavior of the parasitized pill bugs is altered so that they seek out exposed areas, where they are quickly eaten by birds.

Cockroaches parasitized with *Moniliformis moniliformis* are attracted to light and become hyperactive. The roaches are less likely to remain hidden and therefore more likely to be caught by rats, the final host for the parasite.

Scientists are becoming increasingly aware of the possibility of a host's behavior being altered by a parasitic symbiont. From an evolutionary perspective, the life cycles of symbionts and their hosts may show a parallel development.

C. Social Symbioses

Among insects and birds, symbioses occur not only between individuals of different species but also between individuals and societies and between different societies.

In insect societies each individual has only a limited role in relation to the society as a whole. Because of this limitation, other species of insects can easily intrude into the society and this leads the way to symbiotic interactions. A similar situation exists with birds whose young mature slowly and do not form close bonds with their parents. These parents are easily deceived when eggs are laid in their nests by other birds and will incubate the foreign eggs and care for the young along with their own. Among other birds, such as ducks, and among mammals, social symbiosis is rare because the parents form close recognition bonds with their young at an early stage and repel all apparent intruders.

Edward O. Wilson of Harvard University is the leading authority on social symbioses. His studies of insect societies and their complex internal and external interactions have greatly increased our understanding of the subject. Wilson recognizes different types of social symbiosis among animals, and groups them into three major categories: social parasitism, social commensalism, and social mutualism. We consider only these three major groups and examples of them.

1. SOCIAL PARASITISM. Many genera of parasitic ants, most of which live in North and South America and Europe, have evolved symbiotic relationships ranging from the loosely organized to the highly integrated.

The simplest form of social parasitism is one in which a colony of ants builds a nest next to a colony of a different ant species and steals food from and preys on the foreign workers. Some ants may even occupy the nest of another species and maintain a separate colony within the nest.

A highly evolved form of social parasitism among ants is shown in colonies containing different ant species living together. For example, in several subfamilies a fertile queen may enter a colony of different ants and by some unknown mechanism be accepted by the workers. The host queen is killed, either by its own workers or by the intruder queen. As the host workers die of old age or injury, there are no offspring to replace them and the nest becomes fully colonized by offspring of the new queen. Wilson calls this example *temporary social parasitism*.

Another type of social parasitism among ants involves slavery. Some ant species raid the nests of different ant species, steal pupae, and return with them to their own nest, where the young ants are reared as slaves. The slave ants do all of the work in the nest such as foraging, building, and rearing the brood.

One of the most highly integrated examples of social parasitism involves the parasitic ant *Teleutomyrmex schneideri,* which is found in the Swiss and French Alps. *Teleutomyrmex schneideri* parasitizes a close relative, *Tetramorium caespitum.* The ant parasite lives only in the nests of its host and lacks a worker caste. The parasite queen is much smaller than the host queen and spends most of its time clinging to the back of the host queen or workers. The parasite ants are fed by the host workers. Both queens lay eggs; those of the host produce only workers and not sexual forms, while those of the parasite produce only sexual forms. When the parasite ants mate, the new fertile queens either remain in the same nest or fly away and invade other nests of the host ant. It is thought that *Teleutomyrmex* represents an end point in the various evolutionary pathways of parasitic ants. The morphological adaptations of the ant and loss of the worker caste are traits of other highly evolved parasitic ant species.

Parasitic wasps behave in a manner similar to parasitic ants. The queen wasp penetrates the nests of other wasps, kills the host queen, and assumes her role. Some wasps are permanent residents in the nests of other wasps and, like ants, have evolved to a point where they do not have a worker caste and depend on the host workers to care for them. A similar type of behavior has been reported for parasitic bumblebees.

Parasite nestlings may
eject host eggs or
attack host nestlings

Cowbird inspecting host nest

Figure 12-3. Brood parasitism of cuckoo and cow-birds. (Adapted from Wilson, E. O. 1975. Sociobiology. Belknap Press, Harvard University Press, Cambridge, Mass.)

The origin of parasitic ants is of great interest to scientists. In 1909, Carlo Emery developed a hypothesis, known as *Emery's rule,* that states that parasitic ants and slave-making ants evolved from closely related species that now have become their hosts. Wilson suggests that parasitic ant species and their host species developed from a common parent species through geographical isolation and genetic differences. When the range of the two new species later overlapped, one became a parasite on the other. Wilson feels that Emery's rule applies also to parasitic wasps and bumblebees.

Termites do not exhibit the same type of social parasitism as other social insects because termite queens form new nests after fertilization rather than returning to their old nests or attempting to invade established nests.

Some beetles are accepted into an ant or termite colony because they produce substances that attract the host insects. The substances are similar to *pheromones,* which the host normally produces and have a calming effect on the hosts after they lick these substances. Wilson calls these compounds *appeasement substances* because they help the beetle to become adopted into an ant or termite population. After the beetle obtains entry into a nest, it feeds on the host larvae and by stroking worker ants with its antennae can stimulate them to regurgitate food. Many of the beetles that live with ants and termites have evolved morphological features, such as a swollen abdomen, that mimics those of their hosts. Loss of eyes and wings are also common changes the beetles have undergone.

Social parasitism is also present among birds, especially cuckoos. About 50 species of cuckoos are obligate brood parasites, which means that they depend on other birds to hatch their eggs and rear their young. The cuckoos lay their eggs among those of the host birds, and when the eggs hatch, the young cuckoos destroy the host's eggs and kill any young of the host (Fig. 12.3). This association is so highly evolved that the eggs and young of the cuckoo resemble those of the host bird in color and size. One of the best-studied social parasites among birds is the European cuckoo (*Cuculus canorus*). This species consists of different populations, each having a different host bird, whose eggs are mimicked. How the cuckoo populations maintain their specific relationships with their host birds

Figure 12-4. Ants licking honeydew from aphids. In this mutualistic symbiosis, the ants obtain food from the aphids, which they protect from predators.

is not clear, particularly since the different cuckoo populations interbreed. Cuckoos have evolved other adaptations, in addition to egg mimicry, for their parasitic habit. Their eggs have thick shells that resist damage when the eggs are dropped into host nests and the birds have extendible cloacae, which enables them to deposit their eggs into small nests. Another remarkable adaptation is the ability of some cuckoos to fly like a hawk and in so doing to distract the host bird from its nest so that the cuckoo may deposit its eggs undisturbed.

2. SOCIAL COMMENSALISM. The nests of social insects are home to many different scavengers such as mites, beetles, and millipedes. The host insects either ignore these commensals or accept them as part of the colony. Some silverfish and millipedes live with army ants, participate in their forays, and even share their prey.

Among insect-eating birds, several species may flock together in order to better avoid predators and be more efficient in foraging for food. Similar groups of mixed species have been reported among marine fishes, such as dolphins and whales, and also African mammals. Mixed groups of animals often congregate together to increase their efficiency of early warning of predators.

3. SOCIAL MUTUALISM. Mutualistic symbioses are common between ants and various insects such as aphids, scales, mealybugs, and treehoppers. The ants tend these insects like cattle, milking them for food and protecting them from predators (Fig. 12.4).

Aphids penetrate the food-transporting system (*phloem*) of plants with their stylets and feed on the plant sap, which is rich in sugar and minerals. The aphids digest only a small amount of the plant sap they take in and excrete the remainder in a form called *honeydew*. Aphids that are tended by ants excrete the honeydew slowly so that it accumulates as drops around the anus. Mutualism between ant and aphid has evolved to such a high degree that their life cycles have become coordinated and the aphid has lost various structures that they normally have to defend themselves, such as a hard exoskeleton and jumping legs. When an ant is ready to feed, it strokes an aphid with its antennae and stimulates it to exude honeydew. Some ants carry the aphids they tend back to the nests each evening, in order to protect them from predators, and return them to the plants each morning.

To ensure a steady source of food, worker ants carry aphids and other honeydew insects to different plants if the original ones are disturbed or uprooted. Some ant species maintain aphid eggs in their nests during the winter and when the aphids hatch in the spring carry them to nearby plants. The queens of some ant species even carry scale insects during their nuptial flights. Bees and wasps also keep honeydew insects.

Some butterfly species have caterpillars that are tended by ants. The large blue butterfly *Maculinea arion* cannot complete its life cycle without help from red ants. The ants keep the caterpillars in their nests until they pupate and hatch, and during this time they protect the caterpillars and feed on their sugary excretions. When the chrysalis hatches, the butterfly leaves the ant's nest and flies away.

Some South American ant species appear to have a mutualistic relationship with each other. The ants form adjacent nests, share common trails, and collect honeydew from the same population of treehoppers. The different ants are friendly to each other and seem to thrive because of the association.

The giant cowbird (*Scaphidura oryzivora*) of Central and South America is a brood parasite that lays its eggs in the nests of oropendolas, which are members of the grackle family. Oropendolas have a natural enemy, the parasitic botfly (*Philornis*), which infests their nests and kills their young. Studies have shown that oropendola nests containing young cowbirds have fewer botflies than nests without cowbirds. Young cowbirds instinctively attack and kill maggots of the botflies and remove them not only from the nests but also from the young host birds. Cowbirds, unlike cuckoos, tolerate the eggs and young of the host birds. Thus, an association that began as one of strict parasitism has evolved into mutualism, since by defending itself the cowbird also protects the host.

D. Summary and Perspectives

Behavior in a symbiotic association is influenced by external and internal factors. Symbiosis commonly modifies the behavior of the associating partners.

Cleaning symbioses are common in tropical, marine waters, where specialized fish and shrimp clean other fish of external parasites and damaged tissues. In return for this service, the cleaners obtain food and protection from predators. Cleaners and hosts communicate with each other by means of various behavioral patterns such as posing or by means of bright colors and unusual color patterns. Some common cleaners include wrasses such as the senorita, the neon goby, remoras, Pederson shrimp, and the boxer shrimp. Experiments have shown that cleaners have an important function in tropical ecosystems by controlling the spread of parasites. Some fish have evolved color patterns similar to those of cleaner fish. This disguise protects the fish from predators and also allows them to get close to prey fish. Cleaning symbioses are present also among land animals.

Clown fish live in a mutualistic association with giant anemones that grow in tropical, coral reef communities. The fish acquire immunity to the anemone by means of an acclimation process during which the fish coats itself with mucus from the anemone. The mucus disguises the fish in such a way that it is not recognized as foreign by the anemone. Clown fish attract other fish to the anemone and share in the prey caught by its host.

Some animal parasites such as nematodes, tapeworms, flukes, and acanthocephalan worms modify the behavior of their intermediate hosts in a manner that makes them more likely to be preyed on by the final hosts.

Social symbioses occur among insect societies and birds and include social commensalism, social mutualism, and social parasitism. In social symbioses each individual has only a limited role. Social parasitism is highly evolved among ants and may involve simple predation, slavery, or the integration of a foreign queen into an existing ant colony. According to Emery's rule, parasitic and slave-making ants evolved from closely related species that now serve as their hosts. Interestingly, a similar situation may exist among parasitic red algae, many of which occur only on other closely related red algae.

Certain beetles become accepted into ant or termite colonies because of "appeasement substances" produced by the beetles that appear to calm the hosts. The beetles feed on host larvae and food that is regurgitated by worker ants.

Cuckoos are highly developed social parasites that lay their eggs in the nests of other birds. When the eggs hatch, the young cuckoos kill the host's young and are raised by the host. Adaptations of the cuckoo include mimicking the eggs of the host, extendible cloacae, and eggs with thick shells that resist breaking when they are dropped into the foreign nest.

Silverfish and millipedes are commensals in the nest of social insects. Ants form mutualistic associations with aphids, scales, and mealybugs, "milking" them for the food (honeydew) they obtain from plants. Ants tend aphids like cattle, protecting them and carrying them from plant to plant. Ants have a similar relationship with the caterpillars of some butterflies.

Cowbirds and oropendolas have a mutualistic relationship in which the young cowbirds kill botfly maggots that are serious parasites of young oropendolas, and thereby protect their foster nest and host.

Review Questions

1. Why are cleaning symbioses more common in tropical than in temperate waters?
2. What do cleaning fish remove from their hosts?
3. Describe the experiments that were used to test the importance of cleaner fish.
4. How do clown fish develop their immunity against the poisonous nematocysts of the anemone?
5. Give several examples of how parasites modify the behavior of their intermediate hosts.
6. Define *temporary social parasitism*.
7. Describe the relationship of the parasitic ant *Teleutomyrmex schneideri* and its closely related host, *Tetramorium caespitum*.
8. Define *Emery's rule*.
9. List some adaptations of the cuckoo bird for its parasitic habit.
10. Give an example of social commensalism.
11. Characterize the relationship between ants and aphids.
12. How do cowbirds benefit their oropendola hosts?

Further Reading

Gilbert, E. L. 1982. The coevolution of butterfly and the vine. Sci. Am. 247 (August): 110–121. (The vine *Passiflora* has evolved morphological features on its leaves that resemble butterfly eggs and discourage female insects from laying their eggs on the leaves.)

Hausfater, G., and R. Sutherland. 1984. Little things that tick off baboons. Nat. Hist. 93 (February): 54–61. (Describes how parasites greatly influence the social behavior of baboons.)

Hölldobler, B., and E. O. Wilson. 1983. The evolution of communal nest-weaving in ants. Am. Sci. 71:490–499. (Cooperation among weaver ants.)

Limbaugh, C. 1961. Cleaning symbiosis. Sci. Am. 205 (August): 42–49.

Love, M. 1980. The alien strategy: Not all parasites trust to luck. Nat. Hist. 89 (May): 30–32.

Lubbock, R. 1981. The clownfish/anemone symbiosis: A problem of cellular recognition. Parasitology (UK) 82: 159–173. (Examines how the associating animals manage to survive in a potentially lethal environment.)

Moore, J. 1984. Parasites that change the behavior of their host. Sci. Am. 250 (May): 108–115.

Rosenthal, G. A. 1983. A seed-eating beetle's adaptations to a poisonous seed. Sci. Am. 249 (November): 164–171. (Beetle larvae feed on legume seeds that contain a highly toxic amino acid.)

Troyer, K. 1984. Microbes, herbivory, and the evolution of social behavior. J. Theor. Biol. 106:157–169. (Suggests how social systems in a variety of animals, from termites to dinosaurs, may have evolved from herbivorous life styles.)

Wilson, E. O. 1985. The sociogenesis of insect colonies. Science 228:1489–1495.

Zuk, M. 1984. A charming resistance to parasites. Nat. Hist. 93 (April): 28–34. (Supports Hamilton's theory that male birds with high parasite levels have a more brilliant plumage and musical ability than uninfected birds.)

Bibliography

Alcock, J. 1984. Animal behavior: An evolutionary approach. 3d ed. Sinauer, Sunderland, Mass. 596 pp.

Allaby, M. 1982. Animal artisans. Alfred A. Knopf, New York. 192 pp. (A popular account of animal behavior.)

Brian, M. V. 1983. Social insects: Ecology and behavioural biology. Chapman and Hall, London. 377 pp.

Bristow, C. M. 1985. Sugar nannies. Nat. His. 94 (September): 63–68. (Mutualism between ants and treehoppers.)

Davenport, D. 1966. The experimental analysis of behavior. In: Symbiosis, Vol. 1, 381–429, ed. S. M. Henry. Academic Press, New York.

Feder, H. M. 1966. Cleaning symbiosis in the marine environment. In: Symbiosis, Vol. 1, 327–380, ed. S. M. Henry. Academic Press, New York.

Mariscal, R. N. 1971. Experimental studies on the protection of anemone fishes from sea anemones. In: Aspects of the biology of symbiosis, 283–315, ed. T. C. Cheng. University Park Press, Baltimore.

Payne, R. B. 1977. The ecology of brood parasitism in birds. Ann. Rev. Ecol. Syst. 8:1–28.

Wilson, E. O. 1975. Sociobiology: The new synthesis. The Belknap Press, Harvard University Press, Cambridge, Mass. 697 pp.

CHAPTER 13

Symbiosis and Evolution

A. Introduction

Although Darwin's theory of organic evolution has provided an intellectual framework for understanding the origin of species and the unity of life forms, the theory has never fully satisfied evolutionary biologists. In Darwin's view, new species arise following reproductive isolation and the gradual accumulation of favorable variations. Randomness in mutation, natural selection, as well as chance have traditionally been considered by many biologists as significant factors in the evolutionary process. Today, evolutionary ideas are being refashioned in the language of punctuated equilibria, genetic polymorphism, cladistics, and sociobiology. To this list of new approaches, one can add the concepts of *coevolution* and symbiosis. In recent years, ecologists have become convinced that evolutionary processes cannot be understood fully until the close interactions between organisms are understood.

Coevolution is a kind of evolutionary "arms race" between organisms living together in intimate associations. Adaptations that give one organism an advantage are countered by neutralizing adaptations of the other organism. For example, evolution of virulence in a pathogenic organism is often matched by a concomitant evolution of resistance of the host species. Reciprocal genetic change, often occurring simultaneously in interacting populations, is the basis of coevolution. Traits of one species evolve in response to those of another species, whose traits, in turn, were influenced by the first species. Evolution of the eukaryotic cell has involved a series of genetic interactions between ancient prokaryotes. Scientists have recently discovered that an exchange of DNA has taken place between the nucleus, mitochondria, and chloroplasts of eukaryotic cells, attesting to a complex evolutionary process. Coevolution that takes place between two different species is called *pairwise coevolution*. On a larger scale, one or more species may evolve a trait in response to traits of several other interacting species, a phenomenon called *diffuse coevolution*. For example, a plant species may develop toxins or other defenses in response to herbivory, and various animal populations evolve means to overcome these defenses. The plant species then develop new defenses and the evolutionary process is repeated. In some interacting species a *coevolutionary equilibrium* appears to take place.

The theory of coevolution includes *cospeciation* as well as *coadaptation* among interacting organisms. Long-term, close contacts between organisms and the reciprocal influences that take place between them often result in the formation of new species. The phylogeny or evolutionary histories of organisms involved in symbioses are closely related to each other. The many relationships between insects and plants, in particular those involved in pollination, are clear examples of the adaptation and speciation that can occur in symbiotic associations. The lichen symbiosis is an example of how free-living fungi (ascomycetes) become transformed into new species during the course of symbiotic associations with algae. Parasitologists have often observed a close relationship between the phylogeny of parasites and their host species. For example, *Enterobius,* a genus of pinworms that infect primates, has a phylogeny and geographical distribution that closely parallel those of its primate hosts. Evolutionary insights have been made on the taxonomic relationships of species of bats, squirrels, and pocket gophers from the studies of their ectoparasites such as lice and fleas. Ecologists have applied the following rules to better understand the evolution of host-parasite interactions.

Farenholz's rule states that the phylogeny of many parasites mirrors that of their hosts. Since most parasites are obligate symbionts, it is assumed that parasites of present-day hosts had ancestors that were parasites of earlier hosts.

Szidat's rule is that the more primitive the host, the more primitive its parasites.

Eichler's rule states that since the host and parasite share a parallel evolution, parasites can be used to understand the phylogeny and origin of host species. Large host groups will have more genera of parasites than small host groups. Parasitic symbionts may evolve in two ways. Speciation may occur if some parasitic species become established in different hosts, or if the parasites invade new tissues or organs in the same host species.

B. Genetic Interactions in Symbiotic Systems

Genetic strategies for reciprocal coadaptation have been examined for a number of host-symbiont relationships. The strategies include gene-for-gene relationships, epistatic interactions, and polymorphic variation.

1. GENE-FOR-GENE RELATIONSHIPS. The gene-for-gene hypothesis was proposed first by Harold Flor and was based on his studies with flax (*Linum usitatissimum*) and the rust fungus *Melampsora lini*. Flor found that two independent genes for resistance in the host plant were matched with two genes for virulence in the fungal pathogen. The resistance gene is inherited as a dominant trait while virulence is inherited as a recessive trait. The host plant and its fungal pathogen have thus evolved a complementary genetic system in which each gene determining a host response is matched by a gene for a certain behavior in the pathogen. Since 1942, gene-for-gene coevolution has been documented or strongly suggested in more than three dozen cultivated plants and the fungi that infect them; these include (a) wheat-*Puccinia graminis tritici,* (b) wheat-*Puccinia recondita,* (c) barley-*Ustilago hordei,* (d) barley-*Erysiphe graminis hordei,* (e) wheat-*Erysiphe graminis*

tritici, and (f) apple-*Venturia inaequalis.* Gene-for-gene interactions have also been reported between the snail *Biomphalaria glabrata* and the blood fluke *Schistosoma mansoni;* resistance in the snail results from a single, dominant gene, which is matched with an avirulent gene of the fluke. Resistance or susceptibility of the individual snails is controlled by the relative frequencies of genes. Gene-for-gene interactions are present in animal-plant symbioses such as the cyst nematode parasitism of potato, barley, and soybean and aphid infestation of rushberry and alfalfa. Resistance to the Hessian fly *Mayetiola destructor* in wheat plants is governed by genes that are matched by corresponding genes for virulence in the insect pest.

Scientists suspect that there may be a clustering of genes in host chromosomes that confer resistance to more than one pathogen. For example, barley chromosome number 5 has genes that control resistance to the powdery mildew fungus *Erysiphe graminis hordei* and also to the rust fungi, *Puccinia hordei* and *Puccinia striiformis.*

Gene-for-gene coevolution does not explain all the types of host-parasite interactions. Gurmel Sidhu has described the complex interplay of two or more symbionts associated with a host plant. For example, he has shown that the plant nematode *Meloidogyne incognita* alters the resistance of tomato plants to the wilt fungus *Fusarium,* which alone cannot parasitize the plant. Similar interactions between the fungal pathogens *Fusarium* and *Verticillium* on tomato plants have been described. Sidhu has explained these observations in terms of the concept of *epistatic parasitism.* The disease expression by the hypostatic parasite depends on the establishment of the epistatic parasite.

2. GENETIC POLYMORPHISM. Genetic polymorphism is common among species that participate in symbiotic associations, particularly those involving host-parasite interactions. J. B. S. Haldane was the first to suggest that parasitic symbionts might be the reason for polymorphism in their hosts. For example, the variety of antibodies produced by a host organism has evolved in response to the many different antigens produced by trypanosome parasites. Multiple genes for resistance and virulence in host-parasite symbioses have resulted in a high degree of genetic polymorphism in the interacting species. Some scientists have suggested that parasitic associations may be the principal cause of protein polymorphism within a population. Stephen J. Gould and Richard Lewontin have argued that evolutionary changes increase polymorphism within a population. Such a view is a radical departure from Darwin's theory, which states that genetic changes of an organism are subject to selective pressures of the environment and that natural selection conserves one genetic type best suited to a particular set of environmental conditions and eliminates the others.

In humans, malarial disease has been responsible for the selection of three or more types of genetic variation with respect to resistance to the disease. Resistance caused by the sickle-cell hemoglobin gene (HbS) has been studied extensively. Red blood cells containing sickled hemoglobin collapse and assume a sickle-shape under low oxygen tension. The HbS gene is inherited as a recessive gene and individuals who are homozygous for it die early. Individuals heterozygous for that trait survive because they contain a normal hemoglobin gene. In Africa, where malaria is endemic, heterozygous individuals have a survival advantage because *Plasmodium* is unable to develop in the red blood cells. Such an evolutionary strategy on the part of a host is called *heterozygote advantage.*

Duffy blood groups, MN antigens, and glucose-6-phosphate dehydrogenase (G6PDH) deficiency are other biochemical variations in man that have evolved in response to *Plasmodium* infection. Individuals deficient in G6PDH, heterozygous and homozygous individuals, are resistant to the parasite because *Plasmodium,* an obligate intracellular symbiont of red blood cells, depends on the host to supply G6PDH. The gene for G6PDH is highly polymorphic, with more than 50 alleles coding for hundreds of variant forms of the enzymes. Individuals with low levels of this enzyme are susceptible to anemia and are also sensitive to drugs such as sulfanilamides and the antimalarial drug primaquine. The G6PDH deficiency gene occurs with high frequency among black people and people from the Mediterranean region. The absence of Duffy antigen from African people may also be related to resistance to *Plasmodium.* Duffy antigens occur on the surface of red blood cells at the sites to which *Plasmodium vivax* binds in order to invade the cell. The presence of MN

antigens is also thought to be a significant example of genetic polymorphism in response to *Plasmodium* infection.

C. Nature of Host Resistance

The biochemical expression of resistance in host animals and plants has been explored by many scientists. Resistance responses in vertebrates include antibody production, phagocytic activities of macrophages, hypersensitivity and tissue damage, cross-protection, and sequestration of the parasite. In a similar manner, flowering plants respond to pathogens by producing phytoalexins, hypersensitivity, cross-protection, and compartmentalization of the infected tissue. The effectiveness of these host responses constitutes an important reason for the evolutionary success of plants and animals.

1. RESISTANCE IN ANIMALS. Immunology is one of the fastest-developing areas of biology. Recent discoveries have greatly increased our understanding of how humans and other animals resist infection by parasitic symbionts. Immunologists are using techniques of molecular biology to mass-produce monoclonal antibodies and interferons, and to understand the basic mechanisms of autoimmunity and tissue transplantation. During the past few years, there have been remarkable advances in our knowledge of T-cell activities and the role of major histocompatibility complex genes. In order to understand the nature of resistance in man and other animals, we give a brief introduction to the immune mechanisms.

Immunological responses of vertebrates involve organs (thymus, spleen, and bone marrow), lymphocyte cells (T and B cells), and antibody molecules. These components of the immune system are responsible for recognizing, sequestering, and destroying invading parasites. Resistance to an infection may be nonspecific, or *innate,* for example, where parasites are prevented from entering the body tissues by barriers such as the skin. Mucus secreted by epithelial cells also is a protective barrier because it prevents infectious agents from attaching to the epithelial cells. Again, secretions such as tears, saliva, nasal fluids, and milk contain antibacterial substances. *Specific acquired immunity* is another type of host resistance that involves the production of antibodies and a complex interplay of T and B cells. The acquired immune response results in the recovery of the host from a disease and is followed by the host's becoming endowed with a specific memory, which responds vigorously to an infection by the same parasite.

Antibodies are *immunoglobulins* (Ig) that react specifically with target molecules called *antigens*. Chemical bonding of the host's antibody molecules to a parasite's antigens constitutes one important phase of a host's defenses. Each Ig molecule is composed of four peptide chains: two identical heavy chains are linked by disulfide bonding to two identical light chains. There are an estimated one billion variations of Ig molecules in humans. Each type of Ig molecule is a product of a single clone of B cells. Immunologists recognize five classes of Ig molecules: IgG, IgA, IgM, IgD, and IgE.

Immunoglobulin G (IgG) is the most abundant antibody in the interstitial tissues and plays a major role in destroying bacterial toxins and enhancing the phagocytosis of bacteria by lymphocytes. Since IgG can cross the placental barrier, it provides the first level of immunological defense against invading organisms during the first few weeks of an infant's life.

Immunoglobulin A (IgA) is generally found in saliva, tears, nasal fluids, sweat, and secretions produced in the lining of the trachea, lungs, and urogenital and digestive tracts. IgA plays a vital role in defense against skin-inhabiting microorganisms, and strongly inhibits the attachment of microbes to the surface of mucosal epithelial cells.

Immunoglobulin M (IgM) antibodies are large molecules that are efficient agglutinating and cytotoxic agents. These antibodies are confined mainly to the bloodstream and are an effective defense against blood parasites.

Immunoglobulin D (IgD) antibodies are unique because of their susceptibility to proteolytic breakdown and their short life span. Most IgD molecules and IgM antibodies are confined to the surface of lymphocytes. Some scientists suspect that IgD and IgM molecules interact to form receptor sites for antigens.

Immunoglobulin E (IgE) molecules remain bound mostly to the surfaces of large tissue cells (mast cells) and occasionally are found in the serum. IgE molecules protect the external mucosal surfaces of the body by causing an inflammatory reaction in response to pathogenic agents. Parasitologists have often noted high levels of IgE following infection by helminthic parasites. Release of IgE from the mast cells is also responsible for the symptoms caused by allergic reactions, such as hay fever.

The adaptive immune response in a vertebrate host has the following stages: lymphocytes are first activated, and undergo a phase of *clonal proliferation;* some then become *effector cells* while the remainder constitute *memory cells.* The secondary immune response in a host originates with the activation of the memory cells. B-lymphocytes mature in the bone marrow of mammals and become the antibody-producing cells. This constitutes the *humoral immunity.* T-lymphocytes mature under the influence of the thymus and form the basis of *cellular immunity* (cell-mediated immunity). Cellular immunity, the principal host response against intracellular symbionts, is a complicated multistep process. The reader should consult an immunology text for details. In brief, cellular immunity involves the following steps. Different surface antigens of a parasite are recognized by the T and B cells. Macrophages bind to the antigen molecules and carry them to helper T cells, which in turn transfer the antigens to B-lymphocytes. The activated B cells then proliferate and mature to manufacture antibodies. Some activated T cells become cytotoxic and search out foreign cells and cause them to lyse. In addition, some of the activated T and B cells produce diffusible factors called *lymphokines* that activate macrophages and help to localize the immune responses to the infected areas.

Interferons are antiviral molecules that are secreted by virus-infected host cells. Interferons attach to neighboring healthy cells and make them resistant to infection by other viruses. The phenomenon of viral interference or cross-protection has been noted in bacteria, plants, and animals. By continuously changing the structure of their surface antigens, viruses such as the influenza virus render ineffective previously manufactured antibodies. In addition to humoral antibodies, the host also uses sensitized cytotoxic T cells to destroy virus-infected host cells. Examples of such responses to viral infection include mumps, measles, herpes, pox, and rabies. Children with T-cell immune deficiency are unable to recover from viral infections.

Parasitologists have often described the phenomenon of *premunition,* which is the immunity of a host to reinfection following recovery from disease. The exact mechanism of premunition is not understood. Humoral antibodies are effective against the forms of parasites that occur in the host's bloodstream. Some parasites, however, have evolved novel strategies to counter a host's immune system. A few well-known examples of such strategies are described below.

Malarial parasites (sporozoites), when injected by a mosquito into the host bloodstream, are protected from the host's antibodies by a unique surface protein, the *circumsporozoite protein*. This protective protein presents numerous false sites for antibody molecules to bind to without affecting the sporozoites.

Trypanasomes circumvent the host's humoral antibodies by changing the structure of their surface antigens with each cycle of reproduction. When the host produces new antibodies, the parasite also changes its antigens.

Some parasitic symbionts such as *Trypanosoma cruzi, Leishmania,* and *Toxoplasma* escape the host's antibodies by taking up residence in macrophages. Adults of *Schistosoma mansoni,* while living in the host blood vessels, disguise themselves by covering their surfaces with antigens obtained from the red blood cells. This mimicry successfully masks the parasite and the host's humoral antibodies fail to recognize it as foreign.

Sequestration is beneficial to both host and parasite. The invading organism becomes isolated and "stands apart" from the host. Parasites also avoid the host immune response by living in tissue or cells of immunologically isolated zones. For example, larvae of *Trichinella spiralis* reside in muscle cells, where they are protected from the host's antibodies.

2. RESISTANCE IN PLANTS

a. Phytoalexins. The biochemical basis of plant resistance to fungal parasitism is not understood entirely. When a fungus infects a plant, it stimulates the plant to produce compounds, called phytoalexins, that inhibit growth of the fungus. Phytoalexins are produced in response either to complex carbohydrates that are present in the walls of the invading fungus or to compounds excreted by the pathogen. The fungal compounds are called *elicitors* because they stimulate the plant cell to synthesize enzymes that bring about the production of phytoalexins. If a

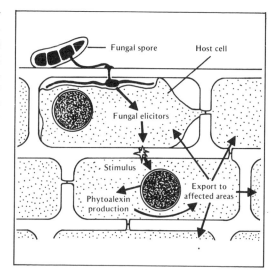

Figure 13-1. Schematic model for phytoalexin synthesis by host cells near the infected cell. (Adapted from Sequeira, L. 1984. Cross protection and induced resistance: Their potential for plant disease control. Trends in Biotechnology 2:25–29.)

fungus lacks specific elicitors, then the infected plant cannot form phytoalexins and, therefore, has no defense against the pathogen. Although the relationship between phytoalexins and elicitors appears simple, it is extremely complex (Fig. 13.1). Many different compounds and conditions stimulate phytoalexin production and there is no clear pattern to the findings of many studies on this subject. There is also a continuing controversy among scientists over the general effectiveness of phytoalexins in determining plant resistance to pathogens. It is assumed that phytoalexins are produced by most angiosperms, but in fact only about a dozen families have been confirmed as having these compounds. One species of plant may produce a large number of different phytoalexins; for example, the potato plant produces about 30 different types of these compounds, presumably as a safeguard against a pathogen developing a tolerance to any particular phytoalexin. Interactions between fungus and host may also involve specific chemical-binding sites on the host, the absence of which makes infection impossible.

b. Hypersensitive Response. The *hypersensitive response* is a common defensive strategy of plants against pathogens such as fungi, bacteria, and viruses. Plant cells that are infected die quickly and often produce a dark pigment. These cells stimulate neighboring cells to respond in a similar manner and a necrotic lesion surrounds the infection site. The pathogen often dies along with the infected cells. This rapid response by the plant prevents the spread of the pathogen and thus limits the infection to localized areas of the plant.

c. *Cross-protection.* Instead of causing severe disease some strains of plant viruses confer immunity against more virulent strains of the same virus on a host plant. This phenomenon is called *cross-protection.* The viruses are modified strains of pathogenic forms and their presence stimulates the host to produce unknown compounds that inhibit the replication of the pathogenic viruses. The resistance may spread from an infected host cell to other uninfected cells. Cross-protection has economic applications. Tomato and citrus plants can be protected from disease-causing viruses by inoculating them with less-virulent strains of the same virus. For example, a common practice used to protect tomatoes against the tobacco mosaic virus is to infect seedlings with a mutant form of TMV. The mutant virus slows the growth of the plants somewhat but does not cause the severe symptoms of the normal virus, such as leaf mosaic, growth inhibition, and fruit discoloration. The infected seedlings produce plants that are disease resistant. This method is useful for protecting plants grown in greenhouses, but scientists are hesitant to apply the technique to field crops. Under natural conditions the modified viruses might spread to different plants, causing new diseases, and they might also mutate back to the virulent strains.

d. *Compartmentalization.* A newly discovered form of resistance of trees against invading pathogens is called *compartmentalization.* Infected trees establish boundaries that seal off pathogens from the healthy tissues. Once that has been accomplished the tree produces new growth rings that further isolate the affected area. If the disease is on a branch or part of the root system, the tree will seal off such parts from the rest of the tree and they eventually decay and fall off.

A tree responds in several ways to disease or injury. The first response is the production of antimicrobial chemicals such as fallic and tannic acids, by the infected cells as well as by neighboring healthy cells. These chemicals permeate the cell walls, often discoloring the wood. A second response by the tree is to form a wall or barrier of new parenchyma cells, which are resistant to microbial infection, around the infected area. A third response of the tree is to continue growing and form new tissues.

Compartmentalization is similar to what a tree normally does when it discards leaves or flowers after they have fulfilled their function. A layer of cells seals off these organs from the rest of the plant, effectively creating a wall, and they eventually fall from the plant.

D. Products of Coevolutionary Symbioses

Plants are attacked by viruses, bacteria, fungi, nematodes, insects, and grazing animals. Plants are unable to move away when attacked; thus many have developed adaptations to protect themselves from predatory and parasitic organisms. Morphological features such as spines and thorns and chemical substances harmful to other organisms are examples of such adaptations.

The primary chemical substances of plants are carbohydrates, lipids, and proteins. Secondary plant chemicals include phenylpropanes, acetogenins, terpenoids, steroids, and alkaloids. These secondary chemicals protect plants from being eaten by herbivores. Some insects, however, have evolved mechanisms to detoxify poisonous plant metabolites.

Passiflora adenopoda protects itself by trapping plant-eating insects by means of hooked hairs on the outer surface of the plant. Many species of *Passiflora* also have structures that resemble the eggs of *Heliconius* butterflies. These structures deter female butterflies from laying their eggs on the plant.

Ants have evolved mutualistic associations with a variety of flowering plants. In plants such as *Acacia* the ants live in hollow areas located at the base of thorns (Fig. 13.2). The ants are attracted to the host plant by secretions from glandular nectaries at the base of the leaves. The ants protect the plants from other insects and disperse the plant's seeds, and their excrements provide nutrient to the host plant.

Many evergreen plant species produce steroids that simulate the molting and juvenile hormones of insects. These steroids accelerate larval development and may cause premature

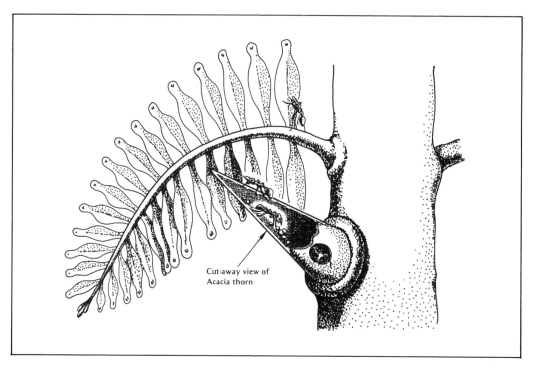

Cut-away view of
Acacia thorn

Figure 13-2. Pseudomyrmex *ants living inside the hollow cavity of thorns of the acacia plant protect the plant from insect predators. The ants feed on gly-* *cogen-rich nodules produced by the plant.* (Adapted from Perry, N. 1983. Symbiosis: Close encounters of the natural kind. Blandford Press, Poole, Dorset.)

death of an insect. Balsam fir (*Abies balsamea*) has evolved an analog of the insect juvenile hormone that arrests larval development and prevents the formation of normal adults. Analogs of insect hormones are the products of fine-tuned coevolutions between plants and insects.

Hypericin is a toxic metabolite secreted from glands of the flowering plant genus *Hypericum.* The chemical, when ingested by animals, produces skin irritation, blindness, and leads to starvation; consequently, the plant is generally avoided by most grazing animals. Beetles of the genus *Chrysolina,* however, can detoxify hypericin and thus have gained access to a food supply that is generally unavailable to other animals.

All members of the mustard family, Cruciferae, produce oils that are harmful to animals and are also effective antibiotics. Insects such as the cabbage butterfly, mustard beetle, and cabbage aphid have evolved mechanisms to detoxify the active ingredients of these oils but in the process have become dependent on the substances for their growth and development.

E. Ecological and Evolutionary Perspectives on Symbiosis

1. THE SYMBIOTIC CONTINUUM. During the past decade, the terms *symbiosis* and *mutualism* have assumed a much broader significance. Throughout this text we have examined the *mechanisms* of symbiosis between interacting species. Evolutionary ecologists, however, have a different perspective of symbiosis and have emphasized the *outcomes* of symbiotic interactions, which they measure in terms of *potential fitness.* For ecologists, symbiosis includes not only intimate associations but also interspecific interactions in which the symbionts are not connected physically, for example, a *diffuse mutualism* such as the relationship between flowering plants and their animal pollinators.

Mortimer Starr and other scientists have developed a classification they call the *symbiotic continuum.* This classification of symbiotic categories is based on the potential fitness, or reproductive ability, of the symbionts. Competition and mutualism are at opposite ends of the

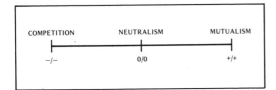

Figure 13-3. The symbiotic continuum classification is based on the potential fitness of interacting species.

continuum and neutralism is in the center (Fig. 13.3). Competition between interacting species produces detrimental outcomes to both species, whereas mutualistic relationships increase the potential fitness of the symbionts. Thus, symbiosis means that two interacting organisms can influence each other's rates of survival and reproduction. Mutualisms are interorganismic interactions that result in improved survival or reproduction of the partners and may involve aids to dispersal, provision of nutrients, energy, shelter, or fertilization. Other continua in Starr's classification deal with the duration of the symbiosis, from transient to permanent; the physical contact between the symbionts, from incidental to close; and nutrition, from saprophytic to biotrophic. For further details of these continua the reader is referred to chapter 1.

2. THEORIES OF MUTUALISM. According to Douglas Boucher and his colleagues, there are four types of models for ecological theories of mutualism. These models are as follows: (1) individual selection model; (2) population dynamics model; (3) model of shifts of interaction; and (4) the "keystone mutualist" model. In the individual selection model, which is usually a "fitness" cost-benefit type, symbiotic associations tend to favor mutualism because the number of competitors that are benefited is less in a mutualistic association. As more needs of an association are met by the combined contributions of the symbionts, the pressure of competition on these symbionts from ecologically similar species will decrease. Population dynamics models begin with the Lotka-Volterra competition theory. More than half a century ago, Alfred Lotka and Vito Volterra used mathematics to understand competition and predator-prey relationships. They devised a model that is now called a phase-plane model and is being used as a prototype to study mutualisms. In this model,

the population densities of two interacting species are plotted along vertical and horizontal axes, thus producing a plane that is separated by lines that form four regions. The regions show different population densities of the two species and correspond to what will happen to the populations; that is, both populations may increase or decrease, or one will increase and the other will decrease. In recent years, the phase-plane model has been applied to over 25 different case histories of mutualism with virtually identical results in the forms of their graphs. The model predicts stable equilibrium, persistence of mutualistic interaction, both species attaining higher population densities in symbiosis than when alone, and decrease of mutual benefit as the population grows larger. Models of shifts of interaction are based on the symbiotic continuum model, which was described above and takes into account fluctuations from one segment of the continuum to another, for example, from competition to mutualism. In the keystone mutualist model, removal of a particular symbiont from an association will produce significant changes in the structure of a community.

3. CATEGORIES OF MUTUALISM. Douglas Boucher argues that mutualism is a major organizing principle in nature and that researchers should be able to test the following predictions.

Mutualists control survival and reproduction; for example, the number of seeds produced in a flower depends on the activity of its mutualist pollinators.

Mutualisms that do not appear to have clear values may, in fact, have definite fitness values. For example, seed dispersal by mutualists is more efficient than by other means such as wind and results in a higher survival value of the seeds.

When mutualistic organisms are either removed from or introduced into a community, the relative abundance of other members of the community is significantly altered.

Interspecific competition in nature often produces *indirect mutualisms,* associations in which species are not in physical contact but positively affect each other's fitness or growth rates; for example, plants with small populations may mimic flowers of other plants, thereby increasing the population size recognized by pollinators.

Mutualistic environments tend to expand because mutualisms can colonize marginal habitats.

Antagonistic interactions stimulate mutualisms to evolve as the antagonists develop defenses against each other.

The biomass of natural as well as artificial communities increases when mutualistic organisms are added to the community, for example, mycorrhizal fungi, and decreases when they are removed.

Sharing of mutualists, as with pollinators, will generally produce indirect mutualisms rather than competition for mutualism.

Many new examples of mutualism in nature remain to be discovered.

Four categories of mutualism in terrestrial ecosystems have been recognized: seed dispersal, pollination, resource harvest, and protection.

a. Seed Dispersal Mutualism. Seed dispersal mutualisms are diffuse ones and involve plants that are generally woody and broad-leaved and many different animals. Seed dispersal occurs in many habitats, from the Arctic to tropical biomes. When an animal swallows a seed and deposits it at another site, the mutualism has taken place. The potential fitness of a seed is increased if it can germinate in a site that is more favorable for growth than where it was produced, that is, away from the parent population. Plant fruits are bait and they protect the seeds as well as provide the mutualists with their reward. Ants disperse seeds by carrying them and frequently burying them. The system provides dispersal for the seeds and food for the ants. There are also nonmutualistic means of seed dispersal such as wind and water and explosive fruits, which forcibly disperse their seeds.

b. Pollination Mutualism. Insects, birds, bats, some rodents, marsupials, and primates make up most of the pollinators. Many of them form diffuse types of associations but others are obligate mutualists. The advantages for the pollinators include food, such as pollen and nectar, and for plants the ability to produce more offspring per gram of pollen produced, more viable offspring, and minimized inbreeding. For further discussion of pollination, the reader should consult chapter 11.

c. Harvest Mutualism. In harvest mutualisms one symbiont gathers or "harvests" nutrients that it converts into a form the other symbiont can use. A unique feature of harvest mutualism is that one organism interacts with a geographically restricted population of the other mutualist. Examples of this type of symbiosis include the gut mutualism of herbivores, fungus-ant gardens, mycorrhizal fungi, and lichens. Alimentary canals of all animals contain complex colonies of microorganisms. Some of these endosymbionts are mutualists or parasites; others are commensalistic. These organisms may form diffuse or obligatory mutualisms. The host pays a low cost for maintenance of the symbionts and receives a high rate of return when the food it ingests is converted into a usable form. Termites and ruminants are examples of *digestive mutualism.* If the animal had to depend on its own ability to digest, it would assimilate less material. Thus, host animals take in large amounts of potential food that cannot be directly assimilated and depend on their mutualistic microorganisms to make available molecules such as vitamins, amino acids, fatty acids, and carbohydrates.

d. Protection Mutualism. One of the best known examples of protection mutualism is the association between ants and the thorns of *Acacia* plants. In this obligate mutualism, ants live in colonies in the hollow chambers of thorns and feed on carbohydrates provided by extrafloral nectaries on the plant. The ants protect the host plant from herbivores (Fig. 13.2).

4. DISTRIBUTION OF MUTUALISM. Although the evidence is incomplete, it appears that mutualistic associations are more common in tropical communities than elsewhere. For example, flowering plants in tropical rain forests and tropical deciduous forests are mostly animal pollinated, whereas the vast northern forests and temperate deciduous forests are mainly wind pollinated. Plant-ant relationships, including those of seed dispersal and protection, are more common in the tropics, as are mutualisms involving algae and coelenterates such as corals, cleaning symbioses, and fungus gardens of insects. The abundance of mutualisms in the tropics may be a reflection of the greater species diversity and productivity in this part of the world. Most tropical soils and waters are poor in nu-

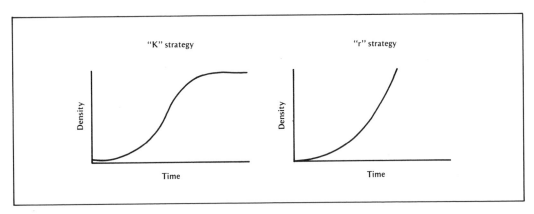

Figure 13-4. Population growth curves for organisms that employ r and K strategies. (Adapted from Esch, G. W., T. C. Hazen, and J. M. Aho. 1977. Parasitism and r- and K- selection. In: Regulation of parasite populations, 9–53, ed. G. W. Esch. Academic Press, New York.)

trients, a condition that favors mutualisms between autotrophic and heterotrphic organisms.

5. r AND K SELECTION: LIFE HISTORY PATTERNS AND SYMBIOSIS. Robert H. MacArthur and Edward O. Wilson in 1967 proposed a concept to elucidate the connection between the abundance of an organism and evolution. According to these scientists, density-dependent selection causes the evolution of high-K traits, whereas r selection is characteristic of an expanding population, which is density independent (Fig. 13.4).

In environments that are constantly changing physically, individuals with high birthrates are often favored by natural selection because their descendants can colonize empty habitats quickly. These organisms usually experience a high degree of mortality among juveniles, and the populations are generally below their saturation levels with little or no competition. The individuals are short-lived and are usually small, fast-growing, and mature early. Organisms such as insects and annual plants have high r values. K-selected species inhabit relatively stable environments in which population levels are consistently high. In these situations, fitness depends more on features that allow individuals to compete for resources, often at the expense of their ability to produce offspring. Populations are at the saturation level most of the time and are regulated by intense competition between and among the species. The individuals are long-lived, grow slowly, and mature late. The most obvious examples of this strategy are found in tropical forests, coral reefs, and specialized environments such as termite intestine or the rumen of herbivores.

Organisms under r selection are often described as *physically controlled,* and the K-selected species are *biologically accommodated.* The r and K selection model offers a broad framework within which to consider the forces of natural selection that regulate symbiotic organisms and their hosts. Gerald Esch and his colleagues have examined the concept of r and K selection within the context of animal parasitism. They have shown that characteristics of morphology, physiology, and life history of parasitic symbionts can be used to determine population density dynamics.

A biological community shows the range of r and K continuum, the r end representing organisms that expend most of their energy on reproduction and little on each individual. K selection produces stabilized symbioses, with efficient use of the environment and with the organisms saturating the environment; competition among individuals is intense. Application of the r and K selection concept to the study of symbiosis is relatively new.

6. EVOLUTION IN A SYMBIOTIC ENVIRONMENT. The evolution of a species in the absence of interspecific interactions is largely determined by forces of natural selection. Ecologists realize, however, that a population of any given species is embedded in the matrix of a community and interactions with other species will produce genetic changes resulting in an evolutionary feedback. Interspecific dynamics such as those involving predator-prey, host-pathogen, and competitive relationships are antagonistic interactions and, therefore, cause deterioration of the evolutionary environment.

Populations of mutualists will bring about an evolutionary change in other mutualists, with the following predictable results: (a) genetic change in the mutualists will be slow; (b) there will be tendency toward a lack of specificity between the interacting populations; and (c) there will be a reduction in the importance of sexual reproduction. Intracellular symbioses that involve two or more species have had a profound influence on the evolution of the symbionts as well as on the environments in which they grow. Examples of such symbioses are those of reef-building corals and foraminifera, which are described in chapter 9. In both cases, symbiosis with unicellular algae has resulted in changes of size and shape of the hosts and an increased number of host species.

One of the unresolved mysteries of evolutionary biology is the role of selection that favors the maintenance of sex in natural populations. Scientists have suggested that antagonistic forces in an environment favor the evolution of sexual reproduction. Conversely, in environments where mutualistic interactions predominate, asexual reproduction is the favored strategy because it results in continued genetic homogeneity of the mutualists, thereby ensuring continuity of the symbiosis.

As a matter of tradition, scientists have considered the environment a product of geological and physico-chemical processes to which living organisms have to adapt or perish. This view has been challenged by James Lovelock and Lynn Margulis, who have proposed the *Gaia hypothesis*. According to this hypothesis, the composition of the atmosphere, sediments, and aquatic environments is controlled by living organisms in their interactions with the environment. Present levels of atmospheric oxygen, carbon dioxide, and nitrogen and temperature ranges are products of life mechanisms. Thus, in the broadest sense, there is a coevolution of life forms and the environmental conditions and each depends on the other for its existence and maintenance.

F. Symbiosis and Human Affairs

According to René Dubos, the late renowned microbiologist, all living things are mutually in-

terdependent. Earlier, the Russian philosopher Peter Kropotkin wrote a book entitled *Mutual Aid: A Factor of Evolution* (1902), which became an alternative explanation to the "struggle for existence" and "survival of the fittest" concepts. Kropotkin felt that cooperation and mutual help, rather than struggle, had been the principal forces of evolutionary change. Cooperation resulted from an instinct of preservation, which gave to species and to individuals a better chance of survival in stressful environments. Even Charles Darwin objected to the simple-minded approach of "devour or be devoured" as a mechanism of selection and evolution. In *The Descent of Man* Darwin wrote that "in numberless animal societies, the struggle between separate individuals for the means of existence disappears. Struggle is replaced by cooperation."

The germ theory of disease, developed during the height of the Darwinian age, provided supportive evidence for the struggle for survival. With much determination, civilized man declared war against microbial infections. Ecological equilibrium between host and microbes was not then a popular view among scientists. As the twenty-first century approaches, man has become a dominate force on earth, but he still struggles with symbionts that cause diseases in himself, his crop plants, and his domestic animals. Human diseases such as leprosy, the black plague of the Middle Ages, typhus, malaria, and yellow fever have had consequences of historic proportions, as have plant disease epidemics such as the potato blight in Ireland.

Tremendous progress has been made in curbing catastrophic diseases. But mass-scale applications of pesticides have polluted our soils, water, and air, and at the same time the pest organisms with millions of years of evolutionary history have successfully resisted man-made challenges to their survival. We have learned several lessons in the past three decades: (a) a symbiont can rapidly develop tolerance to man-made poisons; (b) instability in the ecological equilibrium of the interacting species is an important factor in an outbreak of disease epidemics; (c) genetic uniformity in crop plants and domestic animals is an invitation for epidemics; and (d) a multifaceted approach to pest management is the best means of pest control.

At the present time, distances between nations and continents have shrunk to a few hours of travel time and symbionts move across geographical barriers with ease. Containing epidemics such as those of AIDS or Mediterranean fruit fly infestations of citrus fruit challenges the resourcefulness of scientists.

In agriculture, concepts of symbiosis have been applied to the biological control of insects, increasing nitrogen fixation by symbiotic cyanobacteria, and applications of mycorrhizal symbionts to forest seedlings in order to promote their growth and thus stop soil erosion. Symbiosis provides the conceptual tool for the management of an ecosystem. Its alternative has been the application of chemicals that eliminate symbioses dependent on natural biological interactions. Ever since it became known that *Rhizobium* fixes atmospheric nitrogen within legume root nodules, the application of this symbiosis has become part of modern agricultural practices. Legume crops such as alfalfa have been used as green manure, in crop rotation, and in mixed cropping. Artificial inoculation of legume seeds with strains of *Rhizobium* is a regular practice in many countries. The *Azolla-Anabaena* symbiosis is most active in flooded rice fields, where populations of the symbiotic fern can fix up to three kilograms of atmospheric nitrogen per hectare per day. In rice fields of China and Vietnam, *Azolla* has been used for centuries as a source of nitrogen, with the result that Asian farmers can harvest up to two tons of rice per hectare without using chemical fertilizers. Modern rice varieties need up to 20 kilograms of nitrogen per hectare for each ton of rice produced. Scientists are exploring the possibility of improving the use of this symbiosis by manipulating cultural practices.

Black locust, alder, and *Casuarina* are trees with nitrogen-fixing *Frankia* symbionts. These trees have been used successfully in the reforestation of barren land.

Almost all higher plants form mycorrhizal associations. Mycorrhizas are increasingly being recognized as an important factor in the uptake of nutrients, especially phosphorous, by plants. Donald Marx has demonstrated that pine seedlings inoculated with ectomycorrhizal fungi grow faster and survive better than uninoculated seedlings. Commercial application of vesicular-arbuscular mycorrhizas has yet to become established. These mycorrhizas are also known to increase nitrogen fixation in nodulated legume plants.

The alimentary canals of all animals are unique ecosystems that contain large numbers of interdependent microbial species. Ruminants, with their modified stomachs, have successfully exploited the energy of cellulose and lignin by fermentation symbioses.

Social scientists have found symbiosis to be a useful explanation for a variety of phenomena. For example, Peter Corning has applied the general principles of mutualistic symbiosis to the evolution of human social dominance. According to his synergistic hypothesis, cooperation enhances survival and reproduction of humans. Similarly, psychologists are using symbiosis as a conceptual framework to study interactions between parents and their offspring.

The increasing ability of humans to manipulate natural symbiotic associations for their own advantage and create artificial symbioses through genetic engineering will have far-reaching consequences within the next few decades. Whether all consequences will be positive is difficult to predict. Parasitic symbionts have a long history of survival and despite our best efforts may still invent ways to circumvent our attempts to eradicate them.

G. Summary and Perspectives

The concept of symbiosis and coevolution has attracted much attention from evolutionary biologists and ecologists who are not satisfied with the classical Darwinian approaches to organic evolution. *Pairwise coevolution* occurs between two species, whereas *diffuse coevolution* involves larger numbers of species. Coevolution involves reciprocal genetic changes in the interacting populations and includes coadaptation as well as cospeciation.

Gene-for-gene relationships between parasites and hosts have been found in both plant and animal symbioses. Plants may have clusters of genes that confer resistance to more than one pathogen. In some host-parasite interactions the expression of disease by one parasite depends on the presence of another parasite.

Parasitic symbioses are thought to be the principal causes of genetic polymorphism within a population. Resistance to pathogenic symbionts by vertebrate hosts involves immune mechanisms, cross-protection, and sequestration of the parasites. New advances in the molecular biology of immune responses are offering unique opportunities for combating infectious illnesses. Some parasites have evolved unique ways to protect themselves from the host's immune systems.

Plant responses to parasites include hypersensitivity, cross-protection, and the production of phytoalexins. Plants have evolved morphological and chemical adaptations to protect them from parasitic organisms and herbivores. Some insects have evolved mechanisms to detoxify toxic compounds produced by plants.

By measuring the potential fitness of a species, ecologists are reexamining the role of competition and mutualism in the coevolution of interacting species. The symbiotic continuum is a classification based on the potential fitness of the partners of a symbiotic association. Competition between interacting species produces a decreased fitness, whereas mutualisms result in enhanced potential fitness of both partners.

There are four main models of mutualism: individual selection model, population dynamics model, shifts of interaction model, and keystone mutualist model. Some scientists feel that mutualisms are an important organizing principle in nature and have a significant influence on the environment and community. Mutualisms result in the evolution of stable ecosystems with increased parasexual forms of reproduction. Mutualism in terrestrial ecosystems includes those of seed dispersal, pollination, resource harvest, and protection.

Although the phenomenon of mutualism has been recognized for over one hundred years, it has not been part of the development of modern ecology, despite its prevalence in many biological communities. Only recently have ecologists found the subject of mutualism to be interesting and worthy of serious study. Such interest is welcome, since ecologists, with their theoretical perspectives, bring a new dimension to a subject that has long fascinated descriptive biologists.

Symbiosis can be viewed in terms of r and K selection. Most parasites demonstrate the r-selection strategy; K-selected species are those that occur in stable or specialized environments such as those of tropical forests or the rumen of herbivores.

According to the Gaia hypothesis, the earth's environment is a product of interactions between life and nonlife molecules, and its fragile equilibrium is ultimately controlled by the dynamics of living organisms.

The concept of symbiosis has been applied to human affairs in which cooperation and mutual aid have been the principal forces of evolutionary change. The importance of studying symbiosis is underscored by our increased ability to control pathogenic organisms by understanding the molecular mechanisms of biological interactions. In agriculture, our understanding of symbiotic systems has been of great value in the biological control of insect pests, increasing nitrogen production by natural means, and the artificial inoculation of forest trees and crop plants with mycorrhizal and nitrogen-fixing symbionts.

Review Questions

1. What are the two major subdivisions of coevolution?
2. Why is coevolution considered a type of "arms race"?
3. Describe the gene-for-gene hypothesis.
4. What is meant by *genetic polymorphism?*
5. Describe two types of resistance strategies in plants.
6. Define r and K selection and give the characteristics of each type.
7. Discuss the evolutionary significance of secondary plant chemicals and give a few specific examples.
8. Discuss the various theories of mutualism.
9. Describe the concept of compartmentalization in plant disease symbiosis.
10. Describe the genetic nature of host-parasite relationships in plants.
11. Discuss the immunological basis of resistance in vertebrates and hosts.
12. List areas of study where the concepts of mutualism may be tested experimentally.
13. Define the following terms:
    a. coadaptation
    b. epistatic parasitism
    c. cross-protection
    d. Eichler's rule
    e. symbiotic continuum
    f. Gaia hypothesis
14. Write an essay on the application of symbiology in agriculture, forestry, and human medicine.

Further Reading

Anderson, R. M., and R. M. May. 1982. Coevolution of hosts and parasites. Parasitology (UK) 85:411–426. (Presents mathematical models to explore host-parasite coevolution.)

Boucher, D. H., S. James, and K. H. Keeler. 1982. The ecology of mutualism. Ann. Rev. Ecol. Syst. 13:315–347. (A comprehensive ecological perspective of symbiosis and mutualism.)

Brooks, D. R. 1979. Testing the context and extent of host-parasite coevolution. Systematic Zoology 28:299–307.

Bryant, J. P. 1981. Hare trigger. Nat. Hist. 90 (November): 46–52. (Author suggests how a toxin produced by woody plants may play a role in regulating the ten-year reproductive cycle of the snowshoe hare.)

Combe, R. D., P. L. Ey, and C. R. Jenkins. 1984. Self/non-self recognition in invertebrates. Quart. Rev. Biol. 59:231–255.

Corning, P. A. 1983. The synergism hypothesis: A theory of progressive evolution. McGraw-Hill, New York. 492 pp.

Dawkins, R., and J. R. Krebs. 1979. Arms race between and within species. Proc. R. Soc. Lond. B. 205:489–511. (Considers mutualism to be the result of a mutually exploitative arms race.)

Day, P. R. 1984. Genetics of recognition systems in host-parasite interactions. In: Cellular interactions: Encyclopedia of plant physiology, new series, Vol. 17, 134–147, ed. H. F. Linskens and J. Heslop–Harrison. Springer-Verlag, Berlin.

Ehrlich, P. R., and P. H. Raven. 1967. Butterflies and plants. Sci. Am. 216 (June): 104–113. (A popular article by the scientists who developed the concept of coevolution.)

Esch, G. W., T. C. Hazen, and J. M. Aho. 1977. Parasitism and r- and K- selection. In: Regulation of parasite populations, 9–53, ed. G. W. Esch. Academic Press, New York.

Heslop-Harrison, J. 1978. Cellular recognition systems in plants. Institute of Biology, Studies in Biology, no. 100. University Park Press, Baltimore. 60 pp.

Inouye, D. 1984. The ants and the sunflower. Nat. Hist. 93 (June): 49–52. (Since sunflowers lack chemical defenses, they rely on ants to protect them from herbivorous insects.)

Lackie, A. M. 1980. Invertebrate immunity. Parasitology (Uk) 80:393–412.

Leder, P. 1982. The genetics of antibody diversity. Sci. Am. 246 (May): 102–115.

Mackenzie, C. D. 1984. Sequestration, beneficial to both host and parasite. Parasitology (UK) 88:593–595.

McLaren, J. D. 1984. Disguise as an evasive stratagem of parasitic organisms. Parasitology (UK) 88:597–611.

May, R. M. 1983. Parasitic infections as regulators of animal populations. Am Sci. 71:36–45. (Describes the dynamic relationships between parasites and their hosts with regard to the control of infectious diseases.)

Rosenthal, A. G. 1986. The chemical defenses of higher plants. Sci. Am. 254 (January): 94–99. (An excellent introduction to the newly emerging field of chemical ecology.)

Schlutz, J. C. 1983. Tree tactics. Nat. Hist. 92 (May): 12–25. (Describes how plants defend themselves against herbivory by insects.)

Sequeira, L. 1983. Mechanism of induced resistance in plants. Ann. Rev. Microbiol. 37:57–79. (Explores the role of elicitors and similar compounds in the control of plant diseases.)

Shigo, A. L. 1985. Compartmentalization of decay in trees. Sci. Am. 252 (April): 96–103.

Sidhu, G. 1984. Genetics of plant and animal parasitic systems. BioScience 34:368–373. (Comprehensive review of the genetic nature of host resistance and pathogenic virulence.)

Starr, M. P. 1975. A generalized scheme for classifying organismic associations. In: Symbiosis: Symposia of the Soc. for Experimental Biology, no. 29, 1–20, ed. D. H. Jennings and D. L. Lee. Cambridge Univ. Press, Cambridge.

Wakelin, D. 1978. Genetic control of susceptibility and resistance to parasitic infections. In: Advances in parasitology, 217–308, ed. W. Lumsden, R. Muller, and J. Baker. Academic Press, New York. (Comprehensive review of genetics of animal parasitic systems covering protoctist, helminthic, and arthropod parasites.)

Bibliography

Bailey, J. A., and B. J. Deverall (ed.). 1983. The dynamics of host defense. Academic Press, Sydney. 233 pp.

Bailey, J. A. and J. W. Mansfield. 1982. Phytoalexins. John Wiley, New York. 334 pp.

Boucher, D. H. (ed.). 1985. The biology of mutualism. Croom Helm, London. 388 pp. (Considers ecological and evolutionary aspects of mutualism mainly from a theoretical perspective.)

Clark, B. C. 1979. The evolution of genetic diversity. Proc. R. Soc. Lond. B. 205:453–474.

Cohen, S., and K. S. Warren (eds.) 1982. Immunology of parasitic infection. 2d ed. Blackwell Scientific Publications, Oxford. 864 pp.

Daley, J. M., and I. Uritani (eds.). 1979. Recognition and specificity in plant host-parasite interactions. University Park Press, Baltimore. 355 pp.

Day, P. R. 1974. Genetics of host-parasite interaction. W. H. Freeman, San Francisco. 238 pp.

Dubos, René. 1959. Mirage of health. Perennial Library, Harper and Row, New York. 292 pp.

Ewald, P. W. 1983. Host-parasite relations, vectors, and the evolution of disease severity. Ann. Rev. Ecol. Syst. 14:465–485.

Futuyma, D. J., and M. Slatkin. 1983. Coevolution. Sinauer, Sunderland, Mass. 555 pp.

Gould, S. J., and N. Eldredge. 1977. Punctuated

equilibria: The tempo and mode of evolution reconsidered. Paleobiology 3:115–151.

Gould, S. J., and R. C. Lewontin. 1979. The spandrels of San Marco and the Panglossian paradigm: A critique of the adaptationist programme. Proc. R. Soc. Lond. B. 205:581–598.

Haldane, J. B. S. 1949. Disease and evolution. Ricerca Scient. suppl. 19:68–76.

Hamilton, W. D. 1982. Pathogens as causes of genetic diversity in their host populations. In: Population biology of infectious diseases, 269–296, ed. R. M. Anderson and R. M. May. Springer-Verlag, New York.

Heithaus, E. R., D. C. Culver, and A. J. Beattie. 1980. Models of some ant-plant mutualisms. Am. Nat. 116:347–361.

Howe, H. F. 1984. Constraints on the evolution of mutualism. Am. Nat. 123:764–777.

Kosuge, T., and E. W. Nester. 1984. Plant-microbe interactions. Vol. 1, Molecular and genetic perspectives. Macmillan, New York. 444 pp.

Lovelock, J. E. 1979. Gaia. Oxford Univ. Press, Oxford. 157 pp.

MacArthur, R. H., and E. O. Wilson. 1967. The theory of island biogeography. Princeton Univ. Press, Princeton, N.J. 203 pp.

Margulis, L. 1981. Symbiosis in cell evolution. W. H. Freeman, San Francisco. 419 pp.

Misaghi, I. J. 1982. Physiology and biochemistry of plant-pathogen interactions. Plenum Press, New York. 287 pp.

Nitecki, M. H. (ed.). 1983. Coevolution. Univ. of Chicago Press, Chicago. 392 pp.

Pontin, A. J. 1982. Competition and coexistence of species. Pitman Books, London. 102 pp. (A student's monograph on interspecific competition.)

Rosenblum, L. A., and H. Moltz (eds.). 1983. Varieties of mutualistic interactions in population models. J. Theor. Biol. 74:549–558.

Ruehle, J. A., and D. H. Marx. 1979. Fiber, food, fuel, and fungal symbionts. Science 206:419–422.

Slatkin, M., and J. M. Smith. 1979. Models of coevolution. Quart. Rev. Biol. 54:233–263. (Presents a variety of mathematical models of coevolution.)

Thompson, J. N. 1982. Interaction and coevolution. John Wiley, New York. 179 pp.

Vandermeer, J. H., and D. H. Boucher. 1978. Varieties of mutualistic interactions in population models. J. Theor. Biol. 74:549–558.

APPENDIX

Historical Landmarks in Symbiology

1500 B.C.	Papyrus Ebers discovers the human intestinal parasite *Ascaris lumbricoides* and the tissue parasite *Dracunculus medinensis*.
A.D. *1200*	Albertus Magnus describes the first parasitic symbiont of a green plant—a mistletoe.
1674–81	Antoni van Leeuwenhoek describes microscopic protoctists such as *Emeria* and *Giardia*.
1684	Francesco Redi, father of parasitology, was the first scientist to search for the parasitic symbionts of fish, birds, mammals, and humans and reports the presence of these organisms in intestines, air sacs, and swim bladders. Redi refutes the theory of the spontaneous generation of life in 1668 by demonstrating that maggots develop from the eggs of flies.
1743	Turbevill Needham describes the presence of the nematode *Anguina tritici* in cockled wheat seeds.
1755	Matlieu Tillet, a pioneer experimentalist, proves that bunt disease of cereals is contagious.
1796	Edward Jenner develops the first method of inoculation against an infectious disease (vaccination against smallpox).
1807	Isaac-Bénédict Prévost, the founding father of plant pathology, experimentally proves that a plant disease (bunt disease of wheat) is caused by a fungus.
1835	Agostino Bassi demonstrates that a fungus, *Beauveria bassiana,* causes muscardine disease of silkworm larvae. This was the first experimental proof that microbes cause animal disease.
1836–44	J. Schmidberger and T. Hartig coin the term *ambrosia* and recognize its fungal nature.
1840–44	David Gruby, founder of medical mycology, experimentally shows that fungi cause diseases in man and animals.
1842	Johann J. Steenstrup demonstrates the penetration of a snail's body by cercariae and elucidates the life cycle of a liver fluke. His idea of alternation of generations greatly accelerated research on the life histories of flukes, tapeworms, and nematodes.
1853	H. Anton de Bary unravels the complex life cycle of wheat stem rust fungus and lays the foundation for the experimental study of plant diseases. In 1879, de Bary coins the term *symbiosis.*
1857–60	Louis Pasteur discovers fermentation as a microbial process and produces artificial immunity to anthrax, chicken pox, cholera, and rabies.
1858	J. G. Kuhn writes one of the early textbooks on the theme that pathogens cause disease (the germ theory of disease).
1858	Rudolf Virchow writes a classic book, *Cellular Pathology,* in which disease processes are viewed as abnormalities at the cellular level.

1860	Claude Bernard formulates the concept of homeostasis. In later decades the internal environment of the host is recognized as a specialized habitat for symbionts.
1867	Simon Schwendener proposes the dual hypothesis to explain the nature of lichens, and states that lichens are fungi that live parasitically on algal hosts.
1875	William G. Farlow, after completing studies under de Bary, begins the study of fungi in the United States.
1876	P.-J. van Beneden coins the terms *commensalism* and *mutualism*.
1876–84	Robert Koch proves that in anthrax disease a specific symbiont produces a specific disease and develops criteria for establishing microbes as disease agents (Koch's postulates).
1878	Thomas J. Burrill was the first scientist to show that bacteria cause plant disease by demonstrating the cause of fire blight of pears.
1878	Patrick Manson observes the development of *Wuchereria bancrofti* in the body of female mosquitoes and demonstrates the importance of the intermediate host in the life history of a pathogenic symbiont.
1878	M. Woronin describes the nature of legume nodules.
1880	Alphonse Laveran describes the malarial organism *Plasmodium malariae* in the red blood cells of humans; he wins a Nobel Prize in 1907.
1881–82	Rudolph Leuckart and Algernon Thomas describe the life cycle of a digenetic trematode *Fasciola hepatica*.
1883	A. F. W. Schimper suggests that chloroplasts may have evolved from blue-green bacteria.
1884	Elie Metschnikoff discovers the phenomenon of phagocytosis and pioneers studies on cellular immunity; he wins a Nobel Prize in 1908.
1885	A. B. Frank coins the term *mycorrhiza* to describe the fungus growth around the roots of woody plants.
1888	Hermann Hellriegel, H. Wilfarth, and Martinus W. Beijerinck discover the role of root nodule bacteria in the legume symbiosis and their significance in the nitrogen cycle.
1888	Roland Thaxter begins his lifelong studies on the Laboulbeniales.
1890	S. Kitasato and Emil von Behring discover antibodies for diphtheria and tetanus antitoxins.
1891	Theobald Smith successfully transmits the cattle fever organism by using ticks as vectors.
1891–1903	Dimitri Ivanovski describes the first plant virus disease, tobacco mosaic virus, and demonstrates its filterability. He made microscopic observations and described crystalline inclusions of the virus in diseased tobacco leaves.
1894	Richard Altmann reports on the discovery of mitochondria.
1895	David Bruce discovers the role of the tsetse fly in the transmission of African sleeping sickness.
1897	Friedrich Loeffler and Paul Frosch discover the first animal virus, the causal agent of foot-and-mouth disease.
1897	Ronald Ross discovers the sexual stages of the malarial organism in mosquitoes and was awarded a Nobel Prize in 1902.
1898	Martinus W. Beijerinck experimentally isolates the tobacco mosaic virus and describes it as *contagium vivum fluidum* because of its exclusively intracellular mode of reproduction.
1905	K. C. Mereschkovsky proposes the bacterial origin of mitochondria.
1915	Frederick W. Twort discovers bacteriophages.
1921	Paul Buchner, a pioneer on research in mutualism, writes a classic textbook, *Endosymbiosis of Animals with Plant Microorganisms*.
1921	Elias Melin synthesizes a mycorrhizal association and describes its physiological relationships.
1923–25	Lemuel R. Cleveland explores the relationship of symbiotic intestinal flagellates with their termite and cockroach hosts.

1924 W. Goetsch reports on the mutualistic association between hydra and unicellular algae.

1927 Ivan E. Wallin presents evidence for the bacterial nature of mitochondria and the role such bacteria may have played in the origin of species.

1928 Frederick Griffith discovers the transformation of pneumococci from avirulent to virulent strains following exposure to extracts of killed virulent bacteria. The nature of the "transformation factor" became a subject of intense research activity.

1929 Alexander Fleming reports on the discovery of the antibiotic penicillin.

1933 Charles Drechsler, a pioneer in the study of nematode-trapping fungi in the United States, begins a long and productive career.

1935–37 Wendell M. Stanley isolates and crystallizes the tobacco mosaic virus, which earned him a Nobel Prize in 1946.

1938 T. M. Sonneborn reports on the "killer factor" in *Paramecium.*

1944 Oswald T. Avery, Colin M. MacLeod, and Maclyn McCarty identify Griffith's transforming factor as being nucleic acid with little or no protein. Nucleic acid is later recognized as the hereditary material that made up genes.

1950 Andrew Lwoff and A. Gutman demonstrate that a lysogenic strain of *Bacillus megatherium* carries a noninfectious form of virus (prophage) and that ultraviolet light induces the prophage to produce active viral particles.

1952 Joshua Lederberg coins the term *plasmid* to describe extranuclear genetic structures that can reproduce independently.

1952 Norton D. Zinder and Joshua Lederberg discover transduction, the transfer of genes by viruses.

1954 A. K. Sarkisov publishes the first major textbook on mycotoxicoses and ergotism.

1955 Harold Flor proposes the gene-for-gene hypothesis to explain the nature of the host-parasite relationship of the flax rust, *Melampsora lini.*

1955–56 Heinz Fraenkel-Conrat successfully reconstitutes an active tobacco mosaic virus from protein and nucleic acid components derived from separate sources and demonstrates that RNA is the infective component of the virus.

1957 A. Isaacs and J. Lindemann discover interferon.

1962 Heinz Stolp discovers *Bdellovibrio,* a parasitic symbiont of bacteria.

1962 T. O. Diener discovers the viroid, a new type of infectious agent that consists only of RNA.

1964 Paul R. Ehrlich and Peter H. Raven present the concept of coevolution in an article titled "Butterflies and plants: A study of coevolution."

1970 Clark Reed writes a textbook titled *Parasitism and Symbiology* in which traditional parasitology is considered within the broader scope of symbiosis.

1970 Lynn Margulis uses information from cellular and molecular biology to promote the serial endosymbiotic theory for the origin of the eukaryotic cell.

Glossary

abiotic pathogen A nonbiological entity that produces disease.

acclimation behavior Activity of a clown fish that gains it protection from anemones.

acquired immunity Immunity resulting from host's antibody response.

actinophage Virus that infects actinobacteria.

aeciospore In rust fungi, a binucleate spore that produces a dikaryotic mycelium.

aflatoxins Mycotoxins produced by the fungus *Aspergillus.*

ambrosia fungi Fungi that grow in tunnels of wood-boring beetles.

amoebiasis An infection caused by amoebae.

apicomplexans Parasitic protozoans with complex life cycles and a unique cell structure, the apical complex.

aposymbiotic host A host that is freed of its endosymbionts.

appeasement substances Chemical compounds, produced by some insects, that calm their hosts.

arbutoid mycorrhiza A form of ectomycorrhiza that has a sheath, Hartig net, and root cell penetration.

bacteriophage A virus that parasitizes bacteria.

bacteroids Transformed cells of rhizobia in root cells.

basidiospore Spore produced by basidiomycetes after sexual recombination.

bdellophage Virus that parasitizes the bacterium *Bdellovibrio.*

biotic pathogen Organism that produces disease in other organisms.

biotrophic symbiont Organism that obtains nutrients from a living host.

chemoautotrophs Bacteria that use chemical energy to fix carbon dioxide.

cercaria Larval stage of a fluke parasite.

Chagas' disease Disease of humans and other mammals caused by *Trypanosoma cruzi.*

chlamydospore Asexual spore produced by smuts and other fungi.

circumsporozoite protein Protective protein coat on sporozoite of *Plasmodium.*

cleaning symbioses Associations between marine fishes and smaller fishes and shrimps called cleaners.

commensalism Association in which one symbiont benefits while the other is unaffected.

compartmentalization *See* sequestration.

concomitant immunity *See* premunition.

cross-protection Immunity conferred on a host plant by some viruses.

crowding effect Size of parasites in a host is inversely related to the number of parasites present.

cyanophage Virus that parasitizes cyanobacteria.

cyst A thick-walled stage of many protozoans.

cytocosm A view of the cell as a unique microhabitat.

dermatophytes Fungi that infect the scalp, skin, and nails of vertebrates.

diffuse coevolution Coevolution that occurs between one or more species and several other species.

dikaryon Cell with two genetically different haploid nuclei.

ectosymbiont Symbiont that lives outside of its host.

ectendomycorrhiza A mycorrhiza that lacks an outer sheath but has a Hartig net and penetrates root cells.

ectomycorrhiza A mycorrhiza that forms an outer sheath and a Hartig net but does not penetrate root cells.

endocytobiology A new subdiscipline of biology that involves cell biology and symbiosis.

endosymbiont A symbiont that lives inside its host.

energy parasitism Sexual maturation of a tapeworm caused by the body heat of a host.

entodiniomorphs Mutualistic rumen ciliates with complex cytoskeletons.

entomogenous fungi Fungal parasites of insects.

epistatic parasitism Disease expression by one parasite dependent on the presence of another parasite.

espundia Leishmanial disease of the mouth, nose, and pharynx; endemic to jungles of South America.

ericoid mycorrhiza Mycorrhiza that forms hyphal coils in root cells but does not form a sheath or Hartig net.

facultative symbiont An organism that can live free-living or as a symbiont.

flagellum Organelle used for cell motility in bacteria; *see also* undulipodium.

fossil symbioses Early associations preserved in amber.

Gaia hypothesis The idea that life forms have coevolved with their environment.

gene-for-gene relationship Genes for host resistance are matched by genes for virulence in the parasite.

genetic colonization Introduction of genetic material into a plant host genome by agrobacteria.

giardiasis An intestinal disorder caused by *Giardia lamblia*.

gigantism Weight gained by a host because of a parasitic infection.

glycocalyx Outer mucopolysaccharide layer of a helminthic tegument.

gnotobiology Study of germ-free organisms.

gregarines Extracellular symbionts in the digestive tracts of invertebrates.

Hartig net A hyphal network between root cortical cells formed by mycorrhizal fungi.

haustorium In fungi, a specialized hyphal branch inside a living host cell.

heterokaryosis A condition in some fungi where hyphal cells have two or more genetically different nuclei.

heterotrophic A mode of nutrition where organic molecules are the primary source of energy.

hemiparasites Parasitic plants that can also manufacture food through photosynthesis.

heterocyst In cyanobacteria, a structure in which nitrogen is fixed.

heterozygote advantage Survival strategy against malaria by individuals heterozygous for the sickle cell hemoglobin gene.

holoparasite A plant that obtains all of its nutrients from the host plant.

holotrichs Mutualistic rumen ciliates.

humoral immunity Immunity caused by antibodies in the blood and lymph.

hyperparasitism Parasitism of one parasite on another.

hypersensitive response In plants, the rapid death of infected cells.

hypertrophy Increase in tissue mass resulting from enlarged cells.

hypovirulence Subnormal virulence or reduced pathogenic fitness.

immunoglobulins Antibodies produced by vertebrate hosts in response to foreign antigens.

inguilinism Sharing of living quarters by two or more different animal species.

infection thread A tubular extension of a legume root cell wall that contains rhizobia and grows into the root cortex.

innate immunity Nonspecific resistance to an infection; e.g. skin barrier.

interferon A protein, produced as a result of viral infection, that interferes with viral replication.

K selection A survival strategy of some organisms that is characterized by low mortality, long lifespans, and large populations.

keystone mutualist model A theory of mutualism that predicts that removal of a particular symbiont from an association will significantly change the community structure.

kinetoplast DNA In trypanosomes, DNA that consists of a unique network of circles.

lectins Proteins that may be involved in the early recognition stages between symbionts in the *Rhizobium*-legume symbiosis.

leghemoglobin A red pigment produced in legume root nodules that binds and stores oxygen.

lymphokines Compounds produced by T and B cells that activate macrophages and help localize immune responses.

lysogenic bacteria Bacteria that carry a prophage as part of their chromosome.

merozoite Asexual stage produced by apicomplexans following schizogony.

metacercaria Resistant stage of parasitic flukes.

microfilariae Juvenile stages of filarial nematodes.

microthrix (microtriches) Small projection of the tapeworm tegument.

molecular mimicry Ability of a parasite to produce surface antigens that are similar to those of its host.

monotropoid mycorrhiza A mycorrhiza that forms a thick outer sheath, a Hartig net, and penetrates root cells; occurs in heterotrophic plants such as *Monotropa.*

mutualism An association in which both partners benefit.

mycelium A mass of fungal hyphae.

mycetangium A structure on an insect that contains a fungal colony and is used to infect new substrates.

mycetocytes Insect cells that contain bacteria or yeastlike symbionts.

mycetomes Specialized organs of insects that contain permanent bacterial or fungal symbionts.

mycosymbionts Fungal symbionts.

mycobiont Fungal symbiont of a lichen.

mycoparasitism Parasitism of one fungus on another fungus.

mycophycobioses Obligate associations of marine fungi and large marine algae.

mycoplasmas Bacteria without cell walls; smallest known prokaryotes.

mycoses Animal and human diseases caused by fungi.

mycotoxins Poisons produced by fungi that infect stored grains.

mycoviruses Viruses that infect fungi.

nectaries Nectar-producing glands on flowering plants.

necrotrophic symbiont Organism that obtains nutrients from a dead host.

nematophagous fungi Fungi that prey on amoebae, nematodes, and other small animals.

nodules Gall-like structures on legume and tree roots that house nitrogen-fixing bacteria.

oncogenes Cancer-causing genes in viruses.

onchoceriasis A disease of eyes and skin caused by filarial worms.

ookinetes Motile, worm-like zygotes of *Plasmodium.*

opalinids Protistan symbionts of amphibians.

pairwise coevolution Coevolution that occurs between two species.

parasexuality Genetic recombination in filamentous fungi that involves fusion and segregation of haploid nuclei in a hyphal cell.

parasitism An association in which one organism obtains food at the expense of another organism.

parasitoid Insect that parasitizes immature stages of other insects.

pathogenic fitness Disease-causing ability of a pathogen.

peribacteroid membrane Host-derived plasma membrane that surrounds rhizobia within a root cell.

phagosome Host-derived membrane-bound space that contains intracellular symbiont.

premunition Immunity that lasts as long as the parasite is present in the host; also called concomitant immunity.

periplasmic space Space between bacterial cell wall and plasma membrane; natural habitat of *Bdellovibrio.*

phase plane model Mathematical formulation of population densities of two interacting species.

phoresis An association in which one organism protects or transports another organism.

photobiont Photosynthetic symbiont of a lichen.

phytoalexins Protective chemical compounds produced by some plants in response to fungal infections.

plaques Clear, circular areas in a bacterial lawn resulting from the killing and lysis of bacterial cells by parasites.

plasmid A small piece of DNA that replicates independently of the main chromosome in a bacterial cell.

prions Infectious protein molecules that appear to lack nucleic acids.

prophage Repressed form of a phage present in lysogenic bacteria.

prophage interference Phage that confers immunity in host bacterium to other unrelated phages.

promiscuous DNA DNA that moves between mitochondria, chloroplasts, and nuclei.

protooncogene An altered human gene that can be transformed into an oncogene.

pseudocyst A dormant stage of *Toxoplasma gondii.*

pycniospore Spore of a rust fungus that is produced inside a pycnium.

r-selection A survival strategy of some organisms that is characterized by high birth rates, high mortality, and short life spans.

retroviruses RNA-containing viruses that carry the gene for the enzyme reverse transcriptase and require in their replication a DNA stage.

rumination In ruminants, the regurgitating and chewing of food from the rumen.

rust fungi Obligate parasites of cereal crops and other plants.

saprobic Using dead organic matter as food.

satellite viruses Viruses that replicate only in the presence of other viruses.

schistosomulum Juvenile stage of a blood fluke that penetrates the human skin.

schizogony Asexual reproduction of protozoans.

self-cure phenomenon Expulsion of worms from the stomach of sheep following a challenge infection and subsequent resistance to further infection.

sequestration Isolation of an invading parasite by the host.

slow viruses Viruses that require long periods of incubation.

sporocyst Stage of development of some apicomplexan and fluke symbionts.

stichosomes Secretory cells of *Trichinella spiralis.*

stylet A hollow, needle-like structure used by parasitic nematodes to pierce plant tissues.

symbiotic continuum Classification of symbiotic categories based on the potential fitness of the symbionts.

syconium Inflorescence of a fig plant.

syncytium Multinucleate mass of cytoplasm lacking internal cell membranes.

synergistic hypothesis Cooperation enhances the survival and reproduction of humans.

tachyzoite Merozoite-like stage of *Toxoplasma* that develops in macrophages.

teliospore Spore of rust fungi that overwinters and germinates to produce basidiospores.

ti-plasmid Tumor-inducing plasmid of *Agrobacterium tumefaciens.*

termitaria Termite nests.

toxoplasmosis A prevalent human asymptomatic infection caused by *Toxoplasma gondii.*

transduction Transfer of genes from one bacterium to another by means of a bacteriophage.

trophosome Portion of the tube worm body cavity that contains bacterial symbionts.

trophozoite Metabolically active, motile stage of a protozoan.

undulipodia "Eukaryotic flagella" used for locomotion and feeding and consisting of microtubules in a $9+2$ arrangement.

vesicular-arbuscular mycorrhiza A mycorrhiza that forms arbuscules and vesicles inside root cells but does not form an outer sheath.

viral conversion Development of new genetic properties in a host bacterium because of bacteriophage infection.

viroids Naked, infectious nucleic acids; the smallest viruses.

virions Viral particles.

welcome mat hypothesis Nematode parasites acquired by a host induce resistance of host to later infections.

zooxanthellae Common algal symbionts of marine invertebrates.

Index